Structure and Function of the Musculoskeletal System

SECOND EDITION

James Watkins, PhD

Swansea University, Wales

Human Kinetics

Library of Congress Cataloging-in-Publication Data

Watkins, James, 1946-
   Structure and function of the musculoskeletal system / James Watkins. -- 2nd ed.
      p. ; cm.
   Includes bibliographical references and index.
   ISBN-13: 978-0-7360-7890-0 (hard cover)
   ISBN-10: 0-7360-7890-8 (hard cover)
   1. Musculoskeletal system--Physiology. 2. Musculoskeletal system--Anatomy. 3.
Human mechanics. I. Title.
   [DNLM: 1. Musculoskeletal Physiological Phenomena. 2. Biomechanics. 3.
Musculoskeletal System--anatomy & histology. WE 102 W335s 2010]
   QP301.W34 2010
   612.7--dc22
                                    2009025659

ISBN-10: 0-7360-7890-8 (print)   ISBN-10: 0-7360-8557-2 (Adobe PDF)
ISBN-13: 978-0-7360-7890-0 (print)   ISBN-13: 978-0-7360-8557-1 (Adobe PDF)

Copyright © 2010, 1999 by James Watkins

Acquisitions Editor: John Dickinson, PhD; **Developmental Editor:** Christine M. Drews; **Assistant Editors:** Katherine Maurer and Dena P. Mumm; **Copyeditor:** Julie Anderson; **Proofreader:** Maggie Schwarzentraub; **Indexer:** Michael Ferreira; **Permission Manager:** Dalene Reeder; **Graphic Designer:** Nancy Rasmus; **Graphic Artist:** Dawn Sills; **Cover Designer:** Keith Blomberg; **Illustrator (cover):** Jennifer Gibas; **Photographer (cover):** Jason Allen; **Art Manager:** Kelly Hendren; **Associate Art Manager:** Alan L. Wilborn; **Art Style Development:** Joanne Brummett; **Illustrators:** Katherine Auble, Tim Brummett, Mike Meyer, Andrew Recker, Rachel Rogge; **Printer:** Premier Print Group; **Binder:** Dekker Bookbinding

Printed in the United States of America   10 9 8 7 6 5 4 3 2 1

Human Kinetics
Web site: www.HumanKinetics.com

United States: Human Kinetics
P.O. Box 5076
Champaign, IL 61825-5076
800-747-4457
e-mail: humank@hkusa.com

Canada: Human Kinetics
475 Devonshire Road Unit 100
Windsor, ON N8Y 2L5
800-465-7301 (in Canada only)
e-mail: info@hkcanada.com

Europe: Human Kinetics
107 Bradford Road
Stanningley
Leeds LS28 6AT, United Kingdom
+44 (0) 113 255 5665
e-mail: hk@hkeurope.com

Australia: Human Kinetics
57A Price Avenue
Lower Mitcham, South Australia 5062
08 8372 0999
e-mail: info@hkaustralia.com

New Zealand: Human Kinetics
Division of Sports Distributors NZ Ltd.
P.O. Box 300 226 Albany
North Shore City
Auckland
0064 9 448 1207
e-mail: info@humankinetics.co.nz

E4672

To my mother and father

Mary Watkins 1914-2006
William Watkins 1909-1985

Contents

PART II Musculoskeletal Response and Adaptation to Loading

Preface

Human movement is brought about by the musculoskeletal system—the skeletal muscles, bones, and joints—under the control of the nervous system. The bones of the skeleton are linked together at joints in a way that allows them to move relative to each other. The skeletal muscles pull on the bones to control the movements of the joints and, in doing so, control the movement of the body as a whole. Through coordinated activity of the various muscle groups, the forces generated by our muscles are transmitted by our bones and joints to enable us to maintain an upright or partially upright posture, move from one place to another, and manipulate objects, often simultaneously.

The open-chain arrangement of the bones of the skeleton—two arms and two legs attached independently to the vertebral column—allows us to adopt a wide range of body postures and perform a wide range of movements. However, this movement capability is only possible at the expense of low mechanical advantage of skeletal muscles. Most muscles are attached to bones very close to joints, such that in most postures and movements other than lying down, the muscles have to exert very large forces which, in turn, result in very large forces in joints.

In response to the forces exerted on them, the musculoskeletal components experience strain—they are deformed. Under normal circumstances, the musculoskeletal components adapt their size, shape, and structure, a process referred to as structural adaptation, to more readily withstand the time-averaged strain of everyday physical activity. However, excessive strain will result in injury. Structural adaptation is continuous throughout life, but the capacity for structural adaptation decreases markedly with increase in age after maturity; strain that would normally result in structural adaptation in a young person may result in tissue degeneration and dysfunction in an older person. Consequently, there is an intimate relationship between the structure and function of the musculoskeletal system. The purpose of *Structure and Function of the Musculoskeletal System* is to develop knowledge and understanding of this relationship. The book is primarily a course text for undergraduate students of kinesiology, exercise science, sport science, and physical education, but it also has a great deal to offer health care professionals, in particular, physiotherapists and occupational therapists, who deal with the acute and chronic effects of musculoskeletal pathology.

What's New in the Second Edition?

Changes in the organization and content of the second edition are based largely on feedback from students and teachers over a number of years. The main changes are as follows:

- Revision of all content: Some material was removed and some new material has been added.

- Revision of the content into two parts rather than three: functional anatomy of the musculoskeletal system in part I and response and adaptation of the musculoskeletal system in part II.

- Case studies: A number of case studies are included to illustrate the response and adaptation of the musculoskeletal system to exercise at various ages and to generate discussion of issues pertaining to good practice in the promotion and maintenance of musculoskeletal health.

- Reorganization of elementary biomechanical concepts and principles: In the first edition, all of the fundamental mechanical concepts were contained largely within a single chapter. In the second edition, this chapter has been deleted and the content incorporated more appropriately into other chapters, including a relatively short, new chapter at the start of part II.

• Highlighted introductory figures: In part I, the sequence of illustrations is such that often one figure introduces several subsequent figures. In these cases, the introductory figure has been highlighted with a light-blue burst. This may help readers gain perspective as they refer across several figures at a time when reading the text.

In addition to the preceding changes, all of the learning features in the first edition have been retained—content overview at the start of each chapter, chapter objectives, key points and additional details on some points highlighted within the text, an extensive running glossary, extensive use of illustrations, review questions, references to guide further reading, and an extensive index.

Instructor Resource

This edition includes a new instructor resource, an Image Bank delivered in Microsoft PowerPoint, which contains all of the figures and tables found in this text. Instructors can customize their own presentations with these figures and tables. The Image Bank also includes a blank PowerPoint template and instructions for creating custom presentations. This instructor resource can be found at www.HumanKinetics.com/StructureandFunctionoftheMusculoskeletalSystem.

Organization of the Second Edition

Part I, Functional Anatomy of the Musculoskeletal System, describes how musculoskeletal function—the generation and transmission of forces to control the movement of the body—is reflected in the structure of the musculoskeletal components. Chapter 1, The Musculoskeletal System, describes the composition and function of the musculoskeletal system. Chapter 2, The Skeleton, describes the open-chain arrangement of the bones of the skeleton. Chapter 3, Connective Tissues, differentiates the ordinary connective tissues, in particular, ligaments, tendons, and fascia, and the special connective tissues of cartilage and bone. Chapter 4, The Articular System, explains the differences in types of joints and, in particular, the way joint design reflects a trade-off between stability and flexibility. Chapter 5, Joints of the Axial Skeleton, describes the joints between the vertebrae and the joints of the pelvis. Chapter 6, Joints of the Appendicular Skeleton, describes the joints and joint complexes of the appendicular skeleton. Chapter 7, The Neuromuscular System, describes the relationship between the nervous and muscular systems in terms of force generation and proprioception.

Part II, Musculoskeletal Response and Adaptation to Loading, describes the immediate and long-term effects of loading on the external form and internal architecture of the musculoskeletal system. Chapter 8, Elementary Biomechanics, develops knowledge and understanding of elementary biomechanical concepts and principles. Chapter 9, Forces in Muscles and Joints, focuses on how the open-chain arrangement of the skeleton affects the forces exerted in muscles and joints. Chapter 10, Mechanical Characteristics of Musculoskeletal Components, describes the musculoskeletal system's response to loading. Chapter 11, Structural Adaptation of the Musculoskeletal System, describes how the structure of the musculoskeletal system adapts to the time-averaged loads exerted on it. Chapter

eBook available at HumanKinetics.com

12, Etiology of Musculoskeletal Disorders and Injuries, describes the main groups of risk factors that influence the level of loading on the musculoskeletal system.

I hope this book will encourage you to learn more about the structure and function of the musculoskeletal system. Adequate musculoskeletal functioning is an important determinant of our quality of life, and the information contained in this book is essential for anyone wishing to maintain, or help others to maintain, a healthy musculoskeletal system.

James Watkins

Acknowledgments

I thank my dear wife, Shelagh, for her continuous support throughout the various stages of the writing of the book. I also thank all of the staff at Human Kinetics who contributed to the commissioning and production of the book. I thank my academic colleagues and the large number of undergraduate and graduate students who have helped me, directly and indirectly, over many years, to develop and organize the content of the book.

PART I
Functional Anatomy of the Musculoskeletal System

Your musculoskeletal system—your skeleton and skeletal muscles—account for about half of your total body weight. Your skeleton (12-15% of your total body weight) consists of 206 bones that are linked by more than 200 joints. You have approximately 640 skeletal muscles (30-42% of total body weight), which pull on your bones to control the movement of your joints and enable you to stand up, move around, and manipulate objects. Bone is an ideal support material because it is not only amazingly strong, but also fairly lightweight. Relative to its weight, bone is stronger than concrete and it has a higher strength-to-weight ratio than any other naturally occurring material. Your bones have to be strong to withstand the very large forces that the muscles exert on them during normal everyday movements. If all of your muscles could contract simultaneously in the same direction, they would exert a force in the region of 22 tons.

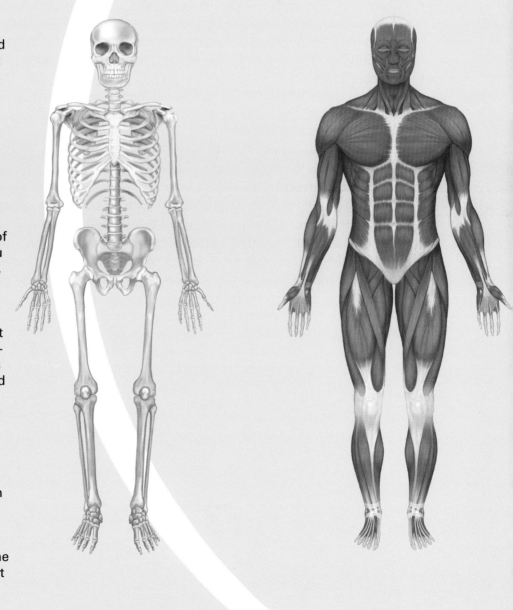

Human movement is brought about by the musculoskeletal system under the control of the nervous system. The muscles pull on the bones to control the movements of the joints and in doing so control the movement of the body as a whole. Part I describes how musculoskeletal function—the generation and transmission of forces to control the movement of the body—is reflected in the structure of the musculoskeletal components.

- Chapter 1, The Musculoskeletal System, describes the basic structural and functional unit of all living organisms—the cell—and how cells are organized into organs and systems. You will learn how one of these systems, the musculoskeletal system, enables you to move in a controlled manner by generating forces in muscles and transmitting the forces across joints.

- Chapter 2, The Skeleton, describes the bones of the skeleton and the open-chain arrangement of the bones. You will learn how the large differences in size, shape, and surface features of the bones reflect the ability of the bones to transmit muscle forces.

- Chapter 3, Connective Tissues, describes the wide variety of connective tissues, which range from areolar tissue—the glue that holds cells together—to ligaments, tendons, cartilage, and bone. You will learn that all connective tissues have the same basic structure but differ in strength and flexibility because of differences in the proportions of the structural components.

- Chapter 4, The Articular System, describes the various types of joints between the bones of the skeleton. The joints, in association with the open-chain arrangement of the bones, facilitate a wide range of body postures. You will learn that all joints transmit forces and allow a certain amount of movement.

- Chapter 5, Joints of the Axial Skeleton, describes the joints between the vertebrae and the joints between the bones of the pelvis. You will learn that compared with the joints in the appendicular skeleton, the joints of the axial skeleton and the shape of the vertebral column provide a high level of shock absorption for the body at the expense of reduced flexibility.

- Chapter 6, Joints of the Appendicular Skeleton, describes the joints in the arms and legs. You will learn that in contrast to the joints of the axial skeleton, the joints of the appendicular skeleton facilitate relative large ranges of movement at the expense of shock absorption.

- Chapter 7, The Neuromuscular System, describes the parts of the nervous and muscular systems that are responsible for bringing about coordinated movement. You will learn that the nervous system constantly monitors and interprets information from the various senses and uses this information to program muscular activity.

The Musculoskeletal System

All living organisms are made up of one or more cells. The human body, like most organisms, is made up of billions of cells organized into complex functional groups including, for example, the musculo-skeletal system. The first part of this chapter describes the four basic types of cells, called tissues, and the organization of tissues into organs and systems. The second part of the chapter describes the composition and function of the musculoskeletal system.

1

OBJECTIVES

After reading this chapter, you should be able to do the following:

1. Describe the four types of tissues.
2. Describe cellular organization in multicellular organisms.
3. Describe the composition and function of the musculoskeletal system.

Unicellular and Multicellular Organisms

The basic structural and functional unit of life is the **cell.** All cells have four main components: cytosol, a nucleus, organelles, and a cell membrane (figure 1.1). Cytosol is a semitransparent fluid consisting of a complex solution of proteins, salts, and sugars. The nucleus and organelles are suspended in the cytosol. The nucleus contains genetic material and controls the **life processes** of the cell, which include movement, growth and development, respiration, circulation, digestion, excretion, and reproduction. The organelles, under the direction of the nucleus, carry out the life processes; each organelle carries out a specific function for the cell as a whole. The cytosol and organelles are usually referred to collectively as cytoplasm.

> **KEY POINT**
>
> The basic structural and functional unit of life is the cell, and all living organisms consist of one or more cells.

The cell membrane, also referred to as the plasma membrane or plasmalemma, encloses the cytoplasm and forms the external boundary of the cell. Like cytosol, the cell membrane consists of a viscous fluid, but it is usually much more viscous than the cytosol that it encloses. The viscous nature of the cell membrane and cytosol enables a cell to change its shape without losing its integrity. During normal functioning, all cells continuously change shape to a certain extent. The amount of change varies

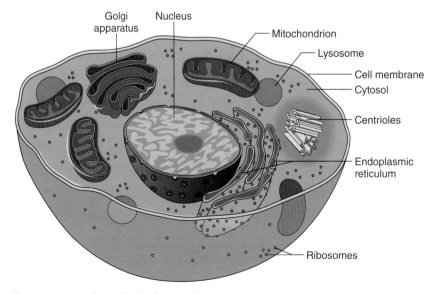

FIGURE 1.1 Cross section of a typical animal cell.

among types of cells. For example, bone cells are likely to experience little change in shape, whereas muscle cells have a highly specialized ability to change shape, which is an essential feature of normal function.

The number of cells that make up an organism reflect its level of evolution. For example, in the lowest forms of animal life such as amoeba and euglena, the entire organism consists of a single cell. These organisms are referred to as **unicellular organisms.** In contrast, most animals consist of many cells and are referred to as **multicellular organisms.** All mammals, birds, and fish are multicellular organisms. The number of cells in multicellular organisms varies considerably. For example, the nematode worm *C. elegans* is 1 mm in length and consists of 959 cells (Kenyon 1988). Most multicellular organisms consist of millions of cells; the human body consists of approximately 10^{14} (one hundred million million) cells (Alberts et al 2002). The size and shape of cells vary considerably among different organisms and within the same organism. Differences in cell size and shape tend to reflect differences in function.

Key Terms

cell   The basic structural and functional unit of all organisms.

life processes   The activities a cell carries out to sustain the life of the cell.

unicellular organism   An organism that consists of a single cell.

multicellular organism   An organism that consists of many cells.

KEY POINT

There are two types of living organisms: unicellular and multicellular.

Cellular Organization in Multicellular Organisms

All living organisms are similar in that they are capable of carrying out all of the essential life processes. In unicellular organisms the life processes are relatively simple. However, in multicellular organisms the cells are organized into complex functional groups that carry out the various life processes for the body as a whole. Multicellular organization can be divided into three structural levels: tissues, organs, and systems.

KEY POINT

All living organisms are capable of carrying out a number of essential life processes. In multicellular organisms the life processes involve a high level of organization and integration between the cells. Cells are organized on three levels: tissues, organs, and systems.

Tissues

In multicellular organisms all of the cells originate from a single cell formed by the fertilization of a female ovum by a male sperm. This cell undergoes rapid cell division to form a ball of cells. Soon after this, the cells begin to differentiate in size, shape, and structure to fulfill different functions in the body of the organism. This process of **cellular differentiation** results in the formation of four types of cells called tissues: epithelia, nerve, muscle, and connective. A **tissue** is a group of cells having the same specialized structure, enabling them to perform a particular function in the body (Freeman and Bracegirdle 1967). The word *tissue* is also used in a general sense to refer to any part of the body: for example, soft tissues, bone tissue, and skin tissue.

Key Terms

cellular differentiation   The specialization of cells into tissues.

tissue   A group of cells having the same specialized structure, enabling them to perform a particular function in the body.

Epithelial Tissue

There are two types of epithelial tissue: covering and glandular. Covering epithelia form the surface layer or layers of cells of all the internal and external free surfaces of the body except the surfaces inside synovial joints. For example, the surfaces of the skin and the lining of the digestive tract, heart chambers, and blood vessels are covering epithelia.

All cells can secrete fluid to a greater or lesser extent, but glandular epithelial cells are specialized for this purpose and as such form the two types of glands found in the body: exocrine and endocrine. Many of the exocrine glands, such as the gastric glands in the lining of the stomach, secrete fluids containing enzymes necessary for the digestion of food. Endocrine glands, such as the pituitary gland at the base of the brain and the adrenals at the upper end of each kidney, secrete hormones that, in association with the nervous system, regulate and coordinate the various body functions.

Nerve Tissue

Nerve cells (neurons) are specialized to conduct electrochemical impulses throughout the body to regulate and coordinate the various body functions. The structure and function of nerve tissue are covered in detail in chapter 7, which deals with the neuromuscular system.

Muscle Tissue

Muscle cells are specialized to contract (i.e., create pulling forces to bring about movement). There are three types of muscle cells: skeletal, visceral, and cardiac. Skeletal muscle is so-called because it generally is attached to the skeleton. It is also called *voluntary muscle* because it is normally under the conscious control of the person. The structure and functions of skeletal muscle are covered in detail in chapter 7.

Visceral or involuntary muscle is found in parts of the body that involve involuntary movement, that is, in body parts not under our conscious control. Visceral muscle is found, for example, in the walls of the alimentary canal and the larger arteries.

Cardiac muscle is found only in the heart. It has characteristics of both skeletal and visceral muscle but differs from them in that it contracts rhythmically throughout life even though the rate of contractions (heart rate) may alter frequently.

Connective Tissue

As its name suggests, connective tissue supports and binds other tissues together. The bones of the skeleton and the fibrous structures holding the bones together at joints are forms of connective tissue. The structure and functions of connective tissue are covered in detail in chapter 3.

Organs and Systems

An **organ** is a combination of tissues designed to carry out a specific bodily function. For example, the heart pumps blood around the body. The structure of the heart consists of

- cardiac muscle cells;
- connective tissue, which binds the muscle cells together;
- epithelial tissue, which lines the chambers of the heart; and
- nerve tissue, which innervates the muscle cells.

Other examples of organs are the lungs, the stomach, and a skeletal muscle (such as the quadriceps) complete with its tendons, which attach the muscle to the skeletal system.

A **system** is a combination of organs working together to carry out a particular function in the body. For example, the cardiovascular system consists of the heart and blood vessels and is responsible for transporting blood around the body. There are 11 separate systems in the human body (Tortora and Anagnostakos 1984):

integumentary system The external covering of the body, that is, the skin and associated structures such as nails.

skeletal system The bones of the skeleton and the structures that form the joints between the bones.

muscular system The skeletal muscles.

nervous system The nerves, organized into central (brain and spinal cord) and peripheral (spinal nerves) components.

endocrine system The glands that secrete hormones, which regulate and coordinate the various body functions in association with the nervous system.

cardiovascular system The heart and blood vessels.

lymphatic system The system of vessels and associated structures that drains and returns fluid leaked from the blood and protects against disease.

respiratory system The lungs and associated passageways.

digestive system The alimentary canal and associated structures that break down food and eliminate solid waste.

urinary system The kidneys, bladder, and associated structures that eliminate nitrogenous waste as urine.

reproductive system The ovaries and associated structures (female) and testes and associated structures (male), which enable the body to produce offspring.

These systems are responsible for carrying out the body's life processes. Whereas all of the life processes involve a certain degree of integration between systems, some processes involve closer integration among systems than do others. For example, the transport of oxygen from the air to all the cells of the body is carried out by the combined activity of the nervous, respiratory, and cardiovascular systems. Similarly, movement of the body is brought about by the combined activity of the nervous, muscular, and skeletal systems. Consequently, for descriptive purposes it is usual to refer to combinations of systems: for example, the cardiorespiratory system, the musculoskeletal system, and the neuromuscular system. Cellular differentiation and organization in multicellular organisms are illustrated in figure 1.2.

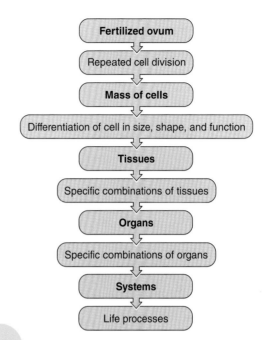

FIGURE 1.2 Cellular differentiation and organization in multicellular organisms.

Key Terms

organ A combination of tissues designed to carry out a specific bodily function.

system A combination of organs working together to carry out a particular bodily function.

KEY POINT

All of the cells of the body originate from a single cell—a fertilized ovum. This cell undergoes rapid cell division and, subsequently, cellular differentiation that results in the formation of different types of tissues that combine to form specialized organs. Organs then work together to form systems to carry out the life processes of multicellular organisms.

Composition and Function of the Musculoskeletal System

Human movement is brought about by the **musculoskeletal system** under the control of the nervous system. The musculoskeletal system consists of the **skeletal system** and the **muscular system** (figure 1.3). The skeletal system consists of the skeleton (the bones) and

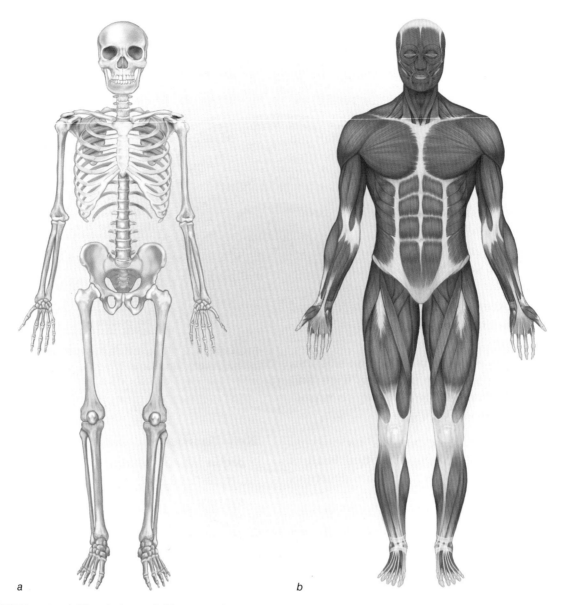

a   b

FIGURE 1.3   *(a)* The skeleton. *(b)* The muscular system.

the structures forming the joints between the bones. Similar to the metal framework that supports a building, the skeletal system gives the body its shape and provides a supporting framework for all the other systems. Bone is an ideal support material because it is not only strong but also fairly lightweight. The adult skeletal system normally has 206 bones and more than 200 joints and accounts for between 12% and 15% of total body weight (McArdle et al 1996).

Skeletal, visceral, and cardiac are the three types of muscle tissue. Visceral muscle is usually considered to be part of the digestive system (in the walls of the digestive tract) and

cardiovascular system (in the walls of arteries). Cardiac muscle, found only in the heart, is part of the cardiovascular system (Tortora and Anagnostakos 1984). The muscular system refers only to the skeletal muscles. There are approximately 640 skeletal muscles. The skeletal muscles account for an average of approximately 34% and 42% of total body weight in young (18-29 years), healthy, untrained adult females and males, respectively, and an average of approximately 30% and 34% of total body weight in elderly (70-88 years), healthy, untrained adult females and males, respectively (Janssen et al 2000).

Musculotendinous Units

Each complete skeletal muscle consists of a large number of long muscle cells bound together by layers of connective tissues (see chapter 3). In most muscles the muscle cells occupy the main belly of the muscle, but the ends of the muscle consist entirely of thickened cords or bands of virtually inextensible connective tissue that anchor the muscle onto the skeleton (figure 1.4). The shape of these connective tissue attachments depends on the shape of the muscle and the area of bone available for attachment. There are two basic shapes: a cord or narrow band called a **tendon** and a broad band called an **aponeurosis** (figure 1.4). A skeletal muscle and its tendons

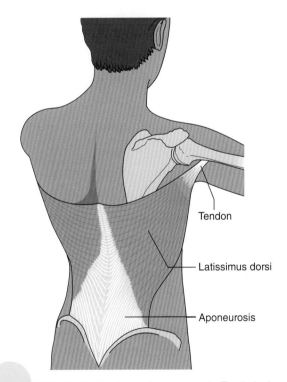

FIGURE 1.4 Tendon and aponeurosis. The latissimus dorsi muscle on each side of the body originates from a broad aponeurosis and inserts onto the humerus via a narrow bandlike tendon.

and aponeuroses are usually referred to as a **musculotendinous unit** (Taylor et al 1990).

Contractility

The most important property of all types of muscle tissue is **contractility,** which is the ability to create a pulling force and, if necessary, change in length (increase or decrease) while maintaining a pulling force. This property is developed to a high level in skeletal muscle tissue. The skeletal muscles pull on the bones to control the movement of the joints and thereby the movement of the body as a whole. By coordinated activity between the various muscle groups, forces generated by the muscles are transmitted by the bones and joints to enable the person to maintain an upright or partially upright posture and bring about voluntary controlled movements.

Antagonistic Muscle Groups

The skeletal muscles are arranged in groups, and each group is attached to the skeleton such that the muscle group crosses over one or more joints. Because a muscle can pull in only one direction, the movement of each joint is controlled by one or more opposing pairs of muscle groups; one group in each pair tends to move a joint in one direction and the opposing group in each pair tends to move the joint in the opposite direction. For this reason, the various pairs of muscle groups are referred to as **antagonistic pairs.** The coordinated activity within and between the antagonistic pairs of muscle groups that cross a joint allow us to control the amount and rate of movement in the joint. For example, when an athlete kicks a ball, the quadriceps and hamstrings work together as an antagonistic pair of muscle groups to control movement at the hip and knee joints (figure 1.5).

The knee joint is designed to rotate in one plane, as shown in figure 1.5. Movements of the knee in this plane are referred to as knee flexion (bending the knee) and knee extension (straightening the knee). However, other joints such as the hip and shoulder are designed to move in more than one plane (see chapter 2). Consequently, two or more antagonistic pairs of muscle groups are required to control movement

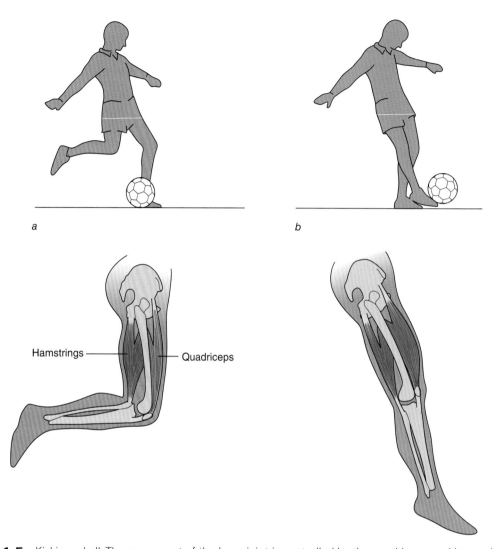

a   b

Hamstrings — — Quadriceps

FIGURE 1.5 Kicking a ball: The movement of the knee joint is controlled by the quadriceps and hamstrings.

in these joints. For example, the quadriceps and hamstrings control hip flexion (bending the hip) and hip extension (straightening the hip), but another antagonistic pair of muscles (the hip abductors and hip adductors) control the movement of the hip in the plane at right angles to the plane of flexion and extension.

Key Terms

musculoskeletal system   The combined system consisting of the skeletal system and the muscular system.

skeletal system   The bones and the structures that form the joints between the bones.

muscular system   The skeletal muscles.

tendon   A cord or narrow band of connective tissue that attaches skeletal muscle to bone.

aponeurosis   A broad band of connective tissue that attaches a skeletal muscle to bone.

musculotendinous unit   A skeletal muscle and its tendons or aponeuroses.

contractility   The ability of muscle tissue to create a pulling force.

antagonistic pair   A pair of musculotendinous units that control a joint's movement in a particular plane.

Each skeletal muscle is attached to the skeleton such that the muscle crosses one or more joints; when a muscle contracts it tends to bring about movement in the joints it crosses. The muscles are arranged in antagonistic pairs that coordinate action to control the amount and rate of movement in each joint.

Force, Load, Strain, and Stress

All movements and changes in movement are brought about by the action of forces. The two most common types of force are pulling and pushing. A **force** can be defined as that which tends to deform an object, move an object from rest, or change the way that the object is moving.

A **load** is any force or combination of forces that are applied to an object. There are three types of load: tension, compression, and shear (figure 1.6). Loads tend to deform the objects on which they act. Tension is a pulling (stretching) load that tends to make an object longer and thinner along the line of the force (figure 1.6, *a* and *b*). Compression is a pushing or pressing load that tends to make an object shorter and thicker along the line of the force (figure 1.6, *a* and *c*). A shear load is composed

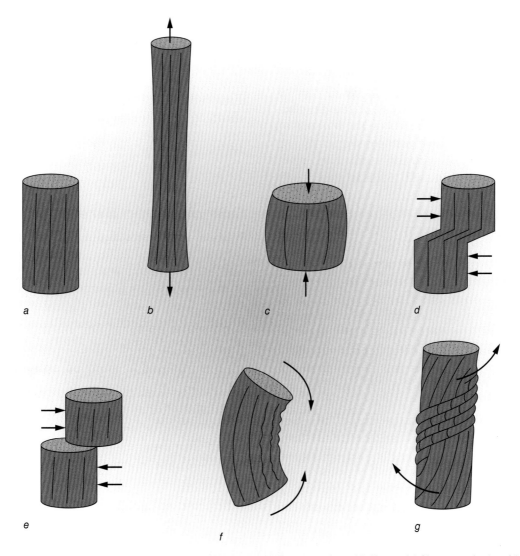

FIGURE 1.6 Types of load. *(a)* Unloaded. *(b)* Tension. *(c)* Compression. *(d)* Shear. *(e)* Shear producing friction. *(f)* Bending. *(g)* Torsion.

of two equal (in magnitude), opposite (in direction), parallel forces that tend to displace one part of an object with respect to an adjacent part along a plane parallel to and between the lines of force (figure 1.6, *a* and *d*). The cutting load produced by scissors and garden shears is a shear load, whereas the cutting load produced by a knife is a compression load. The load that forces one object to slide on another is also a shear load (figure 1.6*e*). The sliding or tendency to slide is resisted by a force called *friction*, which is exerted between and parallel to the two contacting surfaces.

The three types of load frequently occur in combination, especially in bending (figure 1.6, *a* and *f*) and torsion (figure 1.6, *a* and *g*). An object subjected to bending experiences tension on one side and compression on the other. An object subjected to torsion (twisting) simultaneously experiences tension, compression, and shear.

In mechanics, the deformation of an object that occurs in response to a load is referred to as **strain.** For example, when a muscle contracts it exerts a tension load on the tendons at each end of the muscle, and consequently the tendons experience tension strain (i.e., they are very slightly stretched). Similarly, an object subjected to a compression load experiences compression strain, and an object subjected to a shear load experiences shear strain. Strain denotes deformation of the intermolecular bonds that comprise the structure of an object. When an object experiences strain, the intermolecular bonds exert forces that tend to restore the original (unloaded) size and shape of the object. The forces exerted by the intermolecular bonds of an object under strain are referred to as stress. **Stress** is the resistance of the intermolecular bonds to the strain caused by the load.

The stress on an object resulting from a particular load is distributed throughout the whole of the material sustaining the load. However, the level of stress in different regions of the material varies depending on the amount of material sustaining the load in the regions; the more material sustaining the load, the lower the stress. Consequently, stress is measured in terms of the average load on the plane of material sustaining the load at the point of interest.

Tension Stress

Figure 1.7*a* shows a person standing upright with the line of action of body weight slightly in front of the ankle joints. In this posture, stability is maintained by isometric (static) contraction of the ankle plantar flexors as shown in the simple two-segment model in figure 1.7*b*. If the force exerted by the ankle plantar flexors in each leg is 36 kgf (kgf = kilogram force; 1 kgf = the weight of a mass of 1 kilogram) and the cross-sectional area of the Achilles tendon at *P* in figure 1.7*b*, perpendicular to the tension load, is 1.8 cm^2, then the tension stress on the tendon at *P* is 20 kgf/cm^2 (kilograms force per square centimeter), that is,

$$\text{tension stress at } P = \frac{36 \text{ kgf}}{1.8 \text{ cm}^2}$$
$$= 20 \text{ kgf/cm}^2.$$

Compression Stress

When a person is standing upright and barefoot, as in figure 1.7*a*, the ground reaction force (the upward force exerted by the ground on the feet) exerts a compression load on the contact area of the feet (figure 1.8*a*). In an adult the contact area is approximately 260 cm^2 (both feet) (Hennig et al 1994). For a person weighing 70 kgf, the compression stress on the contact area of the feet (on a level floor, contact area perpendicular to the compression load) is 0.27 kgf/cm^2, that is,

$$\text{compression stress} = \frac{70 \text{ kgf}}{260 \text{ cm}^2}$$
$$= 0.27 \text{ kgf/cm}^2.$$

When the person raises her heels off the ground, the contact area is approximately halved (figure 1.8*b*). Because the compression load (body weight) is the same as before, it follows that the compression stress on the reduced contact area is approximately doubled. Compression stress is usually referred to as pressure.

Shear Stress

Many of the joints, especially those in the low back and pelvis, are subjected to shear load during normal everyday activities such as standing and walking. For example, in walking, there is a phase when one leg supports the body while the other leg swings forward (figure 1.9*a*). In this situation the unsupported side of the body tends to move downward relative to the supported side, subjecting the pubic symphysis joint to shear load (figure 1.9*b*). In an adult male, the area of the pubic symphysis in the plane of the shear load is approximately 5 cm². If the shear load at the instant shown in figure 1.9*a* is, for example, 2 kgf, then the shear stress on the joint is 0.4 kgf/cm², that is,

$$\text{shear stress} = \frac{2\,\text{kgf}}{5\,\text{cm}^2}$$
$$= 0.4\,\text{kgf}/\text{cm}^2.$$

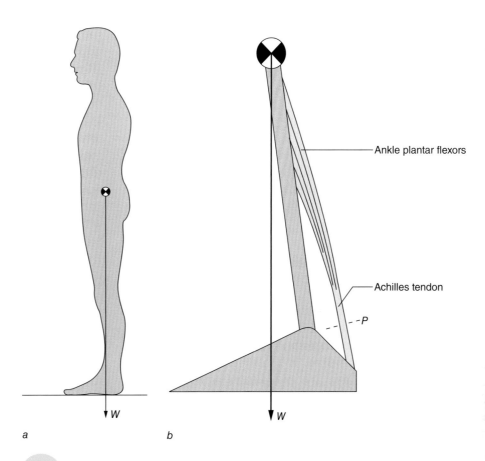

a

b

FIGURE 1.7 Tension load on the Achilles tendon. *W* represents body weight. *P* indicates the plane of the cross section of the Achilles tendon at which the tension stress in the tendon is being assessed.

Ankle plantar flexors

Achilles tendon

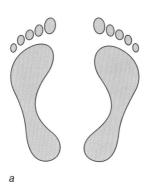

a

b

FIGURE 1.8 Supporting area of the feet. *(a)* Normal upright standing posture. *(b)* Standing upright with the heels off the floor.

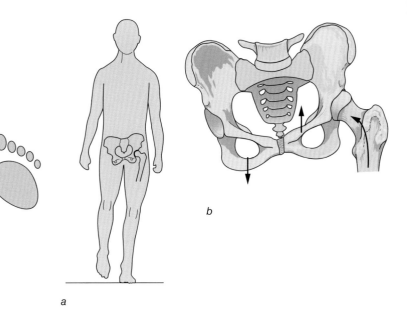

b

a

FIGURE 1.9 Shear load on the pubic symphysis resulting from single-leg support while walking.

13

Key Terms

force That which tends to deform an object, move an object from rest, or change the way that the object is moving.

load Any force or combination of forces that is applied to an object.

strain The deformation of an object that occurs in response to a load.

stress The resistance of the intermolecular bonds of an object to the strain caused by a load.

Musculoskeletal System Function

Posture refers to the orientation of the body segments to each other and is usually applied to static or quasi-static positions such as sitting and standing. When a person is standing upright there are two forces acting on the body: body weight and the ground reaction force (figure 1.10*a*). The combined effect of body weight and the ground reaction force is a compression load (experienced at the feet) that tends to collapse the body into a heap on the ground. This compression load increases with any additional weight carried by the body (figure 1.10*b*). To prevent the body from collapsing while simultaneously bringing about desired movements, the movements of the joints need to be carefully controlled by coordinated activity between muscle groups. For example, when a person is standing upright, the joints of the neck, trunk, and legs must be stabilized by the muscles that control them; otherwise, the body would collapse (figure 1.11). Consequently, the weight of the whole body is transmitted to the floor by the feet, but the weight of individual body segments above the feet (head, arms, trunk, and legs) is transmitted indirectly to the floor by the skeletal chain formed by the bones and joints of the neck, trunk, and legs.

Transmitting body weight to the ground while maintaining an upright body posture illustrates the essential feature of musculoskeletal function, that is, the generation (by the muscles) and transmission (by the bones and joints) of forces. The forces generated and transmitted by the musculoskeletal system are referred to as **internal forces,** and forces that

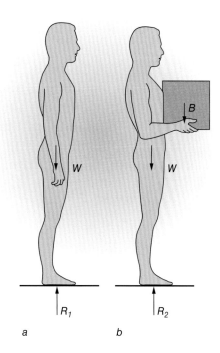

a   *b*

FIGURE 1.10 The forces acting on the body when standing. *(a)* Standing upright and *(b)* standing upright with additional load. W = body weight, R_1 and R_2 = ground reaction forces, B = additional weight.

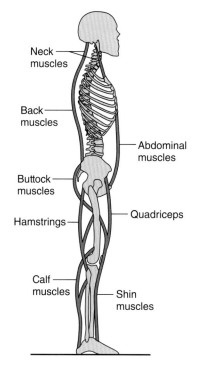

FIGURE 1.11 Location of the main muscle groups that are responsible for maintaining standing posture.

act on the body from external sources, such as body weight, ground reaction force, water resistance, and air resistance, are referred to as **external forces.** The musculoskeletal system generates and transmits internal forces to counteract (in static and quasi-static postures) or overcome (during purposeful movements) the external forces acting on the body in order to maintain upright posture, transport the body, and manipulate objects, often simultaneously.

Key Terms

internal force A force produced or transmitted by the musculoskeletal system.

external force A force exerted on the body by an external agent.

KEY POINT

The musculoskeletal system generates and transmits internal forces to counteract or overcome external forces and bring about controlled movement of the body.

Maintaining Upright Posture

Movements requiring the maintenance of upright posture involve fairly static postures in which the weight of one or more body segments is transmitted indirectly to the floor or other support surface, such as a chair, by other segments. These postures include standing (figure 1.12*a*), sitting (figure 1.12, *b-d*), and balancing activities (figure 1.13*b*). Some upright postures may involve more than one support surface. For example, figure 1.12*c* shows a person sitting in an armchair; his arms are supported by the armrests and his feet rest on the floor. In this case his weight is distributed over four support areas: the backrest, seat, armrests, and floor. Figure 1.12*e* shows a person leaning on a stool. In this case his weight is supported partly by the stool and partly by the floor.

The degree of muscular activity required to maintain a particular body posture depends on the number and size of the support surfaces. Consequently, muscular activity is minimal in the recumbent posture because all of the body segments are supported directly by the support surface (figure 1.13*a*). In contrast, balancing on one hand involves a consider-

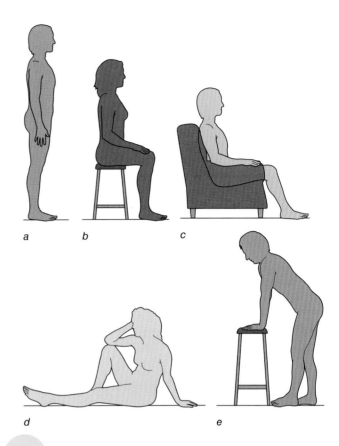

FIGURE 1.12 Standing and sitting postures.

FIGURE 1.13 Lying and handstand postures.

able amount of muscular activity to transmit the entire body weight through the small area beneath the hand (figure 1.13*b*). In this case,

15

the weight of each body segment above the grounded hand is transmitted indirectly to the floor, from segment to segment down through the skeletal chain.

Body Transport

To move from one place to another, the body must push or pull against something to provide the necessary force to drive it in the required direction. In walking and running, forward movement is combined with an upright body posture, and movement is achieved by pushing a foot obliquely downward and backward against the ground. Provided that the foot does not slip, the leg thrust results in a ground reaction force directed obliquely upward and forward. The effect of the ground reaction force is that the body moves forward with an upright posture (figure 1.14).

When one is swimming, the water largely supports body weight (buoyancy force). Consequently, the main function of the musculoskeletal system is to enable the arms and legs to pull and push backward against the water to move the body forward. Because the musculoskeletal system does not have to support the weight of the body in the water, the muscle forces and joint reaction forces (the forces exerted between the articular surfaces in joints) that occur during swimming tend to be considerably less than in land-based activities. For this reason, swimming is often prescribed as part of a rehabilitation program to restore normal muscle function and joint flexibility following injury.

Manipulating Objects

Many movements involve manipulating objects with the hands or the feet. These manipulations may involve fairly forceful pushes and pulls or fine manual dexterity. For example, forceful pushes and pulls are required to push a wheelbarrow, pull on an upright post, carry a heavy suitcase, and throw or kick a ball. Activities involving manual dexterity are usually associated with a fairly static body posture (a stable base of support). Such activities include, for example, driving a car, writing or typing at a desk, turning a door handle, and taking a book off a shelf.

The Human Machine

The skeletal system consists of a number of fairly rigid components (bones) that are joined

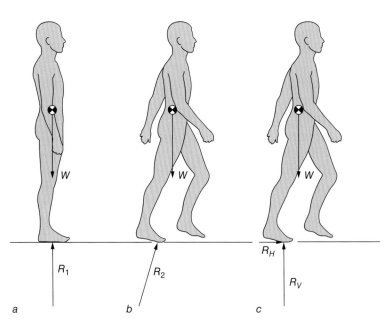

FIGURE 1.14 The ground reaction force in standing and walking. *(a)* Standing. *(b, c)* Push-off in walking. W = body weight, R_1 = ground reaction force when standing, R_2 = ground reaction force during push-off, R_H = horizontal component of R_2, R_V = vertical component of R_2.

together in ways that allow a certain amount of movement in each joint. The muscular system, under the control of the nervous system, controls the movement of the joints and thereby enables the body to apply forces on other objects. Consequently, the musculoskeletal system operates like a **machine**—a powered mechanism that is capable of applying forces to other objects (Dempster 1965). Figure 1.15 shows a simple machine that can be made to apply a gripping force at one end by applying a compression load at the other end.

Key Term

machine A powered mechanism designed to move such that it is capable of applying forces to other objects.

KEY POINT

The musculoskeletal system operates like a machine.

The number of components in a machine, animate or inanimate, reflects the functional complexity of the machine. For example, a pair of scissors, a stapler, and a paper punch all have very few components compared with a car engine. However, the components of an inanimate machine are usually joined together in a closed-chain arrangement that produces simultaneous and predictable movements of the joints between the components; the

movement of one joint determines the movement in all the other joints (as in the simple machine in figure 1.15). This is not the case with the human musculoskeletal system. The human machine enables the person to adopt the most suitable posture for applying forces in any given situation. The body's ability to adopt a wide range of postures stems from the skeleton's open-chain arrangement of bones.

The arms and legs form four peripheral chains, free at their extremities, which attach onto a central chain (see figure 1.3*a*). This open-chain arrangement allows any part of the body to move more or less independently of the rest; movement of one part of the body does not necessarily result in movement in the rest of the body. For example, movements of the arms can accomplish a variety of tasks such as eating, writing, and typing while the rest of the body is in a more or less stationary sitting position.

The human machine has a large number of force-producing components (muscles). Muscle activity is carefully coordinated to use the muscles most efficiently. For example, in a throwing action such as pitching a baseball, the speed of the ball as it leaves the pitcher's hand is the result of the impulse of the force exerted on the ball during the pitching action (the magnitude of the force and the duration of force application) (figure 1.16). The muscles generate the force; the more muscles that are recruited during the pitching action and the greater the length of time that the forces act on the ball, the greater the ball speed. In a well-coordinated pitching action, the forces produced by the individual muscle groups are summated to maximize the impulse of the force exerted on the ball during the pitch (Fleisig et al 1996). The pitching action starts with the rear leg driving the body forward into the step (figure 1.16, *a* and *b)*. The next stage involves foot placement of the lead leg, which provides a firm base for the muscles of the legs and hips to drive the rear hip and the trunk forward (figure 1.16, *b* and *c)*. The muscles of the trunk then drive the throwing arm forward (figure 1.16, *c* and *d)*. Finally, the shoulder, elbow, and wrist of the throwing arm drive the ball forward until release (figure 1.16, *d* and *e)*.

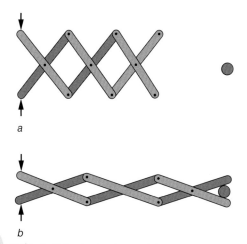

a

b

FIGURE 1.15 A simple machine.

FIGURE 1.16 Pitching a baseball.

In propulsive movements like throwing, the amount of force applied to the implement is limited by the number of muscle groups recruited to drive the implement in the intended direction. If all of an adult's skeletal muscles could contract simultaneously in the same direction, they would exert a force in the region of 22 tons (Elftman 1966). However, the arrangement of the muscles does not permit this theoretical maximum to be achieved. Even in a highly trained and highly motivated athlete, the amount of force generated in propulsive whole-body movements such as jumping and throwing is usually a small fraction of the theoretical maximum. Furthermore, injury to any part of the musculoskeletal system used in the movement reduces this force even further (Nicholas and Marino 1987). For example, injury to an ankle severely affects a pitcher's performance because he is unable to provide a strong, stable base of support for upper body action (figure 1.16). It has been shown that without the initial step, the speed of a baseball pitch is reduced to approximately 80% of normal, and without hip and trunk rotation,

KEY POINT

If all of an adult's skeletal muscles could contract simultaneously in the same direction they would exert a force of about 22 tons. However, the arrangement of the muscles is such that the amount of force generated in propulsive whole-body movements is usually a small fraction of the theoretical maximum.

the speed of the pitch is reduced to about 50% of normal (Miller 1980).

Loading on the Musculoskeletal System

The open-chain arrangement of the bones of the skeleton and the arrangement of the muscles on the skeleton maximize the range of possible body postures. However, the body pays a price for this movement capability: Many of the muscles have very low mechanical advantages (see chapter 9) and usually have to exert much larger forces than the weights of the body segments they control. Furthermore, the size of the joint reaction forces is determined by the size of the muscle forces; the larger the muscle forces, the larger the joint reaction forces. For example, in walking, the peak hip, knee, and ankle joint reaction forces in an adult are normally in the range of 5 to 6, 3 to 8, and 3 to 5 times body weight, respectively (Nigg 1985). The more dynamic the activity, the greater the muscle forces and therefore the greater the joint reaction forces: For example, in fast running (800 m pace), the peak knee and ankle joint reaction forces in an adult are likely to be about 20 and 8 times body weight, respectively (Nigg 1985).

In any body position other than the relaxed recumbent position, the musculoskeletal system is likely to be subjected to considerable loading. In response to the forces exerted on the musculoskeletal components, they experience strain. Under normal circumstances, the musculoskeletal components adapt their size, shape, and structure to the time-averaged forces exerted on them to more readily withstand the strain (Frost 1990, 2003) (see chapter 11). However, when the degree of strain experienced by a particular component exceeds its strength, it becomes injured. Consequently, there is an intimate relationship between the structure and function of the musculoskeletal system.

> **KEY POINT**
>
> The open-chain arrangement of the skeleton enables a person to perform a wide range of postures and movements. However, the body pays a price for this ability: The muscles, bones, and joints are subjected to very high forces in virtually all postures other than lying down.

Summary

The billions of cells that make up the body originate from a single fertilized ovum through a process involving cell division and cell differentiation. The cells are organized into systems for carrying out the essential life processes. The musculoskeletal system consists of the skeletal system and the muscular system. The skeletal system consists of the skeleton and the various structures forming the joints between the bones. The muscular system consists of all the skeletal muscles.

The musculoskeletal system generates and transmits forces to counter the external forces acting on the body and enable it to move in a controlled manner. The open-chain arrangement of the skeleton allows a person to adopt a wide range of postures and movements, but only at the expense of large forces in muscles, bones, and joints. The musculoskeletal components normally continuously adapt their size, shape, and structure to the time-averaged forces exerted on them.

Review Questions

1. Differentiate between cellular differentiation and cellular organization in multicellular organisms.

2. Describe the four basic tissues.

3. Describe what is meant by a combined system.

4. With regard to the musculoskeletal system function, describe the relationship between external and internal forces.

5. Briefly describe the three broad categories of movement brought about by the musculoskeletal system.

6. With reference to recumbent posture and standing posture,

   • explain the difference between direct and indirect transmission of the weight of body segments to the support surface, and

   • describe how direct and indirect transmission of the weight of body segments to the support surface likely affects the degree of activity in the muscles.

7. Describe the main advantage and disadvantage of the open-chain arrangement of the skeleton.

The Skeleton

The skeleton gives the body its shape and provides a strong, protective, and supporting framework for all other systems of the body. With regard to movement, the bones of the skeleton act as levers operated by the skeletal muscles. The levers within the musculoskeletal system vary considerably in terms of mechanical advantage; this is reflected in the wide variety in size and shape of the bones. This chapter describes the bones of the skeleton and, in particular, the features of the bones associated with force transmission and relative motion between bones.

The shapes of some bones and the joints between the bones, especially in the skull, are difficult to describe, and you should read this chapter with models of the bones close by for reference.

2

OBJECTIVES

After reading this chapter, you should be able to do the following:

1. Describe the composition and functions of the skeleton.
2. List the main bones of the skeleton.
3. Identify and describe the main features of the axial skeleton.
4. Distinguish vertebrae from different regions of the vertebral column.
5. Identify and describe the main features of the appendicular skeleton.
6. Distinguish between left and right with regard to the large bones of the appendicular skeleton.

At birth there are approximately 270 regions of the immature skeleton where bone has started to form. During growth and development of the skeleton (see chapter 3), some of these bony regions fuse together so that the mature adult skeleton normally consists of 206 bones. This number can vary; for example, some adults have 11 or 13 pairs of ribs, whereas most adults have 12 pairs.

The skeleton performs three main mechanical functions:

1. It provides a supporting framework for the body.
2. It acts as a system of levers on which the muscles can pull to stabilize and move the body.
3. It protects certain organs. For example, the skull protects the brain, the vertebral column protects the spinal cord, and the rib cage protects the heart and lungs.

Terminology

For descriptive purposes, the bones are usually divided into two main groups: the **axial skeleton** and the **appendicular skeleton** (*axial* = axis, *appendicular* = appendage). The adult axial skeleton consists of 80 bones comprising the skull, vertebral column (backbone), and ribs. The adult appendicular skeleton consists of 126 bones that make up the upper limbs (arms and hands) and the lower limbs (legs and feet) (figure 2.1).

In anatomy, the term **aspect** refers to the appearance of a particular bone (or any other part of the body) from a particular viewpoint. For example, the anterior aspect of the skeleton refers to the features of the skeleton seen from an anterior (frontal) viewpoint (figure

2.1*a*). The lateral aspect (view from the side) (figure 2.1*b*), posterior aspect (view from the back), superior aspect (view from above), and inferior aspect (view from below) are other viewpoints.

The bones vary considerably in size and shape. There are four general categories of shape: long bones, short bones, flat bones, and irregular bones. Some bones fit into more than one category; for example, the small bones of the wrist are categorized as short and irregular. Whereas there are considerable differences in the size and shape of bones, a number of features such as articulating surfaces and points of attachment of tendons are common to many bones. These common features are frequently

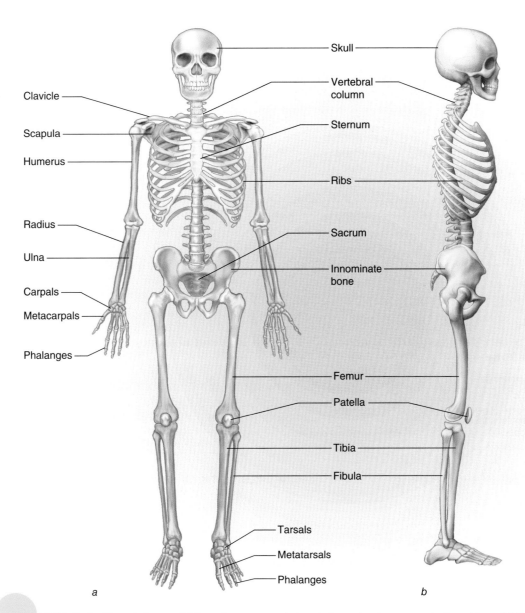

FIGURE 2.1 The skeleton. *(a)* Anterior aspect of the skeleton in the anatomical position. *(b)* Right lateral aspect of the trunk and the right lower limb.

referred to in the description of bones, and it is useful to list them.

Key Terms

axial skeleton   The bones of the skull, vertebral column, and rib cage.

appendicular skeleton   The bones of the upper and lower limbs.

aspect   The appearance of a particular bone (or any other part of the body) from a particular viewpoint.

Common Bone Features

The common features of bones (apart from notch) are illustrated in figure 2.2:

articular surface   Part of a bone that forms a joint with another bone. In a joint, also referred to as an articulation, the articular surfaces articulate with each other; that is, they are joined in one of two ways: (1) the surfaces are joined by fibrous tissue (fibrous joints) or cartilage (cartilaginous joints), which spans

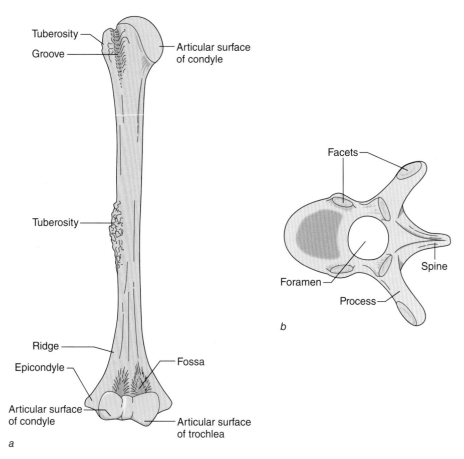

FIGURE 2.2 Common features of bones. *(a)* Anterior aspect of the right humerus. *(b)* Superior aspect of a thoracic vertebra.

the gap between the articular surfaces, and (2) the surfaces are in contact with each other and slide on each other (synovial joints) (see chapter 4, The Articular System).

facet A small fairly flat synovial articular surface. A convex facet on one bone usually articulates with a concave facet on an adjacent bone.

condyle A rounded projection of bone that provides the base for a rounded synovial articular surface. A convex condyle on one bone usually articulates with a concave condyle on an adjacent bone.

trochlea A pulley-shaped condyle.

fossa An oval or circular depression or cavity that can also be an articular surface.

notch An oval depression that is often an articular surface. A notch can also take the

form of a depressed region on the edge of a flat bone.

groove or sulcus An elongated depression (like a trench). One or more tendons usually occupy grooves.

ridge or line An elongated elevation. A ridge is usually the site of attachment of one or more aponeuroses.

crest A broad ridge.

process A projection of bone from the main body usually providing attachment for tendons or ligaments.

spine A smooth process that can be slender or flat.

epicondyle A small process adjacent to a condyle.

tubercle A small roughened process.

tuberosity A large roughened process.

trochanter Another name for a tuberosity used specifically in the description of the thigh bone (femur).

foramen A hole through a bone for the passage of blood vessels and nerves.

> **KEY POINT**
>
> There is considerable variation in size and shape among the bones of the skeleton. The four general categories of bones are long, short, flat, and irregular. The four features common to many bones are
>
> - articular surfaces,
> - areas of attachment of tendons and ligaments,
> - bony grooves that guide tendons, and
> - holes for the passage of blood vessels and nerves.

Those parts of the bone surface that do not have specific features such as articular surfaces are normally fairly smooth. These smooth areas, usually relatively large, are regions where muscles attach directly to the bone. When a muscle is attached to a bone by a tendon, the area of attachment of the tendon to the bone is usually rough and is likely to be referred to as tubercle, tuberosity, or trochanter.

Reference Planes and Spatial Terminology

To describe the spatial orientation of the particular features of a bone, or to describe the position of one bone (or body part) in relation to another, it is necessary to use standard terminology with reference to a standard body posture. In the standard posture, also called the **anatomical position** (see figure 2.1*a)*, the body is upright with the arms by the sides and palms of the hands facing forward. There are three main reference planes: median, coronal, and transverse (figure 2.3).

The **median plane** is a vertical plane that divides the body down the middle into more or less symmetrical left and right portions (figure 2.3). The median plane is also frequently referred to as the **sagittal plane;** the terms *sagittal, paramedian,* and *parasagittal (para* = beside or against) are also used to refer to any plane parallel to the median plane. In this book, the term *sagittal* is used to refer to any plane parallel to the median plane.

The terms *lateral* and *medial* are used to describe the relationship of parts of a bone (or body part) to the median plane. *Lateral* means farther away from the median plane and *medial* means closer to the median plane. For example, in figure 2.1*a* the lateral end of the clavicle (collarbone) articulates with the scapula (shoulder bone) and the medial end of the clavicle articulates with the sternum (breastbone). Similarly, in the anatomical position, the fingers of each hand are medial to the thumbs (and the thumbs are lateral to the fingers).

Any vertical plane perpendicular to the median plane is called a **coronal plane** (or frontal plane) (see figure 2.3). The terms

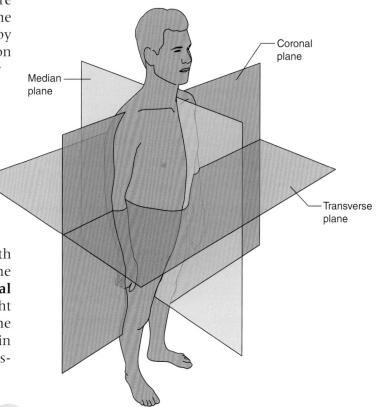

FIGURE 2.3 Reference planes.

anterior (in front) and *posterior* (behind) describe the position of structures with respect to coronal planes. For example, the face forms the anterior part of the skull, the sternum is anterior to the vertebral column (backbone), and the patella (kneecap) is anterior to the lower end of the femur. Similarly, the calcaneus (heel bone) is posterior to the toes, and the fibula (thin long bone in the lower leg) is posterior to the tibia (thicker of the two bones in the lower leg) (see figure 2.1*b*). The terms *ventral* and *dorsal* are synonymous with *anterior* and *posterior,* respectively.

Any plane perpendicular to both the median and coronal planes is called a **transverse plane.** All transverse planes are horizontal (see figure 2.3). The terms *superior* (above or upward) and *inferior* (below or downward) describe the position of structures with respect to transverse planes. For example, as seen in figure 2.1, the ribs are superior to the innominate bones (hip bones) and the patellae (kneecaps) are inferior to the innominate bones. Similarly, the superior end of the right femur (thigh bone) articulates with the right innominate bone to form the right hip joint, and the inferior end of the right femur articulates with the right patella and right tibia to form the right knee joint.

To describe the precise location and orientation of specific features of a particular bone, it is usually necessary to use combinations of the six spatial terms that are applicable to all bones: lateral, medial, anterior, posterior, superior, and inferior. For example, a particular feature may be described as being at the superior lateral part of a bone; another feature may be described as being at the anterior inferior lateral part of the bone.

Some spatial terms apply to some bones but not to others. For example, the terms *proximal* and *distal* are normally only used in reference to the long bones of the limbs. Superior features of these bones (with respect to the anatomical position) are referred to as proximal, whereas inferior features of these bones are referred to as distal. For example, in each arm the proximal end of the humerus (upper arm bone) articulates with its corresponding scapula to form the shoulder joint. The distal end of the humerus articulates with the proximal ends of the radius and ulna (forearm bones) to form the elbow joint. The distal ends of the

Key Terms

anatomical position   Reference body posture for descriptive purposes; the body is upright with the arms by the sides and the palms of the hands facing forward.

median plane   A vertical plane that divides the body down the middle into more or less symmetrical left and right portions.

sagittal plane   Any plane parallel to the median plane.

coronal plane   Any vertical plane perpendicular to the median plane.

transverse plane   Any plane perpendicular to both the median and coronal planes.

Spatial Terminology

Following are terms used to describe the location of a particular feature of a bone or part of the body with respect to the anatomical position:

lateral	On the outside, close to the outside, or toward the outside.
medial	On the inside, close to the inside, or toward the inside.
anterior	At the front, close to the front, or toward the front.
posterior	At the back, close to the back, or toward the back.
superior	At the upper end, close to the upper end, or toward the upper end.
inferior	At the lower end, close to the lower end, or toward the lower end.

radius and ulna articulate with the carpals (wrist bones) to form the wrist joint.

The names of the three reference planes are often used to describe sectional views of bones. For example, figure 2.4 shows a coronal section through the right hip joint. The term *longitudinal section* normally refers to a vertical section, as in figure 2.4: A longitudinal section can be in the median plane, a sagittal plane, a coronal plane, or some other vertical plane. *Cross section* is a general term that can refer to a section in one of the reference planes or to an oblique plane (relative to the reference planes). With respect to a long bone, a cross section usually refers to a section that is perpendicular to the long axis of the bone.

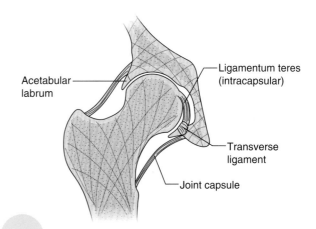

FIGURE 2.4 Coronal section through the right hip joint.

Axial Skeleton

The axial skeleton consists of the skull, the vertebral column, and the rib cage. The skull consists of 29 fairly flat or irregular bones that encase the brain, provide bases for the major sense organs, and form the upper and lower jaws. The vertebral column consists of 26 irregular bones stacked on top of each other. The vertebral column supports the weight of the head, arms, and trunk and provides protection for the spinal cord. The rib cage consists of 25 bones—the sternum (breastbone) and 12 pairs of ribs. The sternum is a rather flat bone. Even though the shafts of the ribs are considerably curved, they are fairly flat in cross section, like a curved strip of paper (see figure 2.20). The rib cage is a flexible structure that protects the heart and lungs and is also very important in the ventilation of the lungs during breathing.

Skull

The 29 bones of the skull consist of 8 cranial bones (cranium), 13 facial bones (face), 6 ear ossicles (3 in each middle ear), the mandible (lower jaw), and the hyoid bone (part of the larynx). The bones of the cranium and face form a single unit that makes up most of the skull (figure 2.5). The cranium encloses the brain and is made up of 8 relatively flat, irregular bones.

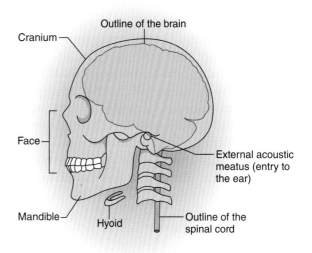

FIGURE 2.5 The main components of the skull in relation to the brain and spinal cord.

Sutures and Fontanels

In the adult skull, the edges of many of the bones are serrated so that the bones interlock closely with each other to form immovable joints. The serrated line of the joints is similar in appearance to a line of stitches, and for this reason each joint is called a suture (see figure 2.6, *b* and *d*). The joints between the bones of the cranium in an infant are also usually called sutures, even though the bones are joined

27

by sheets of fibrous tissue and consequently do not interlock with each other (see figure 2.6, *a* and *c*). Fibrous tissue, described in more detail in chapter 3, is flexible, strong, and, in an infant, moderately elastic.

The flexible linkages allow the bones of the infant's cranium to override each other during birth. This reduces the size of the cranium, which facilitates an easier passage of the baby's head through the birth canal. After birth the normal orientation of the bones is quickly restored. At each of the angles (or corners) of the parietal bones, the fibrous tissue is in the form of a small sheet, called a fontanel (Standring 2004). Because each parietal bone has four angles and the parietal bones join each other superiorly in the median plane, there are six fontanels. The anterior fontanel, which normally closes within the first 18 months, is at the junction of the parietal and frontal bones. At birth the frontal bone is in two halves, joined by the interfrontal or metopic suture; these halves normally fuse together within the first 2 years. The posterior fontanel is at the junction of the parietal and occipital bones and normally closes within the first 2 months. There are two sphenoid fontanels, one on each side of the skull at the junction of the parietal, frontal, sphenoid, and temporal bones. The sphenoid fontanels normally close within the first 3 months. There are also two mastoid fontanels, one on each side of the skull at the junction of the parietal, occipital, and temporal bones. The mastoid fontanels normally close within the first 2 years.

KEY POINT

There are six fontanels. The posterior fontanel normally closes within the first 2 months. The two sphenoid fontanels normally close within the first 3 months. The anterior fontanel normally closes within the first 18 months, and the two mastoid fontanels normally close within the first 2 years.

Mature Skull

The frontal bone forms the anterior and anterior superior part of the cranium including the forehead (figure 2.6, *b* and *d*). The two parietal bones form a large part of the superior and lateral aspects of the cranium. The two temporal bones form a large part of the superior and lateral aspects of the base and sides of the cranium. The sphenoid bone, together with the ethmoid bone and the inferior aspects of the frontal bone, forms the anterior half of the base of the cranium. The occipital bone forms the posterior inferior aspect of the cranium and the major portion of the posterior half of the base of the cranium.

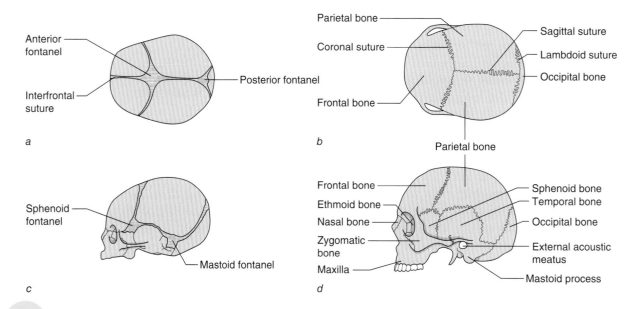

FIGURE 2.6   Superior and left lateral aspects of the cranium and face. *(a, c)* Infant. *(b, d)* Adult.

The temporal and occipital bones have a number of important features. On the inferior lateral aspect of each temporal bone is a funnel-shaped opening that leads on to an open channel called the external acoustic meatus, the external part of the ear (see figure 2.6*d* and figure 2.7). The external acoustic meatus leads to a chamber within the temporal bone called the middle ear. Located within the middle ear are three tiny bones called ossicles, all less than 1 cm in length. The ossicles—the malleus, incus, and stapes—link the lateral and medial walls of the middle-ear chamber. They transmit sound waves from the outer part of the ear to the sound receptors in the inner part of the ear, which is medial to the middle ear.

Just behind the external acoustic meatus is a rounded process projecting downward called the mastoid process. The mastoid process can easily be felt as a bony bump beneath the skin just behind the ear (see figure 2.6*d* and figure 2.7). On the inferior aspect of each temporal bone is a relatively long, slender process arising from the inferior aspect of the external acoustic meatus. This process is called the styloid process and projects medially, forward, and downward (see figure 2.6*d* and figure 2.7). Just in front of the external acoustic meatus there is a concave condyle called the mandibular fossa (figure 2.7) that articulates with the corresponding condyle of the mandible to form the temporomandibular joint. The styloid process provides part of the area of attachment for muscles that control the temporomandibular joint.

The occipital bone has a large hole, the foramen magnum, situated anteriorly (figure 2.7). The foramen is occupied by the start of the spinal cord, which is continuous with the brain (see figure 2.5). Two occipital condyles articulate with the first vertebra (the atlas) and are situated on the anterior lateral aspects of the foramen magnum (figure 2.7). A curved border between the inferior and posterior aspects of the occipital bone is the superior nuchal line (figure 2.7). Anterior and concentric to the superior nuchal line is the inferior nuchal line (figure 2.7). Running between the nuchal lines in the median plane is the external occipital crest. At the posterior end of the external occipital crest is a tuberosity called the external occipital protuberance. The muscles that hold the head upright (and tilt it backward) are attached to the occipital bone between the superior and inferior nuchal lines.

The 13 bones of the face form the middle third of the anterior aspect of the skull (figures 2.5 and 2.8*a)*. The facial bones form the upper jaw, the inferior two thirds of the eye sockets (orbits), and the anterior part of the nasal cavity. The upper jaw, which provides sockets for the upper teeth, is formed almost entirely by the two maxilla bones that join anteriorly in the median plane. The two palatine bones articulate with the posterior aspects of the maxillae to complete the upper jaw (figure 2.7). The eye sockets are formed by parts of the frontal, sphenoid, and ethmoid bones, together with the two lacrimal bones and the two zygomatic (cheek) bones (figure 2.8*a*).

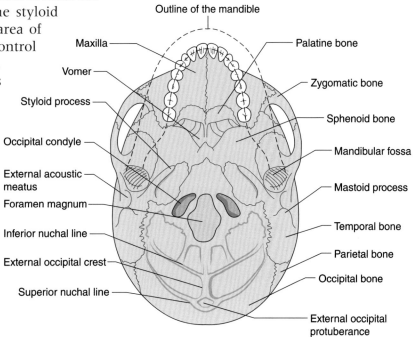

FIGURE 2.7 Inferior aspect of the adult cranium and face.

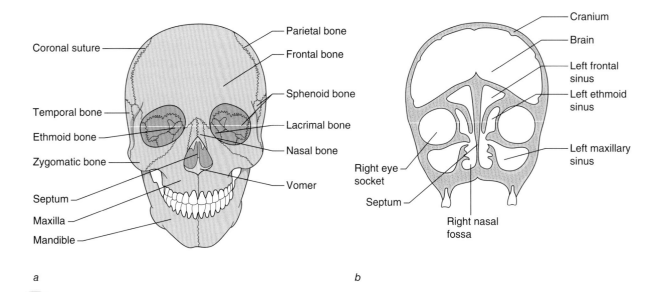

Coronal suture

Parietal bone

Frontal bone

Sphenoid bone

Temporal bone

Ethmoid bone

Lacrimal bone

Zygomatic bone

Nasal bone

Septum

Vomer

Maxilla

Mandible

Cranium

Brain

Left frontal sinus

Left ethmoid sinus

Left maxillary sinus

Right eye socket

Septum

Right nasal fossa

a

b

FIGURE 2.8 *(a)* Anterior aspect of the adult skull. *(b)* Anterior aspect of a coronal section through the eye sockets of the adult skull.

The nasal cavity—the large cavity that stretches backward from the nose to the throat—is formed by bones of the cranium and face (figure 2.8). The anterior part of the nasal cavity, the bony part of the nose, is formed by the maxillae (floor and sides) and the two nasal bones (roof). The posterior part of the nasal cavity is much larger than the anterior part. The ethmoid bone forms the sides and part of the roof, whereas the frontal and sphenoid bones form the remainder of the roof. The sphenoid bone also forms the posterior wall. The vomer bone and the two nasal bones form the floor of the posterior part of the nasal cavity.

The nasal cavity is divided into left and right nasal fossae by a bony-cartilaginous septum in the median plane (figure 2.8b). The posterior part of the septum consists of a bony plate that projects downward from the ethmoid and articulates with the vomer. The anterior part of the septum consists of cartilage. The structure and functions of cartilage are described in detail in chapter 3.

The mandible, or lower jaw, consists of two L-shaped plates of bone that join anteriorly in the median plane (figure 2.8a). The upright part of each half of the mandible is called the ramus (branch), and the horizontal part, which provides sockets for the lower teeth, is called the body (figure 2.9). At the posterior superior

aspect of each ramus is a convex condyle that articulates with the mandibular fossa on its corresponding temporal bone (see figure 2.7). These joints enable the mandible to swing up and down, as in closing and opening the mouth, and to move from side to side. Chewing food involves a combination of these two types of movement.

The hyoid bone is not really part of the skull, but it is convenient to describe this bone in relation to the skull. The hyoid is a U-shaped bone suspended in front of the neck (in front of the fourth cervical vertebra) by ligaments from the styloid processes of the temporal bones (see figure 2.5). The hyoid forms part of the larynx (voice box) and provides attachment for some of the muscles that move the mouth and the tongue.

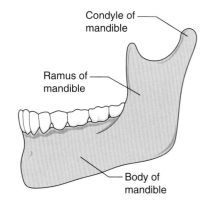

Condyle of mandible

Ramus of mandible

Body of mandible

FIGURE 2.9 Left lateral aspect of the mandible.

Sinuses

The frontal, ethmoid, sphenoid, maxillae, and temporal bones are partially hollow, resulting in cavities called sinuses. The sinuses communicate directly or indirectly with the nasal cavity by means of small channels (see figure 2.8*b*). Like the nasal cavity, the sinuses are filled with air. The frontal, ethmoid, sphenoid, and maxillary sinuses are linked directly to the nasal cavity and are referred to as paranasal sinuses. The sinuses in the temporal bones are linked indirectly to the nasal cavity via the middle ear—the temporal sinuses are linked to the middle ear and the middle ear is linked to the nasal cavity. During respiratory infections the sinuses can become inflamed, resulting in a painful condition called sinusitis. The sinuses make the skull marginally lighter and add resonance to the voice. However, they vary considerably in size among individuals, and their precise function is not known (Standring 2004).

> **KEY POINT**
>
> The skull consists of 29 fairly flat or irregular bones that encase the brain, provide bases for the major sense organs, and form the upper and lower jaws. The main parts of the skull are the cranium (8 bones), face (13 bones), mandible, ear ossicles (3 in each ear), and hyoid.

Vertebral Column

Prior to maturity, the vertebral column—also referred to as the backbone or spine—consists of 33 or 34 irregular bones called vertebrae. The vertebrae are divided into five fairly distinct groups: cervical, thoracic, lumbar, sacral, and coccygeal (figure 2.10, *a* and *b*). The neck consists of 7 cervical vertebrae. The thoracic or chest region consists of 12 thoracic vertebrae that provide articulation for the 12 pairs of ribs. The lower back consists of 5 lumbar vertebrae. The 5 sacral vertebrae form the posterior part of the pelvis; at maturity, the sacral vertebrae fuse together to form the sacrum. The 4 or 5 coccygeal vertebrae are small and represent a vestigial tail. The coccygeal vertebrae normally fuse together at maturity to form the coccyx, or tailbone, which is approximately 3 cm long and is attached to the sacrum by a fibrocartilaginous joint.

> **KEY POINT**
>
> Prior to maturity, the vertebral column consists of 33 or 34 irregular bones called vertebrae, which fuse into 26 bones in the adult vertebral column. The function of the vertebral column is to provide a flexible supporting framework for the head, arms, and trunk.

When viewed from the side, the whole of the vertebral column of a newborn child is concave anteriorly (figure 2.10*c*). Between 3 and 6 months of age the child learns to hold her head upright, and as a result the shape of the cervical region changes from concave anteriorly to convex anteriorly. Similarly, as the child learns to stand and walk—between 10 and 18 months of age—the shape of the lumbar region changes from concave anteriorly to convex anteriorly. The cervical and lumbar curves are referred to as secondary curves because they develop as the child adopts an upright posture. The thoracic and sacrococcygeal curves are called primary curves because they are concave anteriorly throughout life. Figure 2.10*a* shows the shape of the adult vertebral column as viewed from the left lateral aspect.

Structure of a Vertebra

At birth, each vertebra, with the exception of the first two cervical vertebrae, consists of three bony elements united by cartilage (Standring 2004) (figure 2.11*a*). The anterior element, the centrum or body, is basically a block of bone with slightly concave (waisted) sides and fairly flat, kidney-shaped superior and inferior surfaces. The bodies of the vertebrae are mainly responsible for transmitting loads, especially the weight of the head, arms, and trunk. The posterior elements are curved struts that form the two halves of an arch called the vertebral or neural arch. Each half of the arch consists of an anterior portion called the pedicle and a posterior portion called the lamina. The two laminae normally fuse together posteriorly during the first year (figure 2.11*b*). The pedicles normally

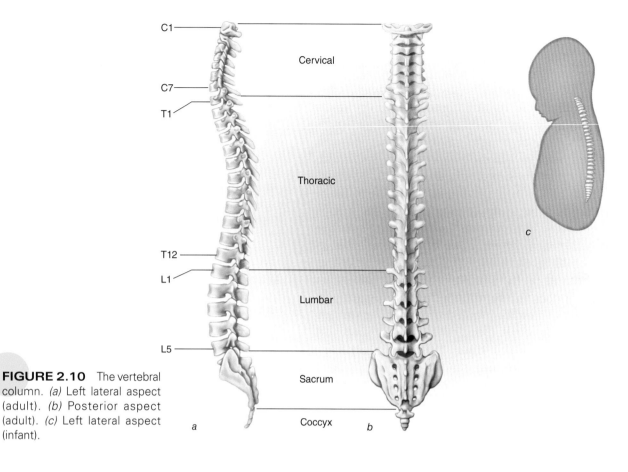

FIGURE 2.10 The vertebral column. *(a)* Left lateral aspect (adult). *(b)* Posterior aspect (adult). *(c)* Left lateral aspect (infant).

C1
Cervical
C7
T1
Thoracic
T12
L1
Lumbar
L5
Sacrum
Coccyx

a
b
c

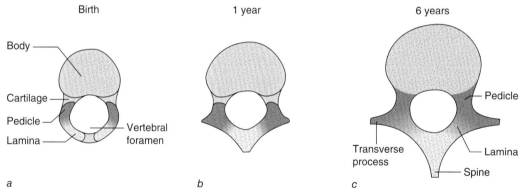

Birth   1 year   6 years

Body
Cartilage
Pedicle
Lamina
Vertebral foramen
Pedicle
Transverse process
Lamina
Spine

a   b   c

FIGURE 2.11 Early stages in the development of a typical vertebra (superior aspect). The vertebral arch consists of the arch of bone, posterior to the body of the vertebra, that consists of both pedicles and both laminae.

fuse with the lateral superior posterior aspects of the body of the vertebra between the third and sixth years (figure 2.11*c*). The hole formed by the arch and the posterior aspect of the body, through which the spinal cord travels, is called the vertebral foramen. The vertebral arch protects the spinal cord.

After fusion of the laminae, seven processes arise from the arch. The spine of the vertebra extends backward from the point of fusion of the laminae. On each side of the arch, three processes arise from the junction of the pedicle and lamina. A transverse process extends laterally, a superior articular process extends upward, and an inferior articular process extends downward (figure 2.12). The spine and transverse processes act as levers that provide areas of attachment for muscles, tendons, and ligaments. With respect to each pair of adjacent vertebrae, the superior articular processes of the lower vertebra articulate by means of facets with the inferior articular processes of the upper vertebra (figure 2.12*a*). These joints are called facet joints or apophyseal joints. In most upright postures the facet joints transmit some load. The load transmitted by these joints decreases with flexion of the trunk (bending forward) and increases with extension of the trunk (bending backward). In addition to load transmission, the orientation of the superior and inferior articular processes somewhat determines the type and range of movement between adjacent vertebrae. The structure and function of facet joints are covered in chapter 5.

Each pair of adjacent vertebrae, except the first two cervical vertebrae, are joined by a tough rubbery disc of fibrocartilage called an intervertebral disc (figure 2.13*a*) to form an intervertebral joint. Movement between adjacent vertebrae occurs because of deformation of the intervertebral discs and sliding in the facet joints (see chapter 5).

On each side of the vertebral arch is a depression in the superior aspect of the pedicle called the superior vertebral notch (see figure 2.12*b*). As the pedicle joins the posterior superior aspect of the body, there is a much larger inferior vertebral notch beneath the pedicle. With respect to each pair of adjacent vertebrae, the inferior vertebral notch of the upper vertebra and the superior vertebral notch of the lower vertebra form a hole called the intervertebral foramen (figure 2.13*a*). A spinal (peripheral) nerve occupies the intervertebral foramen (figure 2.13*b*).

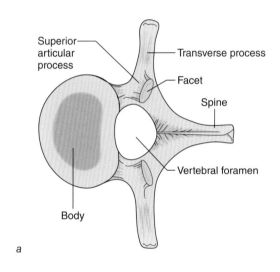

a

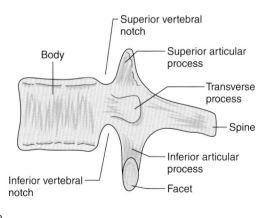

b

FIGURE 2.12 A typical vertebra. *(a)* Superior aspect. *(b)* Left lateral aspect.

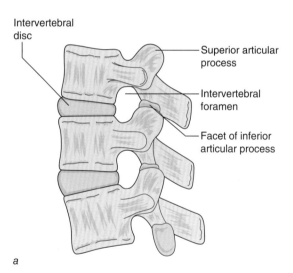

a

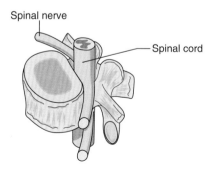

b

FIGURE 2.13 *(a)* Three articulated lumbar vertebrae. *(b)* Relationship of the spinal cord and spinal nerves to a lumbar vertebra.

Distinguishing Features of Vertebrae

The vertebrae gradually increase in size from the second cervical vertebra down to the sacrum. This gradual increase reflects the gradual increase in weight that the vertebrae have to support (see figure 2.10a). The processes of the vertebrae also change in size, shape, and orientation. These changes are fairly gradual within each of the cervical, thoracic, and lumbar regions but tend to be more marked at the junctions between the regions (see figure 2.10, *a* and *b*). The vertebrae in each region of the column have characteristics that distinguish them from vertebrae in other regions.

Cervical   All cervical vertebrae (C1 to C7) have a hole called a transverse foramen in each of their transverse processes. Only cervical vertebrae have this characteristic (figure 2.14). The first cervical vertebra is called the atlas. It has no body and consists of an anterior arch and a posterior arch, which together form a bony ring (figure 2.15a). The facets on the superior articular processes of the atlas articulate with the occipital condyles to form

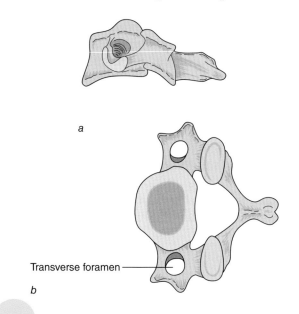

a

Transverse foramen

b

FIGURE 2.14   A typical cervical vertebra. *(a)* Left lateral aspect. *(b)* Superior aspect.

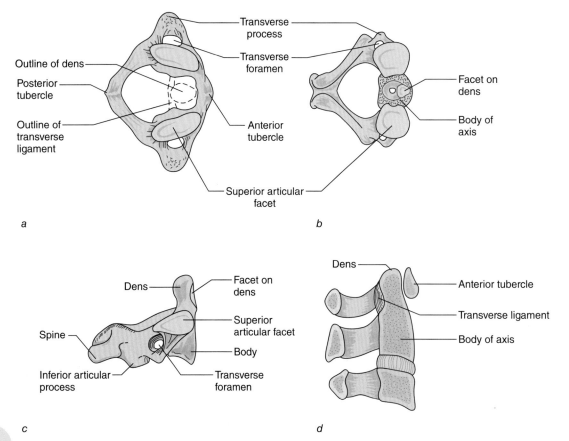

Transverse process
Transverse foramen
Outline of dens
Posterior tubercle
Outline of transverse ligament
Anterior tubercle
Superior articular facet

a

Facet on dens
Body of axis

b

Dens
Facet on dens
Spine
Superior articular facet
Body
Inferior articular process
Transverse foramen

c

Dens
Anterior tubercle
Transverse ligament
Body of axis

d

FIGURE 2.15   The atlas and the axis. *(a)* Superior aspect of the atlas. *(b)* Superior aspect of the axis. *(c)* Right lateral aspect of the axis. *(d)* Median section through the atlas, axis, and third cervical vertebra.

the atlantooccipital joint, linking the skull with the vertebral column. The atlas does not have a spine; the posterior tip of the posterior arch is marked by a posterior tubercle (figure 2.15*a*). The transverse processes of the atlas, as in all the other cervical vertebrae, are short.

The second cervical vertebra is called the axis. The axis has a body and a vertebral arch. Projecting upward from the superior aspect of the body of the axis is a process called the dens or odontoid process (figure 2.15, *b* and *c*). A facet on the anterior aspect of the dens articulates with a facet on the posterior aspect of the anterior arch of the atlas. The dens is held against the anterior arch of the atlas by the transverse ligament of the atlas, which spans the posterior part of the anterior arch of the atlas and runs in a groove on the posterior aspect of the dens (figure 2.15, *a, c,* and *d*). To a certain extent the atlas rotates (in a transverse plane) around the dens, hence the name of the axis.

The dens is the major portion of the body of the atlas that separates from the rest of the body of the atlas during fetal growth and fuses with the axis. There is no intervertebral disc between the atlas and axis. The spine of the axis is fairly short and the tip of the spine is bifid; that is, it has distinct right and left terminal projections (see figure 2.15*b*).

The remaining 5 cervical vertebrae (C3, C4, C5, C6, and C7) are similar in that each consists of a kidney-shaped body and a vertebral arch (see figure 2.14). The spines of C3 to C6 are all bifid and gradually increase in length. The spine of C3 is slightly shorter than that of the axis, and the spine of C6 is slightly longer than that of the axis. The spine of C7 is much longer than that of C6 and can easily be felt as a prominence at the posterior inferior aspect of the neck. Given its unusually long spine, C7 is sometimes referred to as the vertebra prominens.

The facets of the superior and inferior articular processes of the cervical vertebrae articulate in oblique planes that slope downward laterally and posteriorly. The orientation of the facet joints, the short transverse processes, the relatively thick intervertebral discs, and the relatively short spines of C3 to C6 all combine to give a fairly large range of movement in the

cervical region compared with other regions of the column.

Thoracic   Thoracic vertebrae (T1 to T12) can be identified by the presence of facets on the bodies for articulation with the heads (posterior ends) of the ribs. The upper 10 thoracic vertebrae also articulate with the corresponding pairs of ribs by means of facets on the anterior lateral aspects of the transverse processes (figure 2.16). On each side of the body of T1 there is a superior whole facet and an inferior demifacet (half facet); the heads of the uppermost pair of ribs articulate with the whole facets on the sides of T1 (figure 2.16*a*). On each side of the bodies of T2 to T8 there is

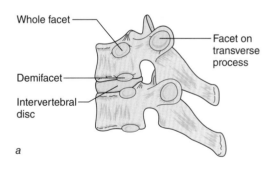

Whole facet

Facet on transverse process

Demifacet

Intervertebral disc

a

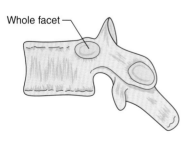

Whole facet

b

FIGURE 2.16   Thoracic vertebrae. *(a)* Left lateral aspect of the first two thoracic vertebrae. *(b)* Left lateral aspect of the tenth thoracic vertebra.

a superior demifacet and an inferior demifacet. The heads of the second pair of ribs articulate, on the corresponding side, with the inferior demifacet of T1 and the superior demifacet of T2 (figure 2.16a). On each side of the body of T9 there is a superior demifacet, and the heads of the third to ninth pairs of ribs articulate with the sides of the bodies of T2 to T9 in the same manner as the second pair of ribs. On each side of the bodies of T10 (figure 2.16b) to T12 there is a whole facet that articulates with the tenth to twelfth pairs of ribs.

The spines of the thoracic vertebrae are fairly long and tend to closely overlap each other, especially in the middle of the region (see figure 2.10a). The transverse processes of the thoracic vertebrae are also fairly long; they gradually decrease in length from T1 to T12. The superior and inferior articular facets articulate in a plane that slopes sharply downward posteriorly. The overlapping spines, relatively thin intervertebral discs, and splinting effect of the ribs result in a smaller overall range of movement in the thoracic region than in the cervical region.

Lumbar The lumbar vertebrae (L1 to L5) have fairly long transverse processes and large, flat spines, rectangular in shape (figure 2.17). The main distinguishing feature of the lumbar vertebrae is the orientation of the facets on the superior and inferior articular processes. The facets on the superior articular processes face medially and posteriorly, and the facets on the inferior articular processes face laterally and anteriorly (figure 2.17). The orientation of the

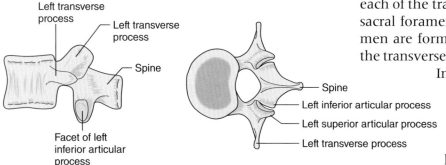

FIGURE 2.17 A typical lumbar vertebra. *(a)* Left lateral aspect. *(b)* Superior aspect.

facet joints severely limits rotation of the lumbar vertebrae about a vertical axis. However, the relatively thick intervertebral discs in the lumbar region ensure a much greater range of movement in other directions.

KEY POINT

Rotation about a vertical axis in the lumbar region is severely restricted because of the orientation of the facet joints. However, rotation in other directions is relatively unrestricted.

Sacrum The sacral vertebrae (S1 to S5) become progressively smaller from S1 through S5. The sacrum is formed by the fusion or partial fusion of the sacral vertebrae. When viewed from the front (or the back), the sacrum is more or less triangular with the apex pointing downward (figure 2.18a). The anterior edge of the upper surface of the first sacral vertebra projects forward and is called the sacral promontory. The anterior aspect of the sacrum is concave, largely because of the orientation of S3, S4, and S5 (figure 2.18b). However, in the anatomical position the large upper portion of the sacrum (S1 and S2) is angled downward and backward, which tends to accentuate the lumbar curve (see figure 2.10a).

The sacrum is sandwiched between the left and right hip bones and thus provides a firm base for the rest of the vertebral column. The anterior aspect of the sacrum is fairly smooth except for four transverse lines resulting from the fusion or partial fusion of the bodies of the sacral vertebrae (figure 2.18a). At the end of each of the transverse lines is a hole called the sacral foramen. The four pairs of sacral foramen are formed by the fusion of the ends of the transverse processes of the sacral vertebrae. In comparison to the fairly smooth anterior surface, the posterior surface of the sacrum is quite rough and provides attachment for a large number of ligaments and aponeuroses. The posterior surface has five ridges (called sacral crests), which run vertically parallel to one another. The fusion of the upper four sacral spines forms the median sacral crest. The inter-

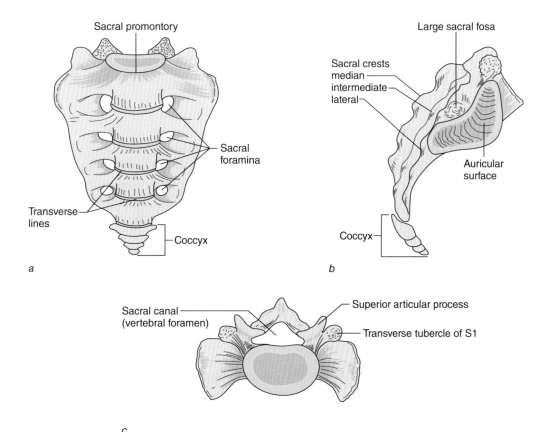

FIGURE 2.18 The sacrum and coccyx. *(a)* Anterior aspect of the sacrum and coccyx. *(b)* Right lateral aspect of the sacrum and coccyx. *(c)* Superior aspect of the sacrum.

mediate sacral crests and lateral crests (two each) are formed by the fusion of the articular processes (except for the superior processes of S1) and the ends of the transverse processes of the sacral vertebrae, respectively (figure 2.18*b*). Processes called sacral tubercles or transverse tubercles, which give the crests an undulating appearance, mark the sites of fusion. In addition to these tubercles there are usually two other tubercles, one on each side, that project posterior laterally from the body of S1 just lateral to the superior articular processes of S1; these tubercles are referred to as the transverse tubercles of S1 (figure 2.18*c*). On each superior lateral aspect of the sacrum is a fairly large C- or L-shaped articular surface called the auricular surface (*auricle* = ear-shaped) (figure 2.18*b*). The auricular surfaces are formed by the lateral expansions of the fused transverse processes of S1, S2, and S3. The auricular surfaces of the sacrum articulate with the hip bones (innominate bones) to form the sacroiliac joints (see chapter 5).

A fairly large oval or circular sacral fossa is adjacent to the posterior border of the angle of each auricular surface. There is usually a smaller sacral fossa at the distal end of each auricular surface. The sacrum and the left and right hip bones form a complete bony ring called the pelvis or pelvic girdle. Consequently, the sacrum is an important part of the vertebral column and the pelvis.

Coccyx The 4 or 5 coccygeal vertebrae normally fuse together at maturity to form the coccyx. The shape and features of the coccyx are similar to those of the sacrum but on a much smaller scale (figure 2.18, *a* and *b*). The coccyx is attached to the sacrum by a fibrocartilaginous joint that allows a small amount of movement. The anterior aspect of the coccyx provides attachment for muscles that support organs in the lower part of the pelvis. The posterior aspect of the coccyx provides attachment for muscles that assist hip extension. Together with the two ischium bones (see figures 2.30 and 2.31), the coccyx forms a weight-bearing tripod when a person is sitting in an upright position.

Rib Cage

The rib cage is roughly the shape of an upright cone partially flattened from front to back (figure 2.19). The rib cage consists of 12 pairs of ribs and the sternum. Because the ribs form the major part of the rib cage (most of the wall of the upright cone), they have a distinct, curved shape (figure 2.20). The heads (posterior ends) of the ribs articulate with the thoracic vertebrae

as previously described. The anterior ends of the upper 10 pairs of ribs are attached to the sternum by pieces of cartilage called costal cartilages (*costa* = rib). The upper 7 pairs of ribs are attached to the sternum by separate costal cartilages and are therefore sometimes referred to as true ribs. The costal cartilages of the eighth, ninth, and tenth pairs of ribs fuse with each other before fusing with the costal cartilages of the seventh ribs (see figure 2.19). Consequently, whereas the upper 7 pairs of ribs have direct cartilaginous attachments to the sternum, the eighth, ninth, and tenth pairs of ribs have an indirect cartilaginous attachment to the sternum. The lower 2 pairs of ribs do not attach onto the sternum at all; the anterior ends of these ribs are free, and these ribs are referred to as floating ribs. Because none of the lower 5 pairs of ribs has a direct cartilaginous or other attachment to the sternum, these ribs are sometimes referred to as false ribs.

Ribs

The upper 10 pairs of ribs are attached to the body and transverse processes of corresponding thoracic vertebrae. In each of these ribs the head is separated from the facet that articulates with the transverse process by a short neck (see figure 2.20). Laterally adjacent to the facet is a tubercle, which provides attachment

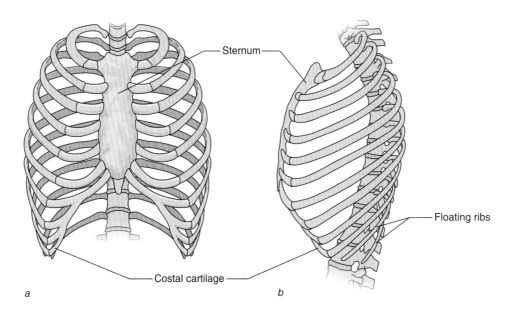

a *b*

FIGURE 2.19 The rib cage. *(a)* Anterior aspect. *(b)* Left lateral aspect.

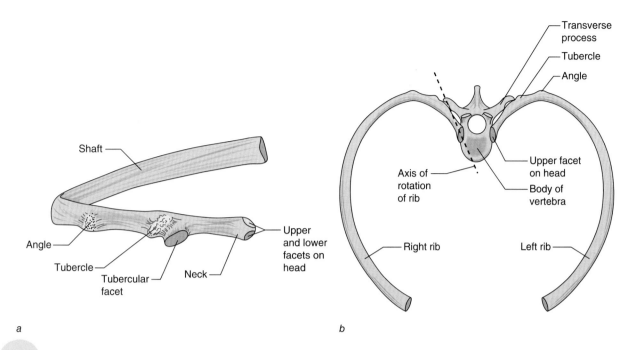

FIGURE 2.20 *(a)* Posterior aspect of a typical left rib. *(b)* Superior aspect of a typical pair of ribs articulating with a thoracic vertebra.

for ligaments that support the vertebral column. Because the facet and tubercle are adjacent to each other, the facet is often called the tubercular facet. Lateral to the tubercle, at about the same distance from the tubercle as the head, the rib bends abruptly forward; this bend in the rib is called the angle of the rib, and it provides attachment for muscles of the back. The part of a rib between the angle and the anterior end is called the shaft.

Sternum

The sternum is a fairly flat bone and consists of three parts: manubrium, body, and xiphoid process (figure 2.21). The manubrium occupies the upper quarter of the sternum. In the center of its superior border is a depression called the jugular notch or suprasternal notch (*supra* = above). On each side of this notch is a facet for articulation with the medial end of the corresponding clavicle (collarbone). On each of the lateral aspects of the manubrium is a facet for articulation with the costal cartilage of the corresponding first rib. At each end of the transverse junction between the

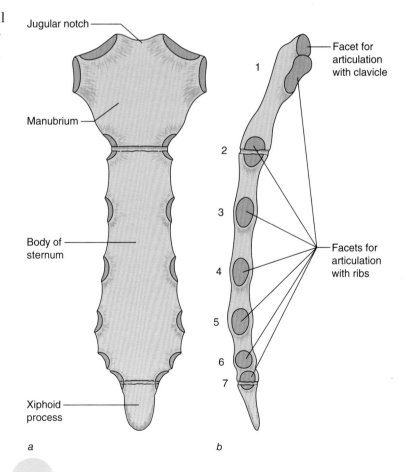

FIGURE 2.21 The sternum. *(a)* Anterior aspect. *(b)* Left lateral aspect.

manubrium and the body of the sternum is a facet for articulation with the costal cartilage of the corresponding second rib. The body of the sternum, which occupies more than half the length of the sternum, also has facets for articulation with the costal cartilages of the third to seventh pairs of ribs. The costal cartilages of the seventh pair of ribs are attached at each end of the transverse junction between the body of the sternum and the xiphoid process. The xiphoid process provides attachment for some of the abdominal muscles; it consists of cartilage that usually becomes transformed into bone by the age of 40 (Tortora and Anagnostakos 1984).

Movement of the Ribs

The spaces between the ribs are called intercostal spaces; muscles largely occupy these spaces. These muscles in association with other muscles of the thorax move the ribs during breathing. During inspiration (breathing in), each rib swings upward and outward about an oblique axis that passes through the costoverte-

bral joints of the rib—the joints between tubercular facet and transverse process and between head and vertebral body or bodies (see figure 2.20b). The upward and outward movement of the ribs is accompanied by a slight upward and forward swing of the body of the sternum about the manubrium. The combined movements of the ribs and sternum decrease the pressure inside the thorax, and air rushes into the lungs. During expiration (breathing out), the ribs and sternum are pulled back down by the elasticity of both the costal cartilages and the lungs themselves and by the pull of various ligaments and muscles that are stretched during inspiration.

> **KEY POINT**
>
> The rib cage consists of the sternum and 12 pairs of ribs. The rib cage is a fairly flexible structure providing protection for the heart and lungs and is also very important in the ventilation of the lungs during breathing.

Appendicular Skeleton

The appendicular skeleton consists of the bones of the upper and lower limbs. In an adult, 32 bones make up each upper limb and 31 bones make up each lower limb; the appendicular skeleton as a whole consists of 126 bones.

Upper Limb

For descriptive purposes, I've divided each upper limb into five regions: shoulder (scapula and clavicle), upper arm (humerus), lower arm (radius and ulna), wrist (eight carpals), and hand (five metacarpals and 14 phalanges) (figure 2.22).

> **KEY POINT**
>
> The upper limbs consist of 64 bones, 32 bones in each limb, and can be divided into five regions: shoulder, upper arm, lower arm, wrist, and hand.

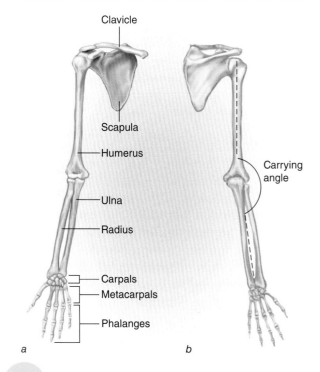

FIGURE 2.22 The right upper limb. *(a)* Anterior aspect. *(b)* Posterior aspect.

Shoulder

The shoulder region consists of the scapula (shoulder blade) and clavicle (collarbone). Together with the manubrium, the scapulae and clavicles of both upper limbs form an incomplete ring of bone called the shoulder girdle (figure 2.23). The arms are suspended from the shoulder girdle.

The medial end of each clavicle articulates with the manubrium to form a sternoclavicular joint, and the lateral end of each clavicle articulates with the acromion process of the corresponding scapula to form an acromioclavicular joint (figures 2.23 and 2.24). The scapulae are not joined to the axial skeleton but are held in position at the lateral superior posterior aspects of the rib cage by muscles. Consequently, each scapula has a considerable range of movement. Most movements of the shoulder region involve movements at the sternoclavicular and acromioclavicular joints (see chapter 6).

From a superior aspect, each clavicle is S shaped—concave anterior laterally and posterior medially (figures 2.23 and 2.24a). In a transverse plane the lateral one third of each clavicle is fairly flat; however, the medial two thirds become progressively thicker and more rounded toward the medial end (figure 2.24b).

In addition to forming the only bony articulations of the upper limb with the axial skeleton, the clavicles act as horizontal struts that maintain the lateral position of the scapulae and give width to the shoulders.

Each scapula consists of a relatively large, flat, triangular portion called the blade, with three prominent features arising from the blade (figure 2.24, c, d, and e). The apex of the blade is called the inferior angle and points directly downward. The large anterior surface of the blade is called the subscapular fossa. A large process called the spine of the scapula arises from an oblique line running laterally and upward across the upper third of the posterior surface of the blade. Like the blade, the spine is fairly flat and triangular in shape; the posterior border of the spine forms a crest that can easily be felt beneath the skin. The superior surface of the spine and the posterior surface of the blade above the spine form a V-shaped trough called the supraspinous fossa. The inferior surface of the spine and the large posterior surface of the blade below the spine form a large area called the infraspinous fossa.

Projecting laterally and slightly upward from the lateral end of the spine is the acromion process, which forms the posterior part of a bony-fibrous arch above the shoulder joint.

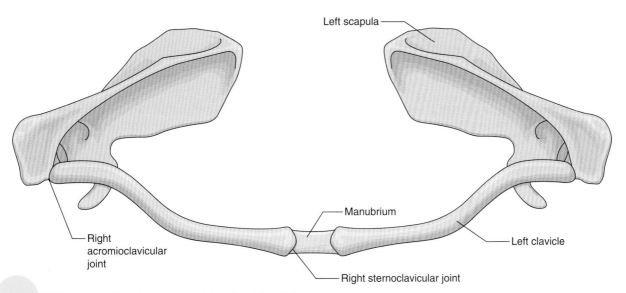

FIGURE 2.23 Superior aspect of the shoulder girdle.

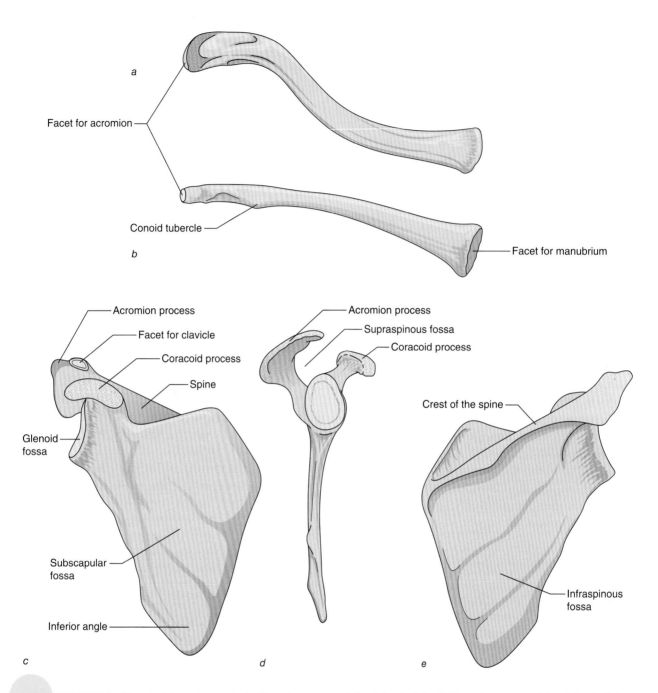

Facet for acromion

a

Conoid tubercle

b

Facet for manubrium

Acromion process

Facet for clavicle

Coracoid process

Spine

Glenoid fossa

Subscapular fossa

Inferior angle

c

Acromion process

Supraspinous fossa

Coracoid process

Crest of the spine

Infraspinous fossa

d

e

FIGURE 2.24 The clavicle and scapula. *(a)* Superior aspect of the right clavicle. *(b)* Anterior aspect of the right clavicle. *(c)* Anterior aspect of the right scapula. *(d)* Lateral aspect of the right scapula. *(e)* Posterior aspect of the right scapula.

The acromion process can also be felt beneath the skin at the tip of the shoulder. On the medial aspect of the acromion process there is a facet for articulation with the lateral end of the clavicle (acromioclavicular joint). At the superior lateral angle of the blade there is a fairly large, oval-shaped, shallow articular surface called the glenoid fossa. In the anatomical position the glenoid fossa faces laterally, slightly forward, and slightly upward. The glenoid fossa forms the shoulder joint (glenohumeral joint) with the head of the humerus.

Arising from the superior anterior part of the base of the glenoid fossa is a fingerlike projection called the coracoid process; this process curves laterally so that its tip is in front of the shoulder joint. The coracoid process, acromion process, crest of the spine, and medial border of the blade all provide areas of attachment for muscles that are mainly concerned with the movement of the scapula and clavicle about the sternoclavicular and acromioclavicular joints and the humerus about the shoulder joint. In contrast, the large anterior and posterior surfaces of the blade (subscapular, supraspinous, and infraspinous fossae) provide attachment for muscles that are mainly concerned with stabilizing the shoulder joint—keeping the glenoid fossa and head of the humerus in close contact.

KEY POINT

The shoulder region consists of the scapula and the clavicle. The medial end of the clavicle articulates with the manubrium to form the sternoclavicular joint linking the upper limb and the axial skeleton. The lateral end of the clavicle articulates with the medial aspect of the acromion process to form the acromioclavicular joint. Together with the manubrium, the scapulae and clavicles of both upper limbs form an incomplete ring called the shoulder girdle.

Upper Arm

There is only one bone in the upper arm, the humerus. The humerus is a typical long bone consisting of a relatively long shaft between two fairly bulbous ends (figure 2.25). The proximal end of the humerus has four main features: head, greater tuberosity, lesser tuberosity, and bicipital groove. The head is an almost perfect hemisphere, which, as mentioned above, articulates with the glenoid fossa to form the shoulder joint. The head faces medially, upward, and backward. The articular surface of the head is much larger than that of the glenoid fossa. This difference in size between the articulating surfaces combined with the shallowness of the glenoid fossa permits a large range of movement in the shoulder joint. Adjacent to the head, occupying the whole of the lateral aspect of the proximal end of the humerus, is the greater tuberosity. Adjacent to the head on the anterior aspect is the much smaller lesser tuberosity, which projects directly forward. Running vertically downward between the two tuberosities is the bicipital groove (intertubercular groove). The head is separated from the two tuberosities by a rather ill-defined anatomical neck, and the proximal end of the humerus as a whole is joined to the main shaft by a short, tapered surgical neck.

The upper two thirds of the shaft of the humerus is more or less cylindrical. The lower one third gradually becomes broader (in the coronal plane) toward the distal end. The surface of the shaft is fairly smooth apart from one roughened area in the middle of the anterior lateral aspect—the deltoid tuberosity—where the deltoid muscle attaches to the bone.

The distal end of the humerus has a cylinder-shaped articular surface consisting of two condyles fused together side by side. The lateral condyle is called the capitulum, and the larger, pulley-shaped medial condyle is called the trochlea. On the anterior aspect just above the capitulum is a small depression called the radial fossa. A similar depression called the coronoid fossa is just above the trochlea. On the posterior aspect there is a relatively large depression just above and continuous with the trochlea—the olecranon fossa. The medial epicondyle, easily felt beneath the skin on the medial aspect of the elbow, projects medially from the trochlea. The smaller lateral epicondyle projects laterally from the capitulum. Extending upward from the lateral epicondyle to the main part of the shaft is a distinct ridge called the lateral supracondylar ridge. A similar ridge, the medial supracondylar ridge, extends upward from the medial epicondyle.

KEY POINT

The humerus is the one bone in the upper arm. The proximal end articulates with the scapula to form the shoulder joint (glenohumeral joint), and the distal end articulates with the radius and ulna to form the elbow joint. The humerus is a typical long bone consisting of a relatively long shaft between two expanded ends; the expanded ends increase the area of articulation in the shoulder and elbow joints and thus decrease the compression stress in these joints.

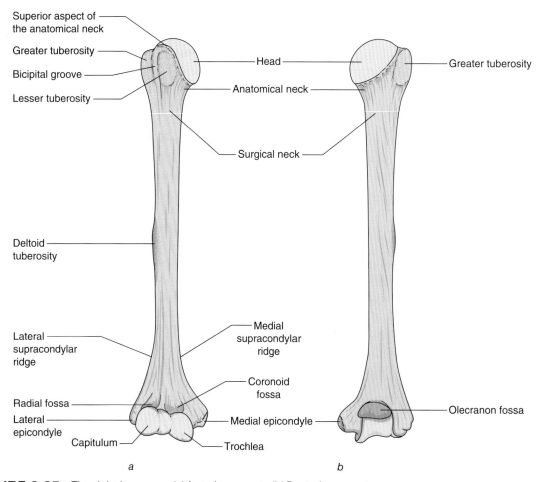

Superior aspect of the anatomical neck

Greater tuberosity

Bicipital groove

Lesser tuberosity

Head

Anatomical neck

Greater tuberosity

Surgical neck

Deltoid tuberosity

Lateral supracondylar ridge

Medial supracondylar ridge

Coronoid fossa

Radial fossa

Lateral epicondyle

Medial epicondyle

Olecranon fossa

Capitulum

Trochlea

a

b

FIGURE 2.25 The right humerus. *(a)* Anterior aspect. *(b)* Posterior aspect.

Lower Arm

There are two long bones in the lower arm (forearm), the radius and the ulna. In the anatomical position, the radius is lateral to the ulna (see figures 2.22 and 2.26). The anterior aspect of the proximal end of the ulna is dominated by a large, pulley-shaped, concave articular surface—the trochlear notch. This notch articulates with the trochlea of the humerus to form part of the elbow joint (figure 2.26). The proximal half of the trochlear notch forms the anterior part of the olecranon (or olecranon process). The tip of the elbow, easily felt beneath the skin, is the posterior superior point of the olecranon. When the elbow is fully extended, the proximal part of the rim of the trochlear notch occupies the olecranon fossa on the posterior aspect of the humerus (see figure 2.22b, figure 2.25b, and figure 2.26b). The distal half of the trochlear notch forms the anterior

superior part of the coronoid process projecting anteriorly from the shaft of the ulna. When the elbow is fully flexed, the coronoid process occupies the coronoid fossa on the anterior aspect of the humerus (figure 2.25a). The anterior inferior aspect of the coronoid process, together with the small part of the shaft with which it is continuous, is usually roughened. This unnamed area is the area of attachment of one of the muscles (brachialis) that flex the elbow joint. Adjacent to and continuous with the inferior lateral edge of the trochlear notch is a smaller articular surface called the radial notch (figure 2.26b).

The shaft of the ulna tapers slightly from the proximal to the distal end. Whereas most of the shaft is fairly smooth, the lower two thirds of the lateral aspect has a rather sharp ridge called the interosseous border of the ulna. The distal end of the ulna has a small drum-shaped head

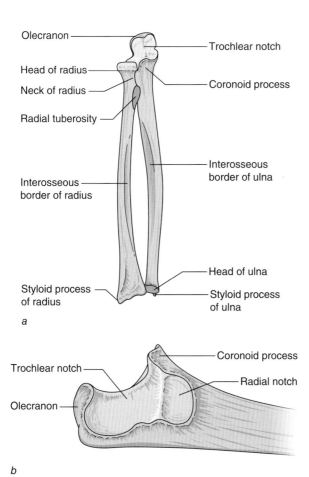

Olecranon

Trochlear notch

Head of radius

Neck of radius

Coronoid process

Radial tuberosity

Interosseous border of ulna

Interosseous border of radius

Head of ulna

Styloid process of radius

Styloid process of ulna

a

Trochlear notch

Coronoid process

Radial notch

Olecranon

b

FIGURE 2.26 The radius and ulna. *(a)* Anterior aspect of the right radius and ulna. *(b)* Lateral aspect of the proximal end of the right ulna.

with a small projection on its posteromedial aspect called the styloid process of the ulna.

The proximal end of the radius consists of a drum-shaped head separated from the main part of the shaft by a short cylindrical neck (figure 2.26*a)*. The circular side of the head and the superior surface of the head form a continuous articular surface. The side articulates with the radial notch on the ulna (figure 2.26*b)* and the superior surface articulates with the capitulum on the humerus. The elbow joint consists of the joint between the trochlea and trochlear notch and the joint between the capitulum and head of the radius. In the elbow joint, the trochlear notch and superior surface of the head of the radius form a virtually continuous articular surface.

A roughened projection at the anterior medial part of the base of the neck is the radial tuberosity. The shaft of the radius is fairly smooth apart from a sharp ridge along the medial aspect called the interosseous border of the radius. In contrast to the ulna, the radius has a distal end that is much thicker than the proximal end. The lateral part of the distal end of the radius forms a small projection called the styloid process of the radius. The inferior aspect of the distal end is dominated by a fairly large, more or less quadrangular, concave articular surface. Adjacent to and continuous with the medial edge of this surface is the ulnar notch, a small articular surface. This notch articulates with the side of the head of the ulna.

In the anatomical position, the radius and ulna lie side by side with their long axes more or less parallel to each other (see figure 2.22*a)*. With the radius and ulna in this position, the lower arm is described as supinated. The radius is able to move with respect to the ulna by means of the proximal and distal joints between the two bones. As the head of the radius rotates within the radial notch, the distal end of the radius moves around the head of the ulna; the radius as a whole rotates about an axis passing through the head of the radius and the head of the ulna (figure 2.27, *a* and *b)*. Consequently, medial rotation of the radius about the ulna from the anatomical position causes the radius to cross over the ulna (figure 2.27*b)*. When the radius crosses over the ulna the lower arm is described as pronated.

The anatomical position is close to the position of extreme supination, and the position in which the radius is fully crossed over the ulna is the position of extreme pronation. When the forearm moves from the position of extreme supination to that of extreme pronation, the hand is rotated about its long axis through approximately 180° (figure 2.27, *c* and *e)*. With the rest of the upper limb in the anatomical position, the position of the forearm in which the plane of the hand occupies a paramedian plane is usually referred to as the neutral position (figure 2.27*d)*.

In the anatomical position the long axes of the upper and lower arms do not coincide, and they form an obtuse angle on the lateral aspect (see figure 2.22*b)*. This angle is called the carrying angle and tends to be in the region

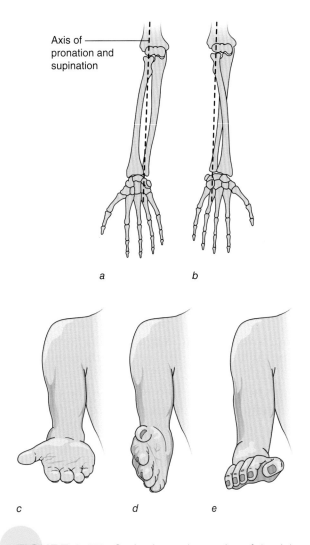

Axis of pronation and supination

a   b

c   d   e

FIGURE 2.27 Supination and pronation of the right forearm. *(a)* Forearm supinated. *(b)* Forearm pronated. *(c)* Maximum supination. *(d)* Neutral position. *(e)* Maximum pronation.

Wrist and Hand

The wrist consists of eight small irregular bones called carpals: trapezoid, trapezium, scaphoid, capitate, lunate, triquetrum, pisiform, hamate (figure 2.28). When articulated, the carpals form the carpus, linking the distal ends of the radius and ulna to the proximal end of the hand (figure 2.28). The carpals are closely packed together; seven of them articulate with three or four other bones from among the other carpals, the radius, the ulna, and the metacarpals of the hand. The carpals, whose names tend to reflect their shapes, are arranged in a proximal row and a distal row. From lateral to medial, the proximal row consists of the scaphoid (boat shaped), lunate (half-moon shaped), triquetrum (triangular), and pisiform (pea shaped). The proximal surfaces of the scaphoid, lunate, and triquetrum form a biconvex (surface rounded outward) elliptical surface that articulates with the biconcave (surface rounded inward) elliptical surface formed by the distal ends of the radius and ulna. The joints between these two elliptical surfaces constitute the wrist joint. The pisiform has one articulation with the anterior medial aspect of the triquetrum. From lateral to medial the distal row of carpals consists of the trapezium (four sided with two parallel sides), trapezoid (four sided), capitate (the central carpal), and hamate (with a distinct hooklike process anteriorly). The series of joints between the proximal and distal rows constitute the midcarpal joint.

of 165° in females and 175° in males. The carrying angle is largely due to the shape of the trochlea of the humerus. The medial end of the trochlea projects downward approximately 6 mm farther than the lateral end, which tilts the ulna outward (Standring 2004). The functional significance of the carrying angle is not clear, but it is thought to increase the precision with which the hand can be controlled in movements involving elbow extension combined with pronation of the forearm (Standring 2004). The significance of the difference in the size of the carrying angle between females and males is unknown, but it may be related to the relatively narrow shoulders, small waist, and broad hips in females compared with males.

The wrist consists of eight small irregular bones called carpals that articulate with each other to form the carpus. The carpals are arranged in a proximal and distal row. The proximal row articulates with the distal ends of the radius and ulna to form the wrist joint. The series of joints between the proximal and distal rows is called the midcarpal joint.

The hand consists of five metacarpals and 14 phalanges (or digits) (figure 2.28). The metacarpals are joined together by soft tissues and form the palm of the hand on the anterior aspect. The metacarpals are miniature long bones, and each metacarpal consists of a base (the proximal end), a shaft, and a head (the distal end). The proximal surfaces of the bases of the metacarpals articulate with the distal row of carpals to form the carpometacarpal joints. The combined ranges of movement in the wrist, midcarpal, and carpometacarpal joints facilitate a large range of movement for the hand as a whole. The bases of the medial four metacarpals articulate with each other side by side. Although the base of the first metacarpal (thumb) is close to that of the second metacarpal (index finger), the bases of these two metacarpals do not usually articulate with each other. The shafts of the metacarpals are fairly flat posteriorly and fairly rounded anteriorly. The head of each metacarpal has a convex condylar articular surface.

Each of the four fingers consists of three phalanges (proximal, middle, and distal), whereas the thumb has only two (proximal, distal). In each finger and the thumb the phalanges become progressively smaller from proximal to distal. Like the metacarpals, each phalanx consists of a base (proximal), a shaft, and a head (distal). The proximal end of the base of each proximal phalanx consists of a concave condyle articulating with the head of

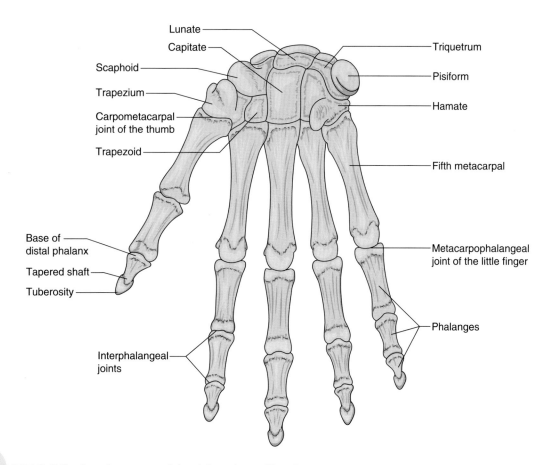

FIGURE 2.28 Anterior aspect of the right wrist and hand.

its corresponding metacarpal to form a meta-carpophalangeal joint. Whereas the shafts of the metacarpals are more or less cylindrical, the shafts of the phalanges are almost semicircular in cross section; the posterior surface of each phalanx is fairly flat and the anterior surface is rounded. The joints between the phalanges, the interphalangeal joints (two joints in each finger and one joint in the thumb), are similar in terms of the shape of the articulating surfaces. The heads of the proximal and middle phalanges all have a pulley-shaped articular surface made up of a lateral condyle and a medial condyle, both of which are convex. Each of these pulley-shaped heads articulates with a biconcave condylar surface on the base of the corresponding middle or distal phalanx. The distal phalanges are quite small, especially those of the fingers. Each distal phalanx has a relatively broad base, a tapered shaft, and a rounded tuberosity at the head.

KEY POINT

The hand consists of five metacarpals and 14 phalanges that are all miniature long bones. Each finger consists of three phalanges, whereas the thumb has only two phalanges. The joints between the phalanges, two joints in each finger and one joint in the thumb, are called interphalangeal joints.

Lower Limb

In the following discussion I've divided each lower limb into four regions: hip (innominate), upper leg (femur and patella), lower leg (tibia and fibula), and foot (seven tarsals, five metacarpals, and 14 phalanges; figure 2.29).

KEY POINT

The lower limbs consist of 62 bones (31 bones in each limb).

Hip

Together with the sacrum, the right and left innominate bones form a complete ring of bone called the pelvis or pelvic girdle (see figure 2.29). Consequently, the innominate bones attach the legs to the axial skeleton. Each

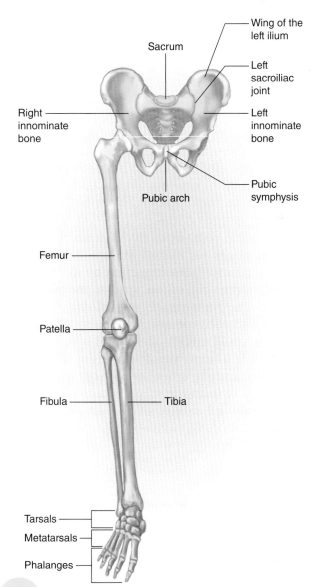

FIGURE 2.29 Anterior aspect of the pelvis and right lower limb.

innominate bone develops from three bones called the ilium, ischium, and pubis, which fuse together at maturity. The region where the three bones fuse together is dominated by a large semispherical concavity called the acetabulum (figure 2.30a). The acetabulum articulates with the head of the femur to form the hip joint. The acetabulum, which faces laterally forward and downward, consists of an outer horseshoe-shaped articular surface called the acetabular rim and a deep central region called the acetabular fossa. The gap between the two ends of the acetabular rim is continuous with the acetabular fossa and is called the acetabular notch. The acetabular

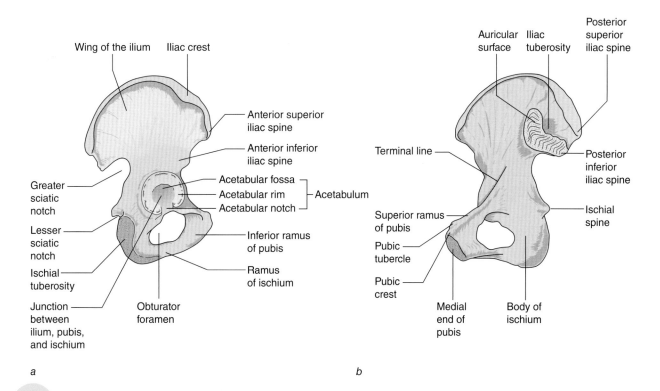

FIGURE 2.30 The right innominate bone. *(a)* Lateral aspect. *(b)* Medial aspect.

fossa and notch are deeper than the acetabular rim would be if it formed a complete articular cup. The acetabular fossa accommodates a ligament that joins the head of the femur to another ligament spanning the acetabular notch (see chapter 6). Consequently, as the head of the femur slides on the acetabular rim, the acetabular fossa prevents the ligament attached to the head of the femur from being crushed.

The ilium comprises the upper two fifths of the acetabulum and the large, more or less flat portion of the innominate bone above the acetabulum. The large flat upper part of the ilium is called the wing of the ilium. The superior border of the wing of the ilium is a broad crest called the iliac crest, felt beneath the skin just above the hip joint. The iliac crest, which has a shallow S shape when viewed from above, provides attachment for muscles comprising the anterior wall of the abdomen.

A projection at the anterior end of the iliac crest is called the anterior superior iliac spine (ASIS). From the ASIS, the anterior border of the ilium runs downward and backward to ter-

minate in a projection called the anterior inferior iliac spine (AIIS). The AIIS lies just above the anterior superior part of the acetabulum and is separated from the ASIS by a notch. At the posterior end of the iliac crest is a projection called the posterior superior iliac spine (PSIS). From the PSIS the posterior border of the ilium runs downward and forward to terminate in a projection called the posterior inferior iliac spine (PIIS). A small notch separates the PSIS and PIIS. Whereas the ASIS, AIIS, and PSIS are usually easy to identify, the PIIS is not usually as well defined. Although normally referred to as spines, these four projections often more closely resemble tubercles or tuberosities.

On the posterior medial aspect of the wing of the ilium there is a large C- or L-shaped auricular surface (figure 2.30*b*) that articulates with the auricular surface on the corresponding side of the sacrum to form the corresponding sacroiliac joint (figure 2.31). Posterior to the angle of the auricular surface is the iliac tuberosity or sacral tuberosity of the ilium. The large lateral and medial surfaces of the wing of the ilium provide attachment for muscles that move the

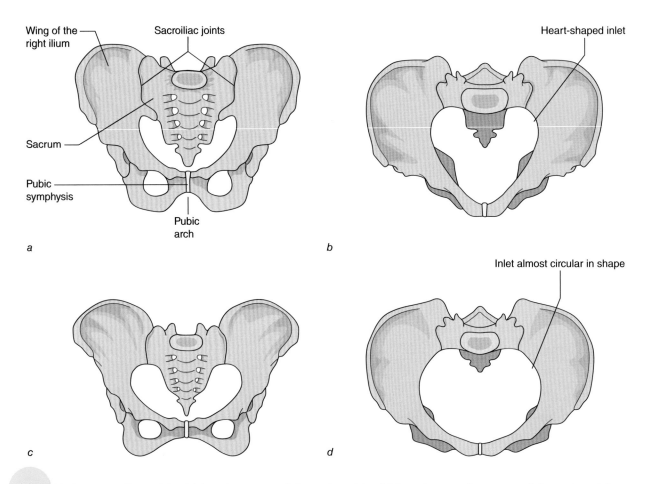

Wing of the right ilium

Sacroiliac joints

Sacrum

Pubic symphysis

Pubic arch

Heart-shaped inlet

Inlet almost circular in shape

a

b

c

d

FIGURE 2.31 The pelvis. *(a)* Anterior aspect of the male pelvis. *(b)* Anterior superior aspect of the male pelvis. *(c)* Anterior aspect of the female pelvis. *(d)* Anterior superior aspect of the female pelvis.

hip joint. The medial surface of the wing also supports the contents of the abdomen.

The ischium, which forms the inferior portion of the innominate bone, consists of the body and the ramus. The body is a more or less vertical pillar that transmits the weight of the trunk, head, and arms to the support surface when the individual is sitting on a chair or stool. The superior part of the body forms the posterior inferior two fifths of the acetabulum. Below the acetabulum the body of the ischium is characterized by the large ischial tuberosity on its lateral and inferior aspects (figure 2.30*a*). Arising from the posterior part of the body, just above the ischial tuberosity, is a process called the ischial spine that projects medially backward. The posterior border of the ilium and ischium between the PIIS and the ischial spine forms the greater sciatic notch. A smaller notch—the lesser sciatic notch—is located between the ischial spine and the ischial tuber-

osity. The ramus of the ischium is a broad, flat process that arises from the base of the body and projects medially, forward and upward.

The pubis forms the anterior inferior portion of the innominate bone. It consists of the body, superior ramus, and inferior ramus. The body forms the anterior inferior one fifth of the acetabulum. The superior ramus extends medially and also slightly forward and downward from the body to join the medial end of the inferior ramus. The junction between the two rami forms a fairly broad, flat region. The medial surface of this junction—the medial surface of the pubis—is elliptical in shape and lies in the median plane. The long axis of the ellipse is inclined at an angle of approximately 45° to the coronal plane. The medial surfaces of the right and left pubic bones are joined in the median plane by a disc of fibrocartilage. This joint is called the pubic symphysis (see figure 2.29).

The inferior ramus of the pubis projects downward and backward laterally to join the anterior end of the ramus of the ischium (see figure 2.30*a*). The inverted V-shaped notch formed by the inferior borders of the right and left inferior pubic rami is called the pubic arch (see figure 2.29*a* and 2.31*a*). On the superior border of each superior ramus is a process called the pubic tubercle situated a short distance from the pubic symphysis. Running between the two pubic tubercles is a ridge, often poorly defined, called the pubic crest. The lateral end of the superior border of the superior ramus of each pubis is continuous with a distinct curved ridge on the medial aspect of the ilium terminating at the anterior inferior margin of the auricular surface of the ilium. This ridge is called the terminal line (see figure 2.30*b*). The pubis and ischium are both essentially V shaped and are joined at their free ends (see figure 2.30*a*). Consequently, when fused together the two bones create a large hole called the obturator foramen on account of the close proximity of the obturator nerve.

KEY POINT

The innominate bones link the lower limbs to the axial skeleton. Each innominate bone develops from three bones called the ilium, ischium, and pubis, which fuse together at maturity.

Pelvis

Pelvis is a Latin word meaning basin (attributable to the large wings of the ilia, which give the impression of an incomplete bowl when the pelvis is viewed from an anterior superior aspect) (figure 2.31, *a* and *b*). The pelvis is made up of the upper pelvis and the lower pelvis. The upper pelvis consists of the two wings of the ilia and the upper third of the sacrum, which together form just over half of the upper part of the bowl. The anterior part of the bowl is missing, and thus the upper pelvis is sometimes referred to as the false pelvis or the greater pelvis. The upper pelvis provides a base of support for the upper body.

The lower pelvis (also called the true or lesser pelvis) can be described very loosely as an incomplete cylinder; it consists of the ischia

and pubic bones together with the inferior parts of the ilia and the inferior two thirds of the sacrum. The lower pelvis transmits the weight of the upper body to the legs for standing and to the seat of a chair or stool for sitting. The margin between the upper and lower pelvis is called the inlet or pelvic brim. The inlet corresponds to a continuous ridge made up of the pubic crest, the superior borders of the superior rami of the pubic bones, the terminal lines on the medial surfaces of the ilia, and a transverse ridge on the anterior aspect of the sacrum just below the uppermost transverse line.

Whereas the male and the female pelvises have the same basic structure, the shapes of the various parts, especially in the lower pelvis, differ considerably because of the childbearing functions of the female. During childbirth the child passes from the upper to the lower pelvis and then exits the abdomen via the inferior aspect of the lower pelvis. Consequently, the broader the lower pelvis, from side to side and from front to back, the easier it is for the child to pass through the pelvis. The shapes of the male and female pelvises differ in four main ways (see figure 2.31, *b* and *d*):

1. The inlet of the male pelvis is heart shaped, whereas that of the female pelvis is more circular (figure 2.31, *b* and *d*).

2. The pubic bones are more in line with each other in the frontal plane in the female than in the male (figure 2.31, *b* and *d*). Consequently, the angle of the pubic arch is obtuse in the female and acute in the male (figure 2.31, *a* and *c*).

3. The relative distance between the acetabulums is greater in the female than in the male. This results in a relatively greater girth around the hips in the female compared with the male.

4. In the male, the sacrum is curved such that the lower half of the sacrum and the coccyx bend forward. This curvature reduces the front-to-back dimension of the lower pelvis. In the female, the sacrum is relatively straight, which tends to maintain a fairly constant front-to-back dimension in the lower pelvis.

Upper Leg

The upper leg or thigh contains a long bone called the femur and a relatively small bone called the patella (knee cap), which articulates with the lower end of the femur. The femur is the longest and strongest bone in the skeleton. The proximal end of the femur consists of a nearly spherical head, which articulates with the acetabulum to form the hip joint (see figures 2.29 and 2.32). The head is joined obliquely to the shaft by a thick neck that runs laterally downward and backward from the head to the anterior medial region of the

proximal end of the shaft (see also figure 2.1*b*). The neck resembles a truncated cone, partially flattened from front to back, its smaller surface joined to the head and its larger surface joined to the shaft. A large process called the greater trochanter dominates the superior posterior lateral region of the proximal end of the shaft. At the base of the neck on the posterior medial aspect of the shaft is another fairly large process called the lesser trochanter. The greater and lesser trochanters are linked on the posterior aspect by a distinct ridge called the intertrochanteric crest. The line of unison between the neck and the shaft on the anterior aspect is marked by another distinct ridge called the intertrochanteric line.

Like the humerus, the upper two thirds of the shaft of the femur is cylindrical and the lower one third gradually becomes broader (in the coronal plane) toward the distal end. In a paramedian plane, the anterior surface of the femur is slightly convex and the posterior

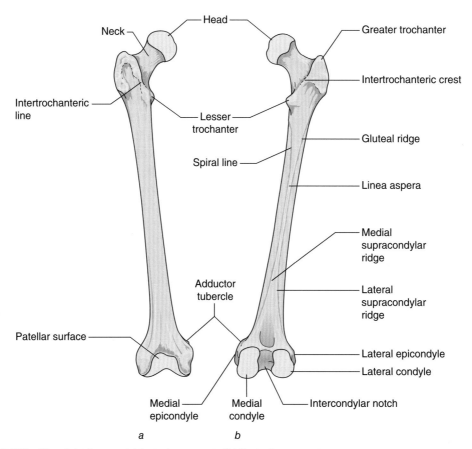

FIGURE 2.32   The right femur. *(a)* Anterior aspect. *(b)* Posterior aspect.

surface is slightly concave (see figure 2.1*b*). The anterior and lateral surfaces are fairly smooth, whereas the posterior surface is dominated by two longitudinal ridges—one lateral and one medial—that run the length of the shaft. The ridges converge and join in the middle one third of the shaft to form the linea aspera (rough line). Above and below the linea aspera the ridges diverge. Above the linea aspera the lateral ridge is called the gluteal ridge, which runs to the greater trochanter. The medial ridge above the linea aspera is the spiral line. The spiral line runs toward the lesser trochanter for half of its length but then curves around the medial part of the shaft in a spiral manner to terminate at the lower end of the intertrochanteric line. Below the linea aspera the lateral and medial ridges are called the lateral supracondylar ridge and medial supracondylar ridge, respectively. At the bottom end of the medial supracondylar ridge is a projection called the adductor tubercle.

The distal end of the femur consists of two large convex condyles, the lateral and medial condyle, fused together side by side anteriorly. The condyles are separated posteriorly by a large notch called the intercondylar notch (or intercondylar fossa). Consequently, the articular surface of the condyles is V shaped. The upper part of the common anterior portion of the articular surface is called the patellar surface. The patellar surface is pulley shaped—depressed in the middle in the sagittal plane—and articulates with the posterior surface of the patella to form the patellofemoral joint (see chapter 6). During extension and flexion of the knee joint, the patella slides up and down on the patellar surface and condyles of the femur.

> ### KEY POINT
>
> The femur and the patella make up the upper leg. The femur is the longest and strongest bone in the skeleton. The proximal end of the femur articulates with the innominate bone to form the hip joint. The distal end articulates with the tibia to form the tibiofemoral joint and with the patella to form the patellofemoral joint.

The patella is a sesamoid (Greek for resembling a sesame seed)—a bone that is partially embedded in a tendon (figure 2.33). A sesamoid bone tends to increase the mechanical efficiency of the associated musculotendinous unit and prevent the tendon from rubbing on an adjacent bone. Consequently, the patella, embedded in the posterior part of the quadriceps tendon, increases the mechanical efficiency of the quadriceps muscle group and prevents the quadriceps tendon from rubbing against the patellar surface of the femur.

The anterior aspect of the patella is rounded superiorly and pointed inferiorly (figure 2.33, *a* and *b*). The whole of the anterior surface and the inferior quarter of the posterior surface are embedded in the quadriceps tendon. The upper three quarters of the posterior surface articulates with the patellar surface of the femur when the knee joint is extended and with the

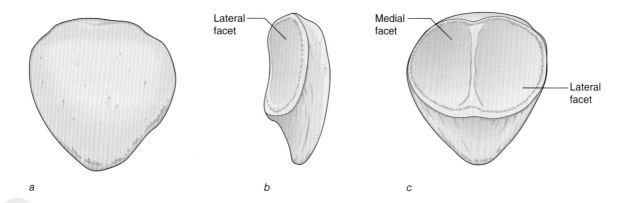

a *b* *c*

FIGURE 2.33   The right patella. *(a)* Anterior aspect. *(b)* Right lateral aspect. *(c)* Posterior aspect.

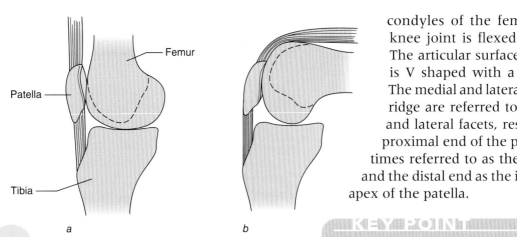

a

b

FIGURE 2.34 Orientation of the patella to the femur in *(a)* knee extension and *(b)* knee flexion.

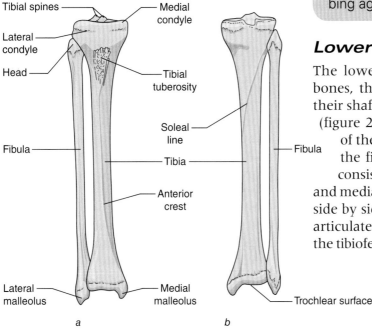

a

b

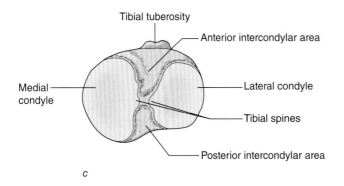

c

FIGURE 2.35 The right tibia and fibula. *(a)* Anterior aspect. *(b)* Posterior aspect. *(c)* Superior aspect of the tibial table.

condyles of the femur when the knee joint is flexed (figure 2.34). The articular surface of the patella is V shaped with a sagittal ridge. The medial and lateral aspects of the ridge are referred to as the medial and lateral facets, respectively. The proximal end of the patella is sometimes referred to as the superior pole and the distal end as the inferior pole or apex of the patella.

KEY POINT

The patella increases the mechanical efficiency of the quadriceps muscle group and prevents the quadriceps tendon rubbing against the femur.

Lower Leg

The lower leg or shank contains two long bones, the tibia and the fibula, aligned with their shafts more or less parallel to each other (figure 2.35, *a* and *b)*. The tibia is the larger of the two bones and is situated medial to the fibula. The proximal end of the tibia consists of two large condyles, the lateral and medial condyles, which are fused together side by side (figure 2.35c). The tibial condyles articulate with the femoral condyles to form the tibiofemoral joint (knee joint). The articular surfaces of the tibial condyles are oval in outline and almost flat. The lateral surface is usually slightly convex and slightly smaller than the medial surface. The medial surface can be slightly convex. The surfaces occupy the same plane more or less horizontally in the anatomical position. This orientation of the condylar surface gives rise to the term *tibial table,* sometimes used to describe the proximal end of the tibia. Between the two condylar surfaces at the center of the tibial table there are two small processes lying side by side; these processes are referred to as the lateral and medial tibial spines (or tibial eminences) or intercondylar tubercles (figure 2.35c). The area of the tibial table in front of the tibial spines and between the anterior aspects of the tibial condyles is referred to as the anterior inter- condylar area. The corresponding area behind the tibial spines is referred to as the posterior intercondylar area.

The shaft of the tibia is fairly smooth apart from a distinct ridge on the upper one third of the posterior aspect of the shaft. This ridge, the soleal line, runs obliquely downward and medially from the inferior posterior aspect of the lateral condyle. The middle two thirds of the shaft is teardrop shaped in cross section; the posterior aspect is rounded, whereas the anterior aspect consists of two fairly flat areas, anterior lateral and anterior medial, which converge anteriorly to form a distinct ridge called the anterior crest. The anterior crest can easily be felt beneath the skin as a ridge running down the bone. The anterior medial surface of the tibia, covered only by skin, is usually referred to as the shin. Above and below the anterior crest, the shaft broadens out toward the proximal and distal ends of the bone. Above the upper end of the anterior crest, on the anterior aspect of the shaft, is a fairly large process called the tibial tuberosity.

On the medial side of the distal end of the tibia there is a downward projection called the medial malleolus. The lateral aspect of the medial malleolus articulates with the medial aspect of the talus to form the medial part of the ankle joint (figure 2.36a). The remainder of the distal end of the tibia is dominated by a large biconcave condylar surface called the trochlear surface of the tibia (see figure 2.35b). The trochlear surface articulates with the supe- rior aspect of the talus to form the main part of the ankle joint. The trochlear surface of the tibia and the articular surface of the medial malleolus are continuous with each other.

The tibia is almost completely responsible for transmitting loads from the upper leg to the foot and vice versa. In contrast, the fibula is a thin, relatively weak bone only margin- ally involved in load transmission between the upper leg and foot. The main functions of the fibula are to help form the ankle joint and to provide additional area for the attachment of muscles that move the ankle and foot. The proximal end of the fibula is called the head. The medial two thirds of the superior aspect of the head articulates with the posterior inferior lateral aspect of the lateral tibial condyle to form the proximal tibiofibular joint.

The shaft of the fibula is characterized by four longitudinal ridges that give rise to four faces of varying width and length along the shaft. The distal end of the fibula is called the lateral malleolus. The medial aspect of the lateral malleolus articulates with the lateral aspect of the talus to form the lateral part of the ankle joint. The medial part of the shaft of the fibula immediately above the articular surface of the lateral malleolus articulates with the lateral part of the distal end of the tibia to form the distal tibiofibular joint.

KEY POINT

The tibia and the fibula are the two long bones in the lower leg. The tibia is much thicker than the fibula and is almost com- pletely responsible for transmitting loads between the upper leg and foot. The fibula helps form the ankle joint and provides additional area of attachment for muscles of the lower leg. The tibia and fibula articu- late with each other at their proximal and distal ends to form the proximal and distal tibiofibular joints, respectively.

Foot

The foot consists of seven tarsals, five meta- tarsals, and 14 phalanges (figure 2.36). When articulated, the tarsals form the tarsus (figure 2.36b). The tarsus corresponds to the carpus in the upper limb, but the tarsals are all much larger than the carpals. Whereas the carpus is not usually considered to be part of the hand, the tarsus forms the posterior half of the foot. The foot articulates with the lower leg at the ankle joint—the joint between the tibia, fibula, and talus.

The talus, the second largest tarsal, has a convex, pulley-shaped articular surface on its superior aspect called the trochlear surface of the talus (figure 2.36c); it articulates with the trochlear surface of the tibia. The trochlear

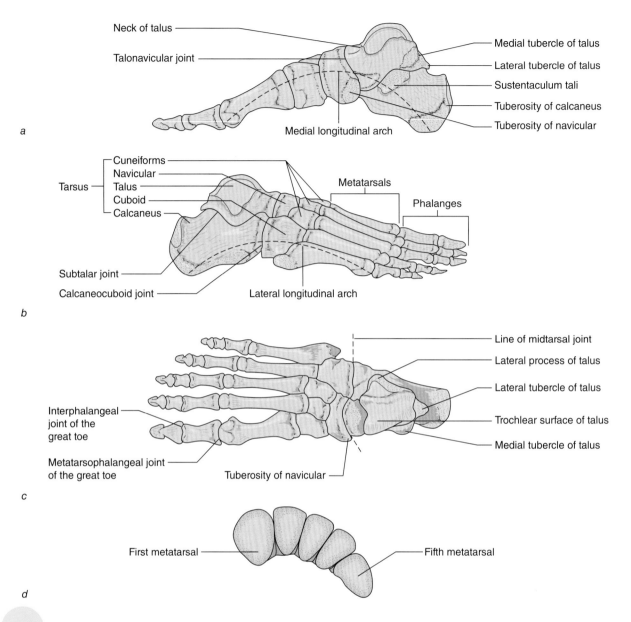

FIGURE 2.36 The right foot. *(a)* Medial aspect. *(b)* Lateral aspect. *(c)* Superior aspect. *(d)* Posterior aspect of vertical section through the proximal ends of the metatarsals.

surface of the talus is continuous with articular surfaces on its lateral and medial aspects, which articulate with the lateral malleolus and medial malleolus, respectively.

The inferior aspect of the talus articulates with the anterior half of the superior aspect of the calcaneus by means of two or three articular facets, which together constitute the subtalar joint (talocalcaneum joint) (see figure 2.36, *a* and *b)*. The anterior aspect of the talus articulates with the posterior aspect of the navicular, on the medial aspect of the foot, to form the talonavicular joint.

The calcaneus—the largest tarsal—is often referred to as the heel bone. The posterior aspect of the calcaneus is characterized by a large tuberosity called the calcaneal tuberosity. The anterior aspect of the calcaneus articulates with the posterior aspect of the cuboid, on the lateral aspect of the foot, to form the calcaneocuboid joint. The calcaneocuboid and talonavicular joints are continuous with each other and constitute the midtarsal joint (see figure 2.36, *a-c)*. The anterior aspect of the navicular articulates with the posterior aspects of the three cuneiforms (medial, middle, lateral), which lie side by side and articulate with each other. The posterior two thirds of the lateral aspect of the lateral cuneiform articulates with the medial surface of the cuboid. The anterior aspects of the cuneiforms articulate with the bases of the first, second, and third metatarsals. The anterior aspect of the cuboid articulates with the bases of the fourth and fifth metatarsals. These joints between the four anterior tarsals and the metatarsals are referred to as the tarsometatarsal joints. The lateral four metatarsals are similar in length but tend to increase in girth from the second through to the fifth. In comparison, the first metatarsal is shorter but has a greater girth than the other four. The short length of the first metatarsal is thought to increase the efficiency of the arches of the feet (Standring 2004).

> **KEY POINT**
>
> The foot consists of seven tarsals, five metatarsals, and 14 phalanges. The tarsals are irregular bones articulating with each other to form the tarsus, which forms the posterior half of the foot. The lower leg articulates with the tarsus at the ankle joint.

The tarsals and metatarsals are arranged in the form of two longitudinal arches (medial and lateral) and a single transverse arch. The medial longitudinal arch is formed by the calcaneus, the talus, the navicular, the three cuneiforms, and the first, second, and third metatarsals (see figure 2.36, *a* and *c*). The lateral longitudinal arch, which is much flatter than the medial arch, is formed by the calcaneus, the cuboid, and the fourth and fifth metatarsals (see figure 2.36, *b* and *c*). In combination, the longitudinal arches form a single arched structure between the posterior inferior aspect of the calcaneus and the heads of the metatarsals. The transverse arch runs across the foot from medial to lateral and is formed by the anterior five tarsals and the bases of the metatarsals. The shape of the arch is attributable to the cuboid, the middle and lateral cuneiforms, and the bases of the middle three metatarsals, which are wedge shaped in coronal section (figure 2.36*d*). The arches, maintained by muscles and ligaments, are normally flexible and resilient, which enables the feet to cushion impacts and provide a stable base of support in weight-bearing activities such as walking, running, jumping, and landing (see chapter 6).

The distribution of phalanges in the foot is similar to that in the hand—two in the great toe (big toe or hallux) and three in each of the other toes. As in the hand, the phalanges of the toes become progressively shorter from proximal to distal. In comparison with the corresponding phalanges of the thumb, the phalanges of the great toe are slightly longer and have a much greater girth. However, the phalanges of the other four toes are much shorter and usually smaller in girth than the corresponding phalanges in the hand. The interphalangeal and metatarsophalangeal joints are similar in structure to their counterparts in the hand.

> **KEY POINT**
>
> The anterior aspect of the tarsus articulates with the proximal ends of the metatarsals to form the tarsometatarsal joints. The distal ends of the metatarsals articulate with the distal phalanges to form the metatarsophalangeal joints.

Summary

The skeleton has three main mechanical functions: to provide a supporting framework for all the other systems of the body; to protect organs such as the brain and spinal cord; and to provide a system of levers, operated by the skeletal muscles, to facilitate force transmission throughout the skeleton.

Bone is a connective tissue, and all connective tissues are concerned to some extent with transmitting forces within and between body systems. The next chapter describes the functional anatomy of the various types of connective tissue.

Review Questions

1. Describe the three main mechanical functions of the skeleton.
2. Describe the three main reference planes and define the spatial terminology associated with the planes.
3. List the bones of the axial skeleton.
4. Describe the following features of the skull:
   - Sutures
   - Fontanels
   - Sinuses
5. Describe the primary and secondary curves of the vertebral column.
6. Describe the components of a typical vertebra.
7. Describe the distinguishing features of cervical vertebrae, thoracic vertebrae, and lumbar vertebrae.
8. Describe the difference between true ribs and false ribs.
9. List the bones of the upper limb.
10. Describe the shoulder girdle.
11. Describe the elbow joint.
12. Describe supination and pronation of the lower arm.
13. List the bones of the lower limb.
14. Describe the differences between the male pelvis and the female pelvis.
15. Describe the tibiofemoral and patellofemoral joints.
16. Describe the arches of the feet.

CONNECTIVE TISSUES

Connective Tissues

All of the cells in the body are joined by connective tissue into progressively larger functional units that are the tissues, organs, and systems. The whole body consists of all the systems joined together. Connective tissue maintains the integrity of the tissues, organs, and systems by providing them with adequate strength and elasticity. Connective tissue also facilitates intercellular exchange of gases and nutrients between cells. These characteristics—passive strength (rather than the active strength produced by muscle), elasticity, facilitation of intercellular exchange—differentiate connective tissue from muscle tissue, nerve tissue, and epithelial tissue. Connective tissue is continuous throughout the body, but its structure gradually changes from one part of a tissue, organ, or system to another depending on the function of the connective tissue at each location. For example, in muscle tissue, nerve tissue, and epithelial tissue, the connective tissue provides a flexible join between the cells and allows the transfer of gases and nutrients between cells. However, as the need for strength increases in, for example, tendons, aponeuroses, and cartilage, the amount of flexibility and intercellular exchange progressively decreases, culminating in bone, the strongest and hardest of the connective tissues.

The purpose of this chapter is to describe the structure and functions of connective tissues.

OBJECTIVES

After reading this chapter, you should be able to do the following:

1. Describe the structure and function of ordinary connective tissues.
2. Describe the structure and functions of the three main types of cartilage.
3. Describe the growth and development of bone.
4. Describe the structure of mature bone.
5. Explain the difference between remodeling and modeling in bone.
6. Describe the effects of aging on bone.

Functions of Connective Tissues

In muscle tissue, nerve tissue, and epithelial tissue, the cells predominate and determine the function of the tissues. In contrast, connective tissues have relatively few cells distributed within a large amount of noncellular material called **matrix** that is produced by the cells. The connective tissues differ from each other, structurally as well as functionally, based on differences in the physical characteristics of the matrix. The matrix ranges from a semiliquid material in one type of connective tissue (areolar tissue) to a very hard solid material in another (bone). Connective tissues have two main functions: mechanical support and intercellular exchange.

KEY POINT

Connective tissues consist of relatively few cells within a large amount of matrix produced by the cells. The connective tissues differ from each other mainly in terms of the physical characteristics of the matrix.

Mechanical Support

All of the connective tissues help maintain or transmit forces by providing variable amounts of strength and elasticity to facilitate a wide range of mechanical functions. These functions include the following:

- Binding together the cells of the body in the various tissues, organs, and systems
- Supporting the organs and holding them in place
- Providing stability and shock absorption in joints
- Providing flexible links between bones in certain types of joints and providing smooth, articulating surfaces between bones in other types of joints
- Transmitting muscle forces

Intercellular Exchange

In multicellular organisms, the cells rely on the circulating body fluids such as blood to supply nutrients, oxygen, and other substances and to carry away waste products such as carbon dioxide. This involves exchange of nutrients, gases, and other substances between the vessels of the circulating body fluids and cells adjacent to the vessels and between cells adjacent to each other. Intercellular exchange ensures that all cells can be supplied with nutrients, gases, and other substances and can excrete waste products, even if the cells do not receive a direct supply of the circulating body fluids.

Connective tissues have two main functions: mechanical support and intercellular exchange.

Classification of Connective Tissues

Although all types of connective tissue are continuous with each other throughout the body, the composition of the matrix gradually changes from one part of an organ or system to another, depending on the function of the connective tissue at each particular location. For example, in a skeletal muscle the individual muscle cells are bound together by connective tissue whose function is to facilitate intercellular exchange and to bind the cells together. In contrast, the belly of the muscle is attached to bone at each end by means of tendons or aponeuroses whose sole function is to provide a strong link between the muscle and the bony attachments (figure 3.1).

Connective tissues are classified according to their function into ordinary and special connective tissues (Standring 2004). Ordinary connective tissues, distributed widely throughout the body, have two main functions:

1. Binding cells together into tissues, organs, and systems
2. Providing mechanical links between bones at joints and between muscles and bones

There are two special connective tissues: cartilage and bone. Cartilage transmits loads across joints efficiently and allows movement between bones at certain joints. The mechanical functions of bone were described in chapter 2; this chapter develops the relationship between these functions and the structure of bone.

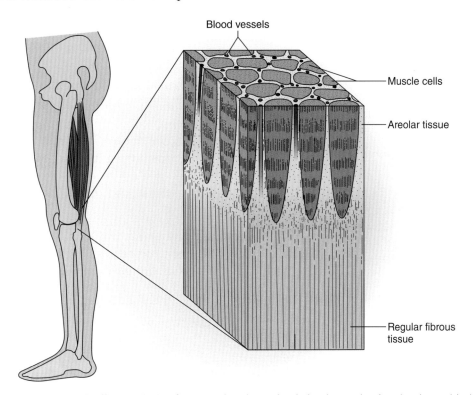

Blood vessels

Muscle cells

Areolar tissue

Regular fibrous tissue

FIGURE 3.1 Location of different kinds of connective tissue in skeletal muscle. Areolar tissue binds the muscle cells together and facilitates intercellular exchange. Regular fibrous tissue provides a mechanical link between muscle and bone.

Ordinary Connective Tissues

The matrix of ordinary connective tissues consists of three components: **elastin** fibers, **collagen** fibers, and **ground substance.** The main difference in structure between the various types of ordinary connective tissue is in the proportion of these basic components in the matrix.

Elastin and Collagen Fibers

Both elastin and collagen fibers are proteins. A protein molecule consists of a long chain of amino acids. In elastin, the molecules are arranged randomly in terms of individual shape and orientation and attachment to each another (Alexander 1975, 2003) (figure 3.2a). When elastin is subjected to tension, the molecules straighten and are then stretched (figure 3.2b). The molecules resist the tension load; that is, they experience tension stress (see

chapter 1), and the greater the tension load, the greater the tension stress. When the tension load is removed (assuming that the molecules have not been stretched to failure), the elastin molecules restore their original orientation and shape. Elastin is, therefore, elastic, hence its name.

An elastin fibril is formed by a number of elastin molecules, and an elastin fiber consists of a number of fibrils grouped together. An elastin fiber is similar in shape, strength, and elasticity to a long, thin rubber band. Elastin fibers can be stretched by about 200% of their resting length before breaking (Nordin and Frankel 2001). They have a yellowish appearance and are often referred to as yellow elastic fibers or yellow fibers.

In contrast to elastin, collagen molecules are arranged in a more regular manner; they

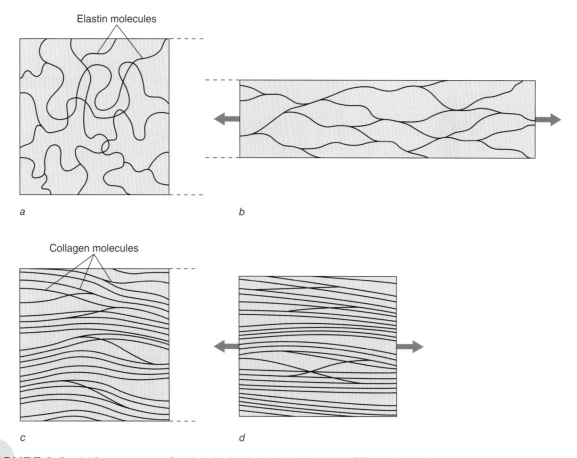

FIGURE 3.2 *(a)* Arrangement of molecules in elastin (unloaded). *(b)* Effect of stretching on elastin molecules. *(c)* Arrangement of molecules in collagen (unloaded). *(d)* Effect of stretching on collagen molecules.

tend to run in the same overall direction and for the most part are aligned parallel to each other (Alexander 1975) (figure 3.2*c*). Like elastin molecules, collagen molecules are attached to each other at various points. When subjected to tension in the direction of their main orientation, collagen molecules, like elastin molecules, straighten and are then stretched. However, for a given tension load, the amount of extension of collagen molecules is very small compared with elastin molecules (figure 3.2*d*). Like elastin molecules, collagen molecules are elastic and return to their resting orientation when the tension load is removed (assuming that the molecules have not been stretched to failure). Each group of closely aligned parallel molecules constitutes an individual collagen fibril, and a collagen fiber consists of a number of fibrils grouped together. A collagen fiber is similar in shape, strength, and elasticity to a shoelace; it is virtually inextensible and, in relation to elastin, is extremely strong. Collagen fibers break after being stretched by approximately 10% of their rest length (Nordin and Frankel 2001). Collagen fibers are white and are often referred to as white collagen fibers or white fibers.

Ground Substance

The ground substance forms the nonfibrous part of the matrix. It is a viscous gel consisting mainly of large carbohydrate molecules (molecules consisting of carbon, hydrogen, and oxygen) and carbohydrate–protein molecular complexes (molecules consisting of carbon, hydrogen, oxygen, and nitrogen) suspended in a relatively large volume of water (Standring 2004, Alexander 1975). The number and type of carbohydrate and carbohydrate–protein substances determine the volume of water. Many of these substances have an affinity for water and thus determine not only the volume of water in the ground substance but also the viscosity of the ground substance. **Viscosity** refers to the resistance of a fluid to flowing; for example, oil is more viscous than water.

In contrast to the elastin and collagen fibers, whose sole function is to provide mechanical support, the ground substance is responsible not only for facilitating intercellular exchange

but also for providing some mechanical support. The gluelike viscosity of the ground substance enables it to bind cells together within the other main tissues (muscle, nerve, and epithelia). In epithelial tissue, the ground substance is the main bonding material between the cells. In muscle and nerve tissue, the cells are bound together by a combination of ground substance and fibers.

The ground substance in ordinary connective tissues is sometimes referred to as tissue fluid or extracellular fluid. It is also referred to as amorphous ground substance (*amorphous* = without definite structure) because it appears, even under a microscope, as a featureless fluid.

Key Terms

matrix The noncellular component of connective tissue.

elastin A protein molecule with high elasticity in the matrix of connective tissue; elastin molecules form elastin fibers, which fail after being stretched approximately 200% of their resting length.

collagen A protein molecule with low elasticity in the matrix of connective tissue; collagen molecules form collagen fibers, which fail after being stretched approximately 10% of their resting length.

ground substance The nonfibrous component of the matrix of ordinary connective tissue.

viscosity The resistance of a fluid to flowing.

KEY POINT

Throughout the body, ordinary connective tissues bind cells together into tissues, organs, and systems and provide mechanical links between bones at joints and between muscles and bones. The matrix of ordinary connective tissues consists of elastin and collagen fibers and ground substance. Elastin fibers provide elasticity and collagen fibers provide strength. The ground substance facilitates intercellular exchange and helps to bind cells within the other main tissues.

Ordinary Connective Tissue Cells

The number and type of cells found in ordinary connective tissues depend on the type of connective tissue and the person's state of health (Standring 2004). When present, the various types of cells are found suspended in the ground substance or, in some cases, attached to the collagen fibers. Six main types of cells are found in ordinary connective tissues:

fibroblasts Usually the most numerous type of cells, fibroblasts are often found attached to the collagen fibers and are responsible for producing the matrix (the ground substance and the elastin and collagen fibers).

macrophages The macrophages are responsible for engulfing and digesting bacteria and other foreign bodies. They also dispose of dead cellular material that occurs as a result of injury or as cells age and die.

plasma cells Plasma cells occur in large numbers in response to infection. They produce antibodies that inactivate and, with the macrophages, destroy harmful bacteria and other substances.

white blood cells The number and type of white blood cells increase in response to infection. They work with the plasma cells and macrophages to identify and destroy harmful bacteria and other substances.

mast cells Mast cells, widespread throughout ordinary connective tissues, are responsible for producing heparin, which prevents the blood plasma from clotting inside blood vessels.

fat cells Fat cells have a variety of functions and occur in great numbers in one particular type of ordinary connective tissue (adipose tissue).

The proportion of elastin fibers, collagen fibers, and ground substance and the number and type of cells within any particular ordinary connective tissue determine its function. Collagen fibers predominate where great strength is required, whereas elastin fibers predominate where considerable elasticity is needed. Similarly, the ground substance tends to predominate where intercellular exchange is of major importance. Under normal circumstances a wide variety of cells are present within ordinary connective tissues. In response to infection, there is an increase in the number of cells responsible for identifying and destroying harmful bacteria.

KEY POINT

The proportion of elastin fibers, collagen fibers, and ground substance and the number and type of cells within any particular ordinary connective tissue determine its function.

Irregular Ordinary Connective Tissues

Ordinary connective tissues are classified into irregular and regular tissues according to the arrangement of the fibrous content of the matrix. In irregular tissues the fibers tend to run in all directions throughout the tissue with no set pattern. In contrast, the fibers in regular tissues tend to be orientated in the same overall direction. There are four types of irregular ordinary connective tissue (loose, adipose, irregular collagenous, irregular elastic) and two types of regular ordinary connective tissue (regular collagenous, regular elastic).

Loose Connective Tissue

Loose connective tissue is the most widely distributed of all the connective tissues. It is the glue that binds cells within the other main tissues (muscle, nerve, and epithelia) and binds these tissues into organs. Loose connective tissue consists of a loose, irregular network of elastin and collagen fibers suspended within a relatively large amount of ground substance (figure 3.3). The large amount of amorphous ground substance gives the impression of a lot of space between the fibers and cells of loose connective tissue. For this reason, loose connective tissue is also referred to as areolar tissue (*areola* = a small open area).

The loose network of elastin and collagen fibers, both of which branch freely, provides moderate elasticity and strength. Consequently, loose connective tissue is well suited to bind cells into tissues and tissues into organs and to provide a supporting framework for nerves and the vessels of the circulating body fluids (blood and lymph). The viscosity of the ground substance is important for binding cells within the other main tissues, and the large amount of ground substance reflects the importance of loose connective tissue in the facilitation of intercellular exchange.

Adipose Connective Tissue

Adipose tissue has a loose network of elastin and collagen fibers similar to loose connective tissue. However, in contrast to loose connective tissue, adipose tissue has little ground substance and a large number of closely packed fat cells. Each fat cell consists of a thin cell membrane surrounding a relatively large globule of fat (figure 3.4). Adipose tissue is widely

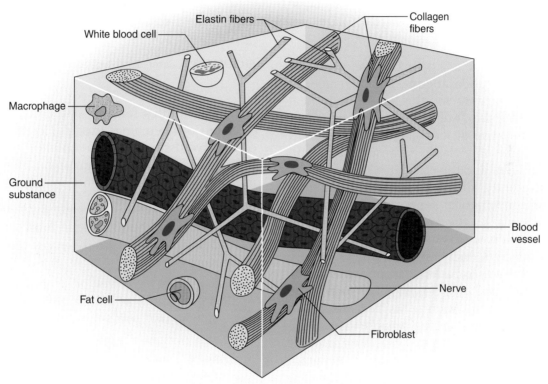

FIGURE 3.3   Loose connective tissue.

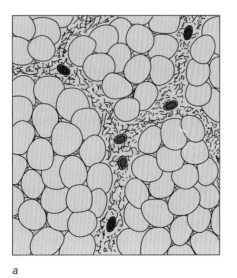

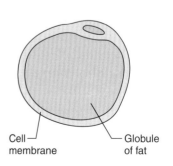

FIGURE 3.4   Adipose connective tissue. *(a)* Groups of fat cells held together by loose connective tissue. *(b)* Cross section of a single fat cell.

a

b

distributed around the body in the following four main locations (McArdle et al 1996):

1. In bone marrow.
2. In association with the various layers of loose connective tissue within certain organs, especially skeletal muscles.
3. As padding around certain organs and joints.
4. As a continuous layer beneath the skin. Skin is sometimes referred to as cutaneous tissue and the layer of fat as the subcutaneous fat layer.

Adipose tissue is a poor conductor of heat and so the subcutaneous fat layer acts as an insulator, reducing the loss of body heat through the skin. Adipose tissue is moderately strong because of its collagen fiber content and considerably elastic because of its elastin fiber content and the elasticity of the fat cells. Consequently, adipose tissue is well suited to provide mechanical support and protection (cushioning) in the form of padding around and between organs such as the heart, lungs, liver, spleen, kidneys, and intestines. Adipose tissue also acts as padding around joints such as the knee and over certain bones such as the heel bone. In addition to its heat insulation and mechanical functions, adipose tissue is the body's main food store. Adipose tissue provides approximately twice as much energy per gram as any other tissue in the body (McArdle et al 1996).

Irregular Collagenous Connective Tissue

The matrix of irregular collagenous connective tissue is dominated by a dense, irregular network of branching, interconnected collagen fiber bundles, together with a few elastin fibers and a relatively small amount of ground substance (figure 3.5). The collagen bundles and their irregular arrangement enable the tissue to resist being stretched in any direction. However, although it is strong, the tissue has a certain amount of elasticity attributable to the wavy orientation of the collagen bundles. When stretched in a particular direction, the

FIGURE 3.5 Irregular collagenous connective tissue.

collagen bundles tend to straighten in the direction of stretching. Irregular collagenous connective tissue is most frequently found as a tough cover around certain organs, where it provides mechanical support and protection. For example, it is found as

- a sheath around skeletal muscles (epimysium) and spinal nerves (epineurium);
- a capsule or envelope around certain organs, such as the kidneys, liver, and spleen, that holds the organs in place;
- the perichondrium of cartilage (discussed later in this chapter); and
- the periosteum of bones (discussed later in this chapter).

Irregular Elastic Connective Tissue

The matrix of irregular elastic connective tissue consists of a dense, irregular network of branching, interconnected elastin fibers, together with a few collagen fibers and a moderate amount of ground substance (figure 3.6). In comparison to irregular collagenous connective tissue, irregular elastic connective tissue is not as strong but is much more elastic. It is found where moderate amounts of strength and elasticity are required in more than one direction, for example, in the walls of arteries and the larger arterioles, the trachea (windpipe), and bronchial tubes. There are few cells in irregular collagenous and irregular elastic connective tissues. The cells that are present are mainly fibroblasts.

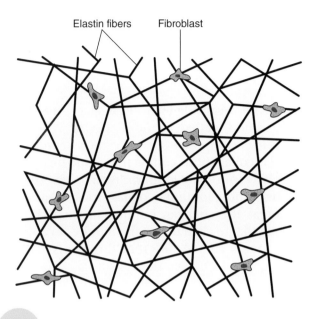

FIGURE 3.6 Irregular elastic connective tissue.

Regular Ordinary Connective Tissues

There are two main types of regular ordinary connective tissue: regular collagenous and regular elastic.

Regular Collagenous Connective Tissue

Regular collagenous connective tissue consists almost entirely of collagen fiber bundles arranged parallel to each other. Usually there are few elastic fibers and little ground substance. The only cells present are fibroblasts arranged in columns between the collagen bundles (figure 3.7). The collagen bundles are gathered together in the form of thick cords, bands, or sheets of various widths. In the unloaded state, the collagen bundles have a slightly wavy orientation. When stretched, the bundles quickly straighten, and the tissue becomes taut. Regular collagenous connective tissue is extremely strong and virtually inextensible. It has three main forms:

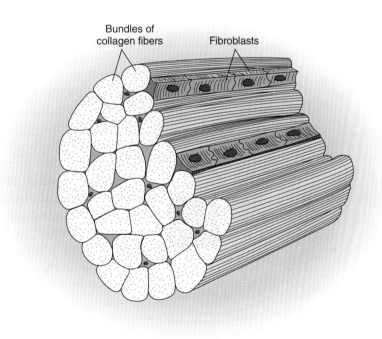

FIGURE 3.7 Regular collagenous connective tissue: part of a tendon.

1. Tendons and aponeuroses: mechanical links between skeletal muscle and bone

2. Ligaments and joint capsules: mechanical links between bones at joints

3. Retinacula: mechanical restraints on tendons that increase the mechanical efficiency of the musculotendinous units or the stability of the associated joints

Tendons and Aponeuroses   Skeletal muscles are attached to other structures, usually bones, by regular collagenous connective tissue in the form of tendons and aponeuroses. A tendon, also sometimes referred to as a sinew, is regular collagenous connective tissue in the form of a cord, as in figure 3.7, or a narrow band, such as the tendon that joins the latissimus dorsi muscle to the humerus (figure 3.8). An aponeurosis is a broad sheet of regular collagenous connective tissue, such as the aponeurosis that joins the latissimus dorsi to the lower half of the vertebral column (figure 3.8).

Ligaments and Joint Capsules   Skeletal muscles provide active links—contractile links—between bones. Ligaments and joint capsules provide passive links—noncontractile links—between bones. In association with

the skeletal muscles, the ligaments and joint capsules bring about normal joint movement. Chapters 4, 5, and 6 describe the various joints of the body. However, at this point it is sufficient to appreciate that each synovial (freely moveable) joint—a joint involving sliding or rolling between the free surfaces of the ends of bones, as in the shoulder and hip—is enclosed within its own joint capsule (figures 3.9 and 3.10).

The joint capsule encloses a space, usually quite small, called the joint cavity. A joint capsule is composed of two or more layers of regular collagenous connective tissue forming a sleeve around the joint, rather like a piece of rubber tubing joining two glass rods together. Whereas the collagen bundles in each layer are parallel to each other, the bundles in adjacent layers run in different directions. This arrangement enables the capsule to strongly resist stretching in a number of directions and therefore helps to maintain joint integrity (figure 3.11).

In all synovial joints the joint capsule is supported by a number of ligaments. These ligaments may be capsular or noncapsular. A **capsular ligament** is a distinct thickening in part of the joint capsule that provides additional strength in one direction. For example, the superior iliofemoral ligament, inferior iliofemoral ligament, and pubofemoral ligament are capsular ligaments that strengthen the anterior aspect of the capsule of the hip joint (see figure 3.10b). A **noncapsular ligament** is a distinct band separate from the joint capsule or only partially attached to it. Noncapsular ligaments may be **extracapsular** (outside the joint cavity) or **intracapsular** (inside the joint cavity). For example, the ligamentum teres of the hip joint is intracapsular, but the lateral ligament, medial ligament, and cruciate ligaments of the knee joint are all extracapsular (figure 3.10a and figure 3.12). Noncapsular ligaments usually consist of a single layer of tissue, but broad ligaments can consist of two or more layers, similar to a joint capsule.

In addition to the noncapsular ligaments associated with synovial joints, other ligaments similar in structure to noncapsular ligaments help stabilize other parts of the skeleton, for example, the ligaments between the clavicle

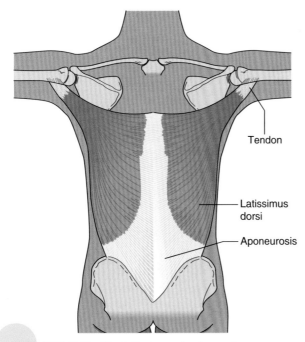

Tendon

Latissimus dorsi

Aponeurosis

FIGURE 3.8   The latissimus dorsi muscles.

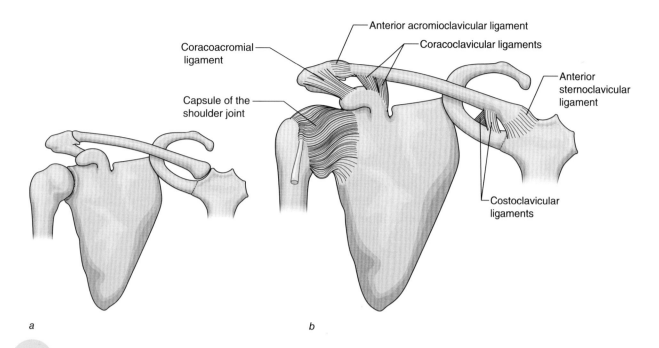

Anterior acromioclavicular ligament

Coracoclavicular ligaments

Coracoacromial ligament

Anterior sternoclavicular ligament

Capsule of the shoulder joint

Costoclavicular ligaments

a

b

FIGURE 3.9 Ligaments of the shoulder girdle. *(a)* Right anterior aspect of the right shoulder girdle and right shoulder joint. *(b)* Shoulder joint capsule and ligaments supporting the shoulder girdle.

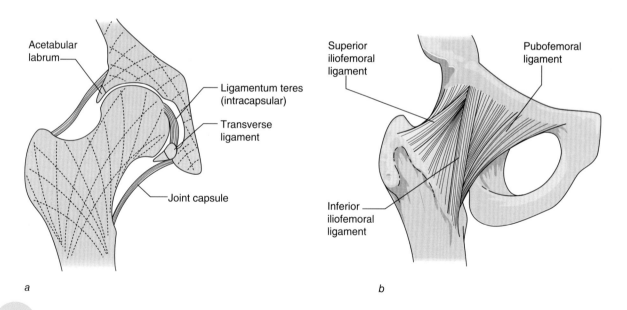

Acetabular labrum

Ligamentum teres (intracapsular)

Transverse ligament

Joint capsule

Superior iliofemoral ligament

Pubofemoral ligament

Inferior iliofemoral ligament

a

b

FIGURE 3.10 Ligaments of the hip joint. *(a)* Coronal section through the right hip joint showing the joint capsule and ligamentum teres. *(b)* Anterior aspect of the right hip joint showing the joint capsule and anterior capsular ligaments.

and scapula and between the clavicle and first rib (see figure 3.9*b*).

Key Terms

capsular ligament A distinct thickening in part of the joint capsule that provides additional strength in one direction.

noncapsular ligament A distinct band separate from the joint capsule or only partially attached to it.

extracapsular ligament A noncapsular ligament outside the joint cavity.

intracapsular ligament A noncapsular ligament inside the joint cavity.

Retinacula A retinaculum is a fairly broad, single-layered sheet of regular collagenous connective tissue that holds in position the tendons of some muscles where they cross certain joints. There are two forms of retinacula: The first type is in the form of a guy rope that restricts the side-to-side movement of a tendon. For example, two retinacula, one on each side of the knee joint, restrain the patella and the quadriceps tendon (figure 3.13). These retinacula help to maintain normal movement between the patella and the femur during flexion and extension of the knee.

The second type of retinacula is in the form of a pulley that restrains or redirects the line of action of the tendons of one or more muscles that cross over a particular joint. For example, extensor retinacula on the anterior aspect of the ankle restrain and redirect the lines of action of the extensor hallucis longus and tibialis anterior muscles, which are aligned with the lower leg (figure 3.14). The extensor retinacula prevent the tendons of the extensor hallucis longus and tibialis anterior muscles from springing away from the ankle joint when the muscles contract. Similarly, the medial

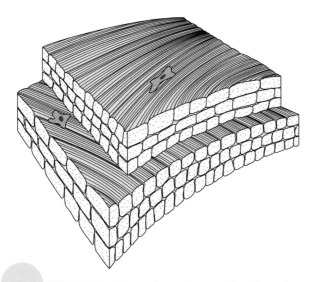

FIGURE 3.11 Three-dimensional section through a two-layered joint capsule.

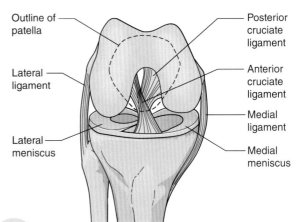

FIGURE 3.12 Extracapsular ligaments of the knee joint. Anterior aspect of the right knee, flexed at 90°, with patella removed to show the cruciate ligaments and femoral condyles slightly raised to show the menisci. The cruciate ligaments are located at the center of the joint but lie outside the joint cavity.

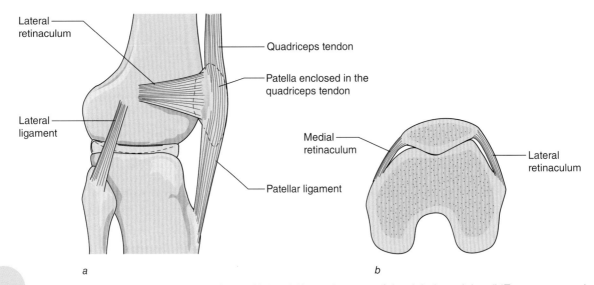

a

b

FIGURE 3.13 Retinacula of the patellofemoral joint. *(a)* Lateral aspect of the right knee joint. *(b)* Transverse section through the patellofemoral joint of the right knee joint.

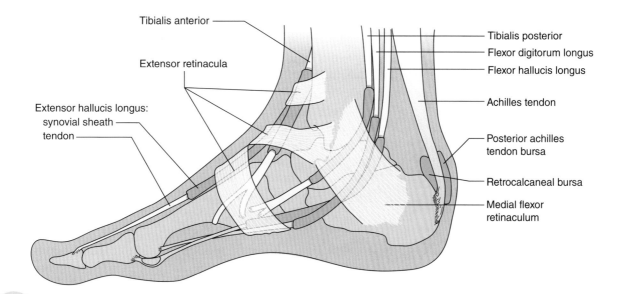

FIGURE 3.14 Retinacula at the ankle.

flexor retinaculum of the ankle joint, in association with the medial malleolus of the tibia, restrains the tendons of the tibialis posterior, flexor digitorum longus, and flexor hallucis longus muscles behind the medial malleolus and prevents the tendons of these muscles from slipping around the medial aspect of the medial malleolus when the muscles contract (figure 3.14). This form of retinaculum considerably increases the mechanical efficiency of the associated muscles.

Regular Elastic Connective Tissue

Regular elastic connective tissue consists largely of elastin fibers arranged parallel to each other. The proportion of collagen fibers and ground substance is usually fairly small. However, the proportion of collagen fibers and ground substance in regular elastic connective tissue is usually greater than the proportion of elastic fibers and ground substance in regular collagenous connective tissue (Akeson et al 1985, Zhang et al 2005). Regular elastic connective tissue is found where moderate amounts of strength and elasticity are required mainly in a single direction. As previously described, most ligaments consist of regular collagenous connective tissue, but a few consist of regular elastic connective tissue. Two of these so-called elastic ligaments (ligamentum

nuchae, ligamentum flavum; see chapter 5) help to stabilize the vertebral column and to allow a certain amount of movement between the vertebrae.

Fibrous Tissue, Elastic Tissue, and Fascia

The matrixes of four of the six main types of ordinary connective tissue are dominated by collagen or elastin fibers: irregular collagenous, irregular elastic, regular collagenous, and regular elastic. Whereas all of these tissues could be described as fibrous, **fibrous tissue** normally refers only to regular or irregular collagenous tissue. **Elastic tissue** normally refers only to regular or irregular elastic tissue.

The term **fascia,** from the Latin for bandage, refers to any type of ordinary connective tissue in the form of a sheet. In this sense, all aponeuroses are fascia. However, fascia most often refers to superficial fascia and deep fascia. Superficial fascia refers to the continuous layer of loose connective tissue that connects the skin to underlying muscle or bone. This layer of loose connective tissue is closely associated with the subcutaneous layer of fat referred to earlier. Deep fascia describes the sheets of irregular collagenous tissue that form sheaths around muscles and groups of muscles, separating them into functional units.

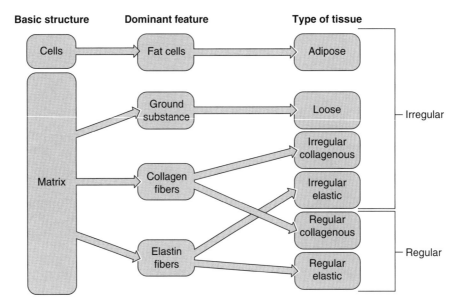

FIGURE 3.15 Ordinary connective tissues.

Figure 3.15 summarizes the six types of ordinary connective tissues in relation to the dominant feature of their matrixes and the arrangement of their fibrous content.

Key Terms

fibrous tissue Regular or irregular collagenous tissue.

elastic tissue Regular or irregular elastic tissue.

fascia Any type of ordinary connective tissue in the form of a sheet.

> **KEY POINT**
>
> The structural differences among types of ordinary connective tissues are in the proportions of elastin, collagen, and ground substance and in the arrangement—irregular or regular—of the fibrous content. There are four types of irregular ordinary connective tissue: loose, adipose, irregular collagenous, and irregular elastic. There are two types of regular ordinary connective tissue: regular collagenous and regular elastic.

Cartilage

The matrix of cartilage is similar to that of fibrous and elastic ordinary connective tissues in that it consists mainly of collagen and elastin fibers embedded in ground substance. However, relative to ordinary connective tissues, the ground substance of cartilage is highly specialized (Caplan 1984, Alexander 1992). It consists of huge carbohydrate–protein molecular complexes, called proteoglycans, suspended in a large amount of water. The large amount of water is attributable to the fact that proteoglycans have a high affinity for water; each proteoglycan complex is capable of attracting to itself a volume of water that is many times

its own weight. Consequently, under normal circumstances, water is the chief constituent of cartilage. The proteoglycans and water produce a highly viscous gel usually referred to as proteoglycan gel. In combination with collagen and elastin, the proteoglycan gel forms a tough, rubbery material, often called gristle, capable of strongly resisting all types of loads, especially bending and twisting. In comparison, fibrous and elastic ordinary connective tissues are only designed to resist tension loads.

Cartilage, like the other connective tissues, is a **composite material**—a material that is stronger than any of the separate substances from which it is made (Alexander 1968, 1992). Consequently, cartilage is stronger than either the fibers (collagen or elastin) or the proteoglycan gel. Wood and bone are other examples of natural composite materials. Fiberglass and the type of rubber from which tires are made are examples of manmade composite materials (Alexander 1968, 1992).

> **KEY POINT**
>
> A cartilage matrix consists of collagen and elastin fibers embedded in proteoglycan gel. In combination with collagen and elastin, the proteoglycan gel forms a tough, rubbery material capable of resisting all forms of loading. The main functions of cartilage are to facilitate load transmission and joint movement.

The only cells found in cartilage are cartilage cells that produce the cartilage matrix. The cells are called chondrocytes (mature cells) or chondroblasts (immature cells); they lie in fluid-filled spaces called lacunae distributed throughout the matrix (figure 3.16). The cells are arranged singly (parent cells) and in groups of two to five cells that originate from a single parent cell. As the cells become mature they

separate from their parent groups and start to produce new groups. Whereas collagen and elastin fibers dominate the matrix of all three main types of cartilage, the region around each lacuna is usually free of fibers. This distinct region—the capsule of the lacuna—consists of proteoglycan gel, which is denser than in other parts of the matrix.

With the exception of the layer of cartilage covering the articulating surfaces of bones in synovial joints, the surface of other cartilages, such as the costal cartilages (chapter 2), is usually covered by a sheath of fibrous tissue called the perichondrium (*peri* = around, *chondrium* = cartilage) (Standring 2004). Some of the fibroblasts of the perichondrium are transformed into chondroblasts, which eventually become chondrocytes. Non-weight-bearing perichondrium normally has a number of blood vessels running through it.

Mature cartilage contains no blood vessels (except in non-weight-bearing perichondrium) or nerves (Nordin and Frankel 2001). This reflects the mechanical functions of cartilage: Blood vessels and nerves would be destroyed by the deformation of cartilage in response to loading. In the absence of a direct blood supply, the cartilage cells depend on intercellular

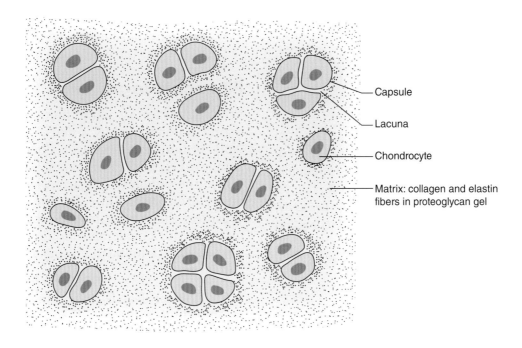

Capsule

Lacuna

Chondrocyte

Matrix: collagen and elastin fibers in proteoglycan gel

FIGURE 3.16 Typical structure of cartilage; specifically, hyaline cartilage is shown here.

exchange via the proteoglycan gel for their nutrition and excretion. For this reason, and the fact that cartilage is often under considerable load, repair of cartilage is slow or may not take place at all (Caplan 1984).

When cartilage is subjected to any type of load, water is forced out of the cartilage and the cartilage deforms. The rate and extent of deformation depend on the size and duration of the load. When the load is removed, the proteoglycan structures restore the original level of water saturation and, consequently, the original size and shape of the cartilage by absorbing water into the cartilage. The ability of a material to gradually deform in response to a load and to gradually restore its original size and shape following unloading is referred to as **viscoelasticity.** In comparison, **elasticity** refers to a material's ability to deform immediately in response to a load and to immediately restore its original size and shape following unloading, like a rubber band. Cartilage tends to behave viscoelastically in response to prolonged loading and elastically in response to sudden impact loads (see chapter 10).

The degree of viscoelasticity, elasticity, and strength of cartilage depends on the proportions of collagen fibers, elastin fibers, and proteoglycan gel in the matrix. There are three types of cartilage: hyaline cartilage, fibrocartilage, and elastic cartilage.

Key Terms

composite material A material that is stronger than any of the separate substances from which it is made.

viscoelasticity The ability of a material to gradually deform in response to loading and, following unloading, to gradually restore its original size and shape.

elasticity The ability of a material to deform immediately in response to loading and, following unloading, to immediately restore its original size and shape.

Hyaline Cartilage

Hyaline cartilage is the most abundant type of cartilage in the body (Tortora 2004). It has a pearly bluish-white tinge, and under a low-power microscope the matrix appears amorphous and translucent (semitransparent), as in figure 3.16. Under a high-power microscope the matrix can be seen to consist of a dense network of very fine collagen fibrils and fibers embedded in proteoglycan gel (Standring 2004). The size, shape, and arrangement of the cells and fibers in hyaline cartilage vary in different parts of the body depending on the function. Most of the skeleton is preformed in hyaline cartilage, and prior to maturity the growth and development of many bones are largely determined by the hyaline cartilage content of the bones (discussed later in chapter). In addition, hyaline cartilage comprises the following:

- Articular cartilage, which forms the smooth, tough, wear-resistant articular surfaces of bones in synovial joints
- The costal cartilages, which link the upper 10 pairs of ribs to the sternum and provide the rib cage with flexibility and elasticity
- Supporting rings within the elastic walls of the trachea (windpipe) and the larger bronchial tubes
- Part of the supporting framework of the larynx (voice box)
- The external flexible part of the nose that forms the major part of the nostrils

Fibrocartilage

In fibrocartilage, or white fibrocartilage, the matrix is dominated by a dense regular network of bundles of collagen fibers arranged parallel to each other in several layers (figure 3.17). The bundles in adjacent layers run in different directions (like the layers in a joint capsule); this structure produces a strong material with a moderate amount of elasticity. Fibrocartilage is found in a number of locations and forms:

- Fibrocartilage is present in complete or incomplete discs interposed between the articular surfaces of some synovial joints including

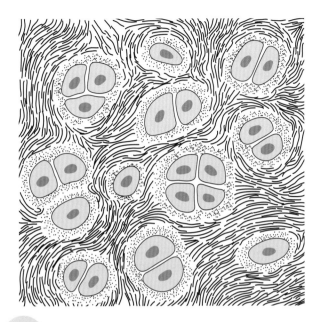

FIGURE 3.17 Fibrocartilage.

the knee joint, the sternoclavicular joint, and the acromioclavicular joint. In these joints the discs improve the congruence (area over which the joint reaction forces are distributed) and stability of the joints (see chapter 4). In addition, the discs deform in response to loading and thereby provide shock absorption.

• Fibrocartilage exists in complete discs that join the bones in certain joints (symphysis joints; see chapter 4). These joints include the pubic symphysis and the joints between the bodies of the vertebrae. In these joints, deformation of the disc in response to loading allows movement between the articulated bones and provides shock absorption.

• Fibrocartilage is present in the lip around the border of each glenoid fossa and acetabulum. The lips increase the areas of articulation and, therefore, the stability of the shoulder and hip joints.

• Fibrocartilage forms the lining of bony grooves, such as the bicipital groove (see figure 2.25), which are occupied by tendons. The grooves act as pulleys that normally increase the mechanical efficiency of the associated muscles.

Elastic Cartilage

In elastic cartilage, or yellow elastic cartilage, the matrix is dominated by a dense network of elastin fibers (figure 3.18). Elastic cartilage provides support with a moderate to high degree of elasticity. It is found mainly in the larynx, the external part of the ear (pinna), and the tube leading from the middle part of the ear to the throat (eustachian or auditory tube).

> **KEY POINT**
>
> There are three main types of cartilage: hyaline cartilage, fibrocartilage, and elastic cartilage. Most of the skeleton is preformed in hyaline cartilage, and bone growth is largely determined by the hyaline cartilage content of the bones. Hyaline cartilage forms the articular surfaces in synovial joints. Fibrocartilage improves congruence and provides shock absorption in some synovial joints and provides shock absorption in symphysis joints. Elastic cartilage helps maintain the shape of certain structures, including the larynx and external part of the ear.

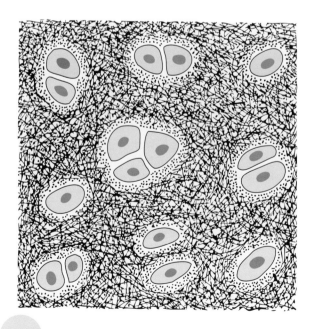

FIGURE 3.18 Elastic cartilage.

Bone

The strongest and most rigid of all the connective tissues is bone. The matrix of bone consists of a dense, layered, regular network of collagen fibers embedded in a hard, solid, ground substance. The ground substance is called bone salt and consists of a combination of calcium phosphate and calcium carbonate with smaller amounts of magnesium, sodium, and chlorine (Alexander 1975, Tortora 2004). In mature bone, bone salt makes up about 70% of the total weight of bone, with collagen making up the remaining 30%. Bone salt is denser than collagen such that the bone salt and collagen both occupy about 50% of the total volume. The composite material made of bone salt and collagen produces a hard, tough, fairly rigid structure. Relative to cast iron, bone has the same tensile strength, is only one third as heavy, and is much more elastic (Ascenzi and Bell 1971, Tortora 2004). The elasticity of bone, although slight relative to cartilage, enables bone to absorb a sudden impact without breaking. The ability of the ends of bones, together with the articular cartilage, to deform in response to loading is also important in maintaining normal transmission of loads across joints.

KEY POINT

The matrix of bone consists of a dense, layered, regular network of collagen fibers embedded in a hard ground substance called bone salt. In combination with each other, the bone salt and collagen produce a very hard, tough material with little, although some, elasticity.

Bone Growth and Development

Around the third week of intrauterine (within the uterus) life, the embryo's skeleton starts to appear in the form of blocks and plates of tissue. The blocks and plates of most of the embryonic skeleton consist of hyaline cartilage. The top of the skull, the clavicles, and

parts of the mandible, however, are formed in a highly vascular form of tissue called fibrous membrane. By the eighth or ninth week of intrauterine life the shapes of the embryonic bones are similar to their eventual adult shape (Standring 2004).

Ossification

Ossification or osteogenesis is the process by which the embryonic skeleton is transformed into bone (*osteo* = bone, *genesis* = creation). The ossification of fibrous membranes is called **intramembranous ossification,** and the ossification of hyaline cartilage is called **intracartilaginous** or **endochondral ossification** (*endo* = within, *chondral* = cartilage). Both forms of ossification are similar and produce the same type of bone tissue. In the following sections, the process of endochondral ossification is described with reference to a typical long bone.

Growth in Girth

An embryonic long bone consists of a block of hyaline cartilage covered in a fibrous perichondrium (figure 3.19*a*). Between the fifth and twelfth weeks of intrauterine life, some fibroblasts in the perichondrium around the middle of the shaft of the cartilage model are transformed into osteoblasts. Osteoblasts are one of three types of bone cells and are responsible for the production of bone. The newly formed osteoblasts invade the hyaline cartilage immediately beneath the perichondrium and start to deposit calcium and other minerals in the matrix. Consequently, the hyaline cartilage is transformed into calcified cartilage. This process of mineralization is called **calcification;** calcified cartilage represents an intermediate stage in the process of ossification of cartilage into bone. Calcification continues until the calcified cartilage is transformed into bone. Consequently, a bony ring or collar is formed around the middle of the shaft of the otherwise cartilaginous model (figure 3.19*b*).

When the perichondrium starts to produce osteoblasts and, in turn, bone, it is called **periosteum.** The first site of bone formation, the middle of the shaft of the cartilage model, is called the **primary center of ossification.** The process of ossification proceeds from the bony collar in two directions: across the shaft from the outside toward the center, and toward the ends of the shaft. By the 36th week, around the time of birth, the bony collar has become a bony cylinder running the length of the shaft but not progressing into the bulbous ends of the bone (figure 3.19*c*). The bony cylinder is thickest at its middle and thinnest at its ends. By this time the remaining hyaline cartilage in the middle of the shaft has been transformed into calcified cartilage. Soon afterward, a second type of bone cells called osteoclasts invade this central portion of calcified cartilage. Whereas osteoblasts produce new bone, osteoclasts remove bone and calcified cartilage.

The osteoclasts start to remove the calcified cartilage in the middle of the shaft, thereby creating a space called the medullary cavity (figure 3.19*d*). The medullary cavity gradually widens and extends toward both ends of the shaft. Simultaneously, the thickness of bone in the shaft gradually increases. The development of the skeleton prior to birth, especially the rate at which ossification occurs, is partly the result of the loading exerted on the skeleton by the developing muscles, which is manifested in increased movement of the fetus.

Eventually all of the calcified cartilage is removed by the combined effects of ossification across the shaft from the outside toward the center and osteoclastic activity from the center outward. By this time, the medullary cavity is occupied by yellow marrow consisting of loose

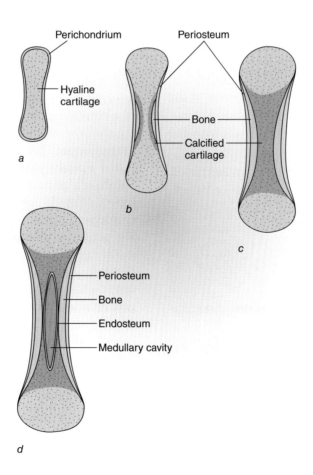

FIGURE 3.19 Early stages in the endochondral ossification of a long bone.

connective tissue containing a large number of blood vessels, fat cells, and immature white blood cells (Standring 2004). A layer of loose connective tissue containing many osteoblasts and a smaller number of osteoclasts lines the medullary cavity. This layer is called the endosteum.

Long bones are designed to resist bending. For a given amount of bone tissue, a hollow shaft is stronger in bending than a solid one (Alexander 1968, 1992), which is the main reason that the shafts of long bones are hollow. Growth in girth of the shaft of a long bone involves formation of new bone on the outside of the shaft by osteoblasts in the periosteum and the removal of bone from the inside of the shaft by osteoclasts in the endosteum. The type of growth produced by the periosteum, which involves laying down new bone on the surface of older bone rather like the addition of rings in a tree, is called **appositional growth.** In mature bone the periosteum consists of irregular fibrous tissue. In addition to producing appositional growth, the periosteum has three other main functions:

1. To provide a protective cover around the shaft of the bone

2. To allow blood vessels to pass into the bone

3. To provide attachments for muscles, tendons, ligaments, and joint capsules

Key Terms

ossification   The process by which the embryonic skeleton is transformed into bone.

intramembranous ossification   Ossification of fibrous membranes.

intracartilaginous ossification   Ossification of hyaline cartilage.

endochondral ossification   Another name for intracartilaginous ossification.

calcification   The process of mineralization of hyaline cartilage into calcified cartilage.

periosteum   The layer of fibrous tissue that covers the nonarticular surfaces of a bone;

responsible for growth in girth of a bone by appositional growth.

primary center of ossification   The first site of bone formation in a bone.

appositional growth   The type of growth in which new tissue is laid down on the surface of existing tissue.

Growth in Length

Around the time of birth a **secondary center of ossification** occurs in the center of each end of a long bone. These new centers of ossification are responsible for the ossification of the ends of the bones; ossification proceeds from the center toward the periphery. After the secondary centers of ossification have been established, the only hyaline cartilage remaining from the original cartilage model is that covering the bulbous ends of the bone and the plates of cartilage separating the ends of the bone from the shaft (figure 3.20). These two regions of hyaline cartilage are continuous with each other and remain so until maturity.

Part of the cartilage covering each end of a bone forms an articular surface and is referred to as **articular cartilage.** Each end of a bone is called an **epiphysis** and the shaft is called the **diaphysis.** The plates of cartilage that separate the epiphyses and diaphysis are called **epiphyseal plates** (figure 3.20). The epiphyseal plates are responsible for growth in length of the bone. During normal growth the epiphyseal plates remain active until the bone has achieved its mature length.

An epiphyseal plate consists of four layers (Tortora 2004) (figure 3.21). The layer adjacent to the epiphysis is called the reserve or germinal layer. In this layer, which anchors the epiphyseal plate to the bone of the epiphysis, the chondrocytes are distributed throughout the matrix usually as single cells or in pairs. The second layer is called the proliferation layer. As its name suggests, it is responsible for chondrogenesis (production of new cartilage). The chondrocytes in this layer undergo fairly rapid cell division, and the cells produce additional matrix that increases the amount of cartilage. Growth in length of the shaft of a bone is the result of chondrogenesis in the proliferation layers of the epiphyseal plates. This type of growth, in which additional new tissue is produced from within the mass of existing tissue, is called **interstitial growth.**

The third layer of the epiphyseal plate is called the hypertrophic layer. In this layer, the chondrocytes are arranged in columns and gradually increase in size, with the larger and more mature cells farthest from the epiphysis. The fourth layer of the epiphyseal plate is called the calcified layer. In this layer the hypertrophied chondrocytes and surrounding matrix are replaced by calcified cartilage. The calcified cartilage interdigitates with the underlying bone, forming a relatively strong bond between epiphysis and diaphysis (see figures 3.20 and 3.21). As new cartilage is formed in the proliferation layer, the calcified cartilage in contact with the underlying bone is itself gradually transformed into bone. The net result of these processes is that the epiphyseal plates, which remain about the same thickness, gradually move farther from the middle of the shaft as the shaft increases in length.

The **metaphysis** is the region where the epiphysis joins the diaphysis; in a growing bone this corresponds to the calcified layer of the epiphyseal plate together with the interdigitating bone (see figures 3.20 and 3.21). The interface between the hypertrophic and calcified layers is sometimes referred to as the tidemark.

When a long bone has achieved its mature length, longitudinal growth in the epiphyseal plates ceases. Shortly afterward, the epiphyseal plates are replaced by bone so that the epiphyses are fused with the shaft. In most long bones one end usually fuses with the shaft before the other end. In the long bones of the arms and legs, fusion of both ends normally

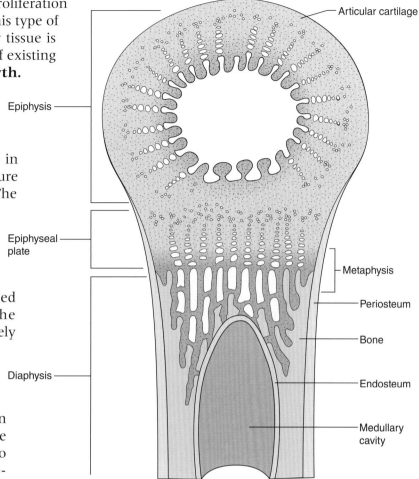

FIGURE 3.20 Longitudinal section through the epiphysis and part of the diaphysis of a typical immature long bone.

Labels: Articular cartilage; Epiphysis; Epiphyseal plate; Diaphysis; Metaphysis; Periosteum; Bone; Endosteum; Medullary cavity

KEY POINT

The adult bony skeleton develops from an embryonic skeleton that forms during the second month of intrauterine life and consists mainly of hyaline cartilage and fibrous membrane. The process by which cartilage and membrane are transformed into bone is called ossification. Growth in girth of bones occurs by appositional growth; growth in length of bones occurs by interstitial growth.

takes place between 14 and 20 years of age (Standring 2004). In some other bones, such as the innominate bones (which consist of three bones—ilium, pubis, ischium—prior to matu-

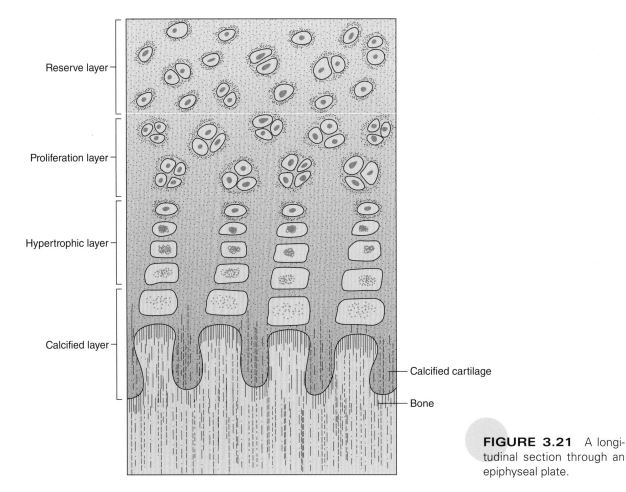

Reserve layer

Proliferation layer

Hypertrophic layer

Calcified layer

Calcified cartilage

Bone

FIGURE 3.21 A longitudinal section through an epiphyseal plate.

rity), fusion takes place usually between 20 and 25 years of age. It follows that the epiphyseal plates of the various bones are vulnerable to injury for a relatively long period. Injury to an epiphyseal plate may, in severe cases, result in one of two types of bone deformity (Pappas 1983, Caine et al 2006):

1. A complete cessation of growth and premature fusion resulting in, for example, a limb length discrepancy

2. An asymmetric cessation of growth across an epiphyseal plate resulting in an angular deformity and joint incongruity

The degree of bone deformity resulting from an epiphyseal plate injury depends on the following factors:

• An individual's physical maturity; the more mature the individual, the lower the likelihood of serious deformity

• The severity of the injury
• Which epiphyseal plate is injured

Following the establishment of epiphyseal plates, subsequent to the establishment of the secondary centers of ossification in the epiphyses, the epiphyseal plates at each end of a long bone usually contribute different amounts to the length of the shaft. For example, the proximal and distal epiphyseal plates of the humerus contribute approximately 80% and 20%, respectively, to the length of the shaft. In contrast, the proximal and distal epiphyseal plates of the femur contribute approximately 30% and 70%, respectively, to the length of the shaft (Pappas 1983) (figure 3.22). Injury to the epiphyseal plate that makes the largest contribution to the total length of a bone is likely to have a greater effect on bone growth than injury to the other epiphyseal plate (Siffert 1987).

Key Terms

secondary center of ossification   The second site of bone formation in a bone.

articular cartilage   The layer of hyaline cartilage that covers each articular surface of a bone (in a synovial joint).

epiphysis   End of a bone separated from the diaphysis by an epiphyseal plate prior to maturity.

diaphysis   The shaft of a bone.

epiphyseal plate   A region of hyaline cartilage that separates an epiphysis from the diaphysis prior to maturity; responsible for growth in length of the bone by interstitial growth.

interstitial growth   The type of growth in which new tissue is produced from within the mass of existing tissue.

metaphysis   The region of a bone where the epiphysis joins the diaphysis; in a grow-

ing bone this corresponds to the calcified layer of the epiphyseal plate together with the interdigitating bone.

The epiphyseal plates are vulnerable to injury because they are the weakest parts of the immature skeleton. For example, ligaments and joint capsules are two to five times stronger than epiphyseal plates (Larson and McMahan 1966, Doschak and Zernicke 2005). When a ligament supporting a particular joint is inserted into the epiphysis (rather than the diaphysis), a load applied to the joint that tends to stretch the ligament is, in a child, more likely to result in a fracture through the epiphyseal plate than in a tear in the ligament. In an adult, the same type of loading would tend to cause a ligament tear because the epiphysis and diaphysis are fused (Pappas 1983) (figure 3.23).

Growth of Epiphyses

Just as the epiphyseal plates are responsible for growth in length of a bone, the hyaline cartilage that covers the end of a bone is responsible for growth of the epiphysis. This cartilage consists of an articular region and a nonarticular region. Like an epiphyseal plate, articular cartilage (and its adjacent nonarticular regions; see figure 3.20) consists of four layers. The only real difference in structure between articular cartilage and an epiphyseal plate is in the arrangement of the fibers in the reserve layer.

In an epiphyseal plate the collagen fibers cross each other obliquely, forming a strong bond between the epiphyseal bone and the reserve layer of the plate. In articular cartilage

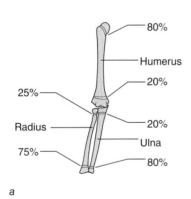

a

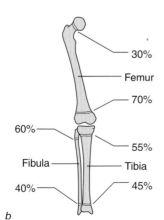

b

FIGURE 3.22   Contributions of the proximal and distal epiphyseal plates to growth in length of the long bones of the *(a)* upper limb and *(b)* lower limb.

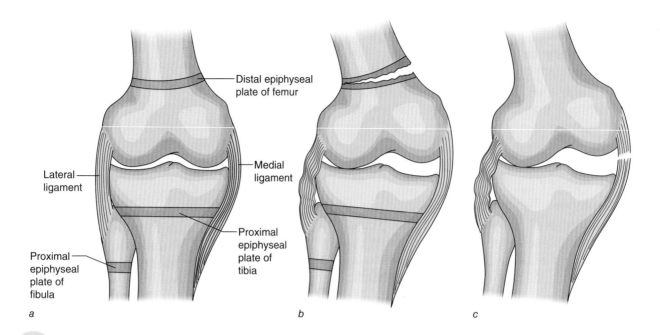

FIGURE 3.23 Effect of degree of skeletal maturity on type of injury. *(a)* Anterior aspect of the right knee joint showing normal alignment of the femur, tibia and fibula. *(b, c)* Excessive abduction of the knee joint. *(b)* In a child this is more likely to result in a fracture through the distal epiphyseal plate of the femur than tearing of the medial ligament. *(c)* After maturity, it is likely to result in partial or complete tearing of the medial ligament.

the reserve layer is the outer layer. Whereas the majority of the layer is similar in structure to the reserve layer of an epiphyseal plate, the outer surface of articular cartilage is cell free and consists of densely packed collagen fibers and fibrils arranged parallel to the articular surface. This arrangement produces a tough, wear-resistant surface.

The type of growth produced by articular cartilage is the same as that produced by an epiphyseal plate—interstitial growth. During the growth period, the rate of ossification of an epiphysis is greater than the rate of growth of the epiphysis. Consequently, the articular cartilage becomes relatively thinner with age (figure 3.24). At maturity, the thickness of articular cartilage is approximately 1 to 7 mm, and this tends to decrease with age because of mechanical wear. Whereas bone growth is largely determined by genetic factors, the mechanical stress experienced by articular cartilage and epiphyseal plates, as a result of movement and the maintenance of an upright posture, also has a major effect on bone growth (see chapter 11).

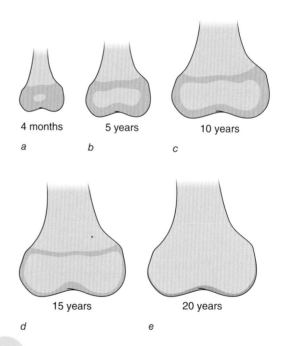

FIGURE 3.24 Successive stages in the ossification of the distal femoral epiphysis.

In a number of bones, such as some of the carpals and tarsals, ossification is completed from a single (primary) center of ossification.

Other bones, such as metacarpals and metatarsals, have a primary center of ossification and only one secondary center of ossification; there is only one epiphysis, at one end of the bone. In all of the large long bones—the bones of the arms and legs—secondary centers of ossification occur in both ends of the bone around the time of birth.

> **KEY POINT**
>
> The articular cartilage and associated nonarticular regions of an epiphysis are responsible for growth of the epiphysis by interstitial growth. During growth and development, the relative thickness of articular cartilage gradually decreases. At maturity it is approximately 1 to 7 mm.

Growth of Apophyses

Secondary centers of ossification occur not only in the epiphyses of long bones but also in some of the rudimentary tuberosities of some bones, including the femurs, innominate bones, and calcaneus bones (figure 3.25). These secondary centers of ossification occur in regions of bone called **apophyses** around 10 to 14 months after birth. Each apophysis grows and ossifies in much the same way as an epiphysis. Apophyses provide areas of attachment for the tendons of powerful muscles such as the quadriceps (tibial tuberosity), hamstrings (ischial tuberosity), and calf muscles (calcaneal tuberosity) (figure 3.26). This form of attachment is different from that of most tendons, which attach directly onto the periosteum.

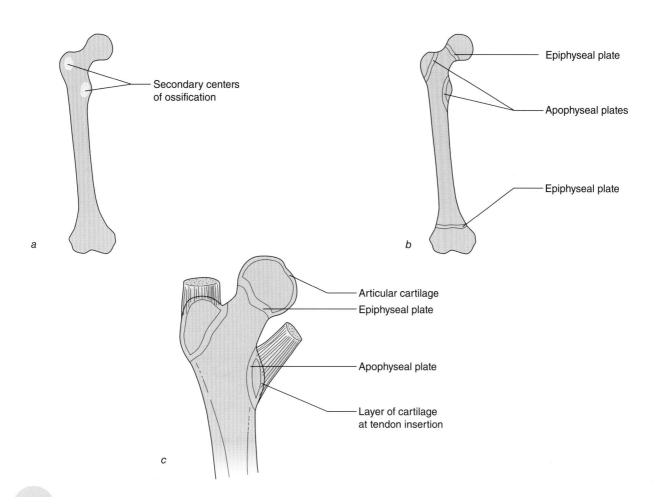

a — Secondary centers of ossification

b — Epiphyseal plate / Apophyseal plates / Epiphyseal plate

c — Articular cartilage / Epiphyseal plate / Apophyseal plate / Layer of cartilage at tendon insertion

FIGURE 3.25 Apophyses of the femur: greater trochanter and lesser trochanter. *(a)* Occurrence of secondary centers of ossification. *(b)* Apophyseal and epiphyseal plates of the femur. *(c)* Growth areas of the head of the femur and the greater and lesser trochanters.

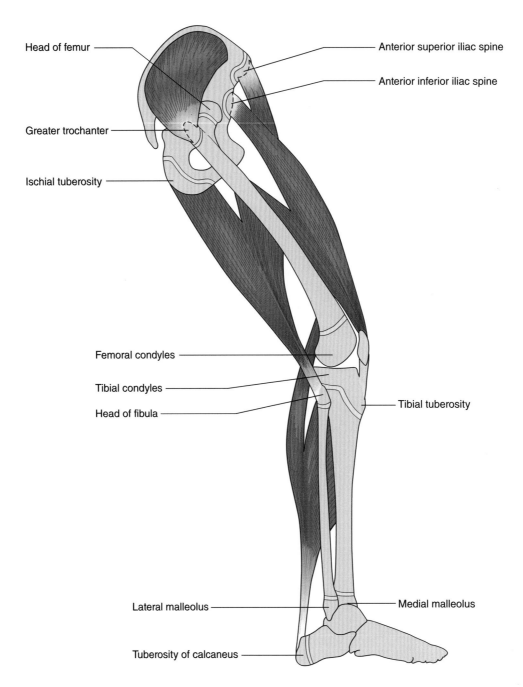

Head of femur

Anterior superior iliac spine

Anterior inferior iliac spine

Greater trochanter

Ischial tuberosity

Femoral condyles

Tibial condyles

Head of fibula

Tibial tuberosity

Lateral malleolus

Medial malleolus

Tuberosity of calcaneus

FIGURE 3.26 Major epiphyses (head of femur, femoral condyles, tibial condyles) and apophyses (anterior superior and anterior inferior iliac spines, greater trochanter, ischial tuberosity, lateral malleolus, medial malleolus, tuberosity of the calcaneus) of the lower limb.

Prior to maturity each apophysis is separated from the rest of the bone by an **apophyseal plate,** which is very similar in structure and function to an epiphyseal plate. Each apophyseal plate is responsible for growth of the bone adjacent to the nonapophyseal side of the plate.

Growth of the apophysis itself is the result of interstitial growth in the layer of cartilage (mixture of hyaline cartilage and fibrocartilage) outside of the apophysis into which the fibers of the tendon insert (see figure 3.25c). At maturity, the apophyses fuse with the rest of the bone.

Key Terms

apophysis A tuberosity separated from the rest of a bone prior to maturity by an apophyseal plate.

apophyseal plate A region of hyaline cartilage that separates an apophysis from the rest of the bone prior to maturity; responsible for growth of the bone adjacent to the nonapophyseal side of the plate.

KEY POINT

Apophyses provide areas of attachment for powerful muscles. Prior to maturity, each apophysis is separated from the rest of the bone by an apophyseal plate that is similar in structure and function to an epiphyseal plate. At maturity, the apophyses fuse with the rest of the bone.

Epiphyses, especially those that form weight-bearing joints, are most frequently subjected to compression loading and therefore are often referred to as pressure epiphyses. In contrast, apophyses are most frequently subjected to tension loading and are often referred to as traction epiphyses. Whereas apophyseal growth plates do not affect growth in bone length, they do affect the alignment and strength of the tendons attached to them. Consequently, injury to apophyseal plates may affect the mechanical characteristics of associated muscles, which, in turn, may affect normal joint function.

Studies of sport-related injuries in children show that the proportion of injuries involving growth plates (epiphyseal and apophyseal) is between 6% and 18% of the total number of injuries (Speer and Braun 1985, Krueger-Franke et al 1992, Gross et al 1994, Caine et al 2006). About 5% of these growth plate injuries result in some type of bone deformity (Larson 1973). On the basis of these figures, the number of growth-plate injuries resulting in bone deformity is in the region of 3 to 9 per thousand. However, this estimate is likely to be conservative because many injuries that occur during free play and sports are not reported or are incorrectly diagnosed (Combs 1994, Caine et al 2006). Case study 1 summarizes a study on the occurrence of physeal injuries in children's and youth sports and provides strategies for preventing physeal injuries.

CASE STUDY 1 PHYSEAL INJURIES IN YOUTH SPORTS

Caine D, DiFiori J, Maffulli N. 2006. Physeal injuries in children's and youth sports: Reasons for concern? *British Journal of Sports Medicine* 40:749-760.

Participation in children's and youth sports is widespread in Western culture. Many children start year-round training and specialization by 9 years of age. Preteens training at regional centers or with high school and club teams in sports such as gymnastics may train more than 20 hr per week. There is increasing concern, especially among physicians, that the frequency and intensity of training and competition experienced by many participants in children's and youth sports are putting them at risk of serious physeal injury (epiphyseal and apophyseal), which may result in permanent skeletal abnormalities. Unfortunately, detailed information, based on the use of standardized recording systems applied over a long period, on the epidemiology of injuries (incidence, location, type, diagnosis, severity) sustained by participants in children's and youth sports is not available for any sport. The purpose of this study was to systematically review the literature on the frequency and characteristics of physeal injuries in children's and youth sports.

The review was undertaken using Medline and SPORTdiscus. More than 150 reports were obtained, largely case reports or case series investigations. The authors used the reports to

(continued)

determine the number, location, type, diagnosis, and severity of physeal injuries that occur in different sports, but it was not possible to calculate incidence of injury because information was not provided on the total number of participants and exposure time of participants in the various sports.

The review indicates that acute physeal injuries (sudden widening or fracture along or through an epiphyseal or apophyseal plate accompanied by considerable pain) and chronic physeal injuries (progressive widening of an epiphyseal or apophyseal plate associated with a progressive increase in pain, especially during exercise) occur frequently in some sports, including football, baseball, gymnastics, basketball, volleyball, judo, weightlifting, soccer, rugby, tennis, cricket, and long-distance running. The main regions of physeal injuries are the shoulder, elbow, wrist, and knee. Most of the injuries are chronic, and most resolve without complication. However, there are several reports of premature partial or complete physeal closure.

The authors express a number of concerns with current practice, in particular, the paucity of epidemiological data on the distribution and determinants of physeal injuries in children's and youth sports and the apparent lack of knowledge on the part of many coaches of children's and youth sports regarding musculoskeletal growth and development in children in general and physeal injuries in particular.

Application

Coaches should consider the following strategies to reduce the incidence of physeal injuries in children's and youth sports:

- Individualize fitness training and practice for athletes experiencing rapid growth. Assess growth rate by monitoring increases in height and limb segment lengths.
- Use a variety of training and practice drills and avoid excessive volume of training.
- For collision sports, ensure that competition is based on physical maturity rather than chronological age.
- Mandate regular medical assessment to ensure early diagnosis and treatment of epiphyseal and apophyseal plate disorders.

Structure of Mature Bone

During the period of growth and development of a long bone, bone tissue is deposited in a manner that maximizes the strength of the bone as a whole. The development of a hollow shaft is just one example of this process. Different regions of the bone are subjected to different types and magnitudes of loading. For example, the epiphyses are mainly subjected to compression loading, whereas the shaft is mainly subjected to bending and twisting loads. For this reason, not only is the shaft hollow, but the density and thickness of bone in the shaft are greater than in the epiphyses because the shaft is subjected to the greatest bending and twisting loads. Dense bone is called **compact**

bone. The thickness of compact bone in the shaft gradually decreases from the middle of the shaft toward the epiphyses (figure 3.27).

In each epiphysis the calcified layer of the articular cartilage merges with the underlying bone. This transitional region is referred to as subchondral bone; it encloses a mass of low-density bone that makes up the remainder of the epiphysis. The low-density bone is in the form of a honeycomb or latticework consisting of thin curved bars of bone attached to each other by interconnecting bars of bone. The bars of bone are called trabeculae and the latticework formed by the trabeculae is called trabecular bone, spongy bone or, most often, **cancellous bone** because of the large number of spaces between the trabeculae. The major-

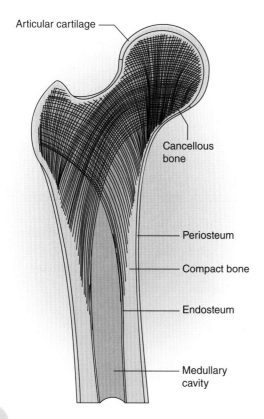

Articular cartilage

Cancellous bone

Periosteum

Compact bone

Endosteum

Medullary cavity

FIGURE 3.27 The structure of a mature long bone: a longitudinal section through the proximal third of the femur.

ity of the trabeculae cross each other at right angles, an arrangement that maximizes the strength of the trabecular bone. The spaces between the trabeculae are filled with red marrow—loose connective tissue containing a large number of blood vessels, some white blood cells and fat cells, and a large number of cells called erythroblasts responsible for producing red blood cells. The spaces in cancellous bone are continuous with the medullary cavity, and therefore the red marrow is continuous with the yellow marrow.

In most joints, especially weight-bearing joints, the loading on the epiphyses is different in different positions of the joint. The change in loading may be in terms of the type, magnitude, or direction of the load or a combination of these characteristics. The trabeculae are arranged to minimize the stress experienced by the epiphyses in all positions of the joint during habitual movements.

The only real structural difference between compact bone and cancellous bone is in the density of the bone. Compact bone is much more dense and therefore much less elastic than cancellous bone. The elasticity of cancellous bone is very important in ensuring congruity in joints during load transmission, thereby minimizing stress within the epiphyses and on the articular cartilages (Ascenzi and Bell 1971, Radin 1984, Standring 2004).

Compact and Cancellous Bone

Compact bone consists of complete and incomplete columns of bone closely packed together (figure 3.28). Each column of bone is called an **osteon** or a haversian system. In long bones the osteons run parallel to the long axis of the bone. Each osteon consists of three to nine concentric rings (or layers) of bone surrounding a central open channel. The concentric rings of bone are called lamellae and the central channel is called a haversian canal. Each lamella basically consists of a single layer of closely packed collagen fibers arranged parallel to each other, with bone salt embedded between the fibers. Whereas the collagen fibers in each lamella are parallel to each other, the orientation of fibers in adjacent lamellae is different (figure 3.29). This arrangement, similar to the layers of collagen fibers in a joint capsule, enables the bone to strongly resist deformation in any direction. In addition to the lamellae found in osteons, a number of circumferential or surface lamellae are found in the outer surface of compact bone; these lamellae encircle the bone and tend to bind the osteons together (figure 3.28b).

Between the lamellae are numerous osteocytes, which are osteoblasts that have become entrapped in the bone. Osteocytes are responsible for repairing bone damage and are also involved in regulating the level of minerals, especially calcium and phosphorus, in the blood (Bailey et al 1986, Parfitt 2002). Each osteocyte lies in a lacuna, a small space, and the lacunae are linked together by tiny channels called canaliculi. The canaliculi run between the lamellae and through the lamellae from one side to the other. This system of canaliculi enables the osteocytes to communicate with each other both physically, by means of projections from the cell bodies into the canaliculi,

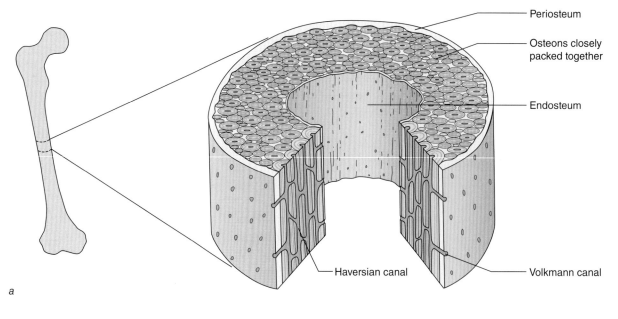

Periosteum

Osteons closely
packed together

Endosteum

Haversian canal

Volkmann canal

a

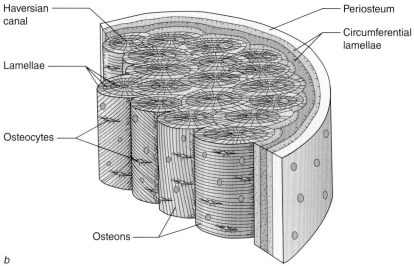

Haversian
canal

Lamellae

Osteocytes

Osteons

Periosteum

Circumferential
lamellae

b

FIGURE 3.28 Structure of compact bone.

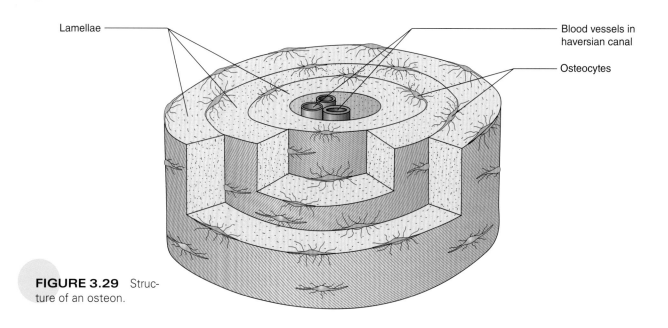

Lamellae

Blood vessels in
haversian canal

Osteocytes

FIGURE 3.29 Structure of an osteon.

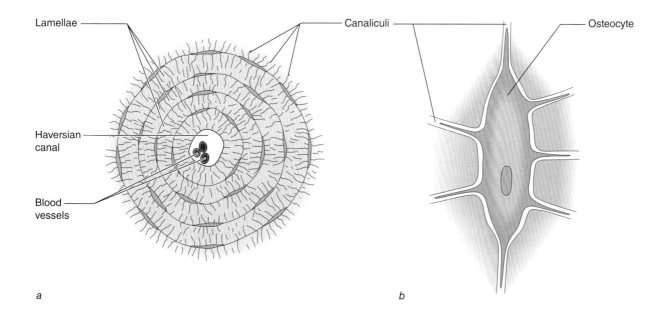

Lamellae — Canaliculi — Osteocyte

Haversian canal

Blood vessels

a   b

FIGURE 3.30 Osteocytes and canaliculi. *(a)* Cross section through an osteon showing circumferential and radial canaliculi. *(b)* An osteon lying in a lacuna with projections of the cell body into the canaliculi.

and chemically, by means of secretions from the cells (see figures 3.29 and 3.30). Communication between the osteocytes is important for coordinating the growth, development, and repair of lamellae and for facilitating intercellular exchange among osteocytes and between the osteocytes and blood vessels (Nijweide et al 2002, Ferretti et al 2003).

The haversian canal at the center of each osteon contains blood vessels and nerves supported by loose connective tissue. The canaliculi are linked to the haversian canals, thereby facilitating intercellular exchange between blood vessels and osteocytes. In addition to being linked together by canaliculi, the haversian canals of adjacent osteons are linked together by channels called volkmann canals, which are similar in size to haversian canals.

Volkmann canals, like haversian canals, contain blood vessels and nerves supported by loose connective tissue. By linking together the periosteum, haversian canals, and endosteum, the volkmann canals form a system of channels that run from the outside of the bone to the medullary cavity (see figure 3.28*a*). The system of haversian and volkmann canals enables blood vessels and nerves to pass along, around, and across the bone.

In compact bone the osteons are very closely packed with little or no space between them. In cancellous bone the osteons are loosely packed with spaces in between them filled with red marrow; the osteons are usually arranged in small groups that form trabeculae. Many of the trabeculae do not contain haversian canals and consist of several lamellae in the form of a narrow strip. Consequently, cancellous bone consists of a mixture of osteonal and nonosteonal bone.

Key Terms

compact bone   Bone in which the osteons are closely packed with little or no space between them.

cancellous bone   Bone in which the osteons are loosely packed with spaces between them filled with red marrow; the osteons are usually arranged in the form of trabeculae.

osteon   A column of bone consisting of three to nine concentric layers of bone surrounding a haversian canal.

In a mature long bone, the diaphysis consists of a cylinder of compact bone thickest in the middle and tapered toward each end. Each epiphysis consists of a relatively thin outer layer of subchondral bone enclosing a mass of cancellous bone. The cancellous bone extends into the diaphysis and tapers toward the middle. The articular surfaces are covered by articular cartilage.

Modeling and Remodeling in Bone

The processes of growth, development, and maintenance of the bones of the skeleton are carried out by the interaction of three subprocesses: the expression of the skeletal genotype, modeling, and remodeling.

- The expression of the **skeletal genotype** refers to the process of genetically programmed change in the external form (size and shape) and internal architecture of the bones.

- **Modeling** refers to the changes in the expression of the skeletal genotype that occur as a result of environmental factors such as nutrition and, in particular, the mechanical strains imposed by normal habitual activity.

- **Remodeling** refers to the coordination of osteoblastic and osteoclastic activity responsible for the actual changes in external form and internal architecture of the bones, including repair of bones (figure 3.31).

Because bone is continuously being absorbed from some places (by osteoclasts) and deposited in others (by osteoblasts), the process of remodeling is sometimes referred to as turnover. Prior to maturity, all bones are in a continual state of change in external form and internal architecture. After skeletal maturity is achieved (approximately 20-25 years of age), modeling of external form decreases to negligible proportions, but modeling of internal architecture continues throughout life (Frost 1979, Bailey 1995, Frost 2003).

Key Terms

skeletal genotype The process of genetically programmed change in the external form (size and shape) and internal architecture of the bones.

modeling Changes in the expression of the skeletal genotype that occur as a result of environmental influences.

remodeling Coordination of osteoblastic and osteoclastic activity resulting in changes in external form and internal architecture of the bones, including repair of bones.

Growth, development, and maintenance of the bones of the skeleton are determined by the interaction of three processes: the expression of the skeletal genotype, modeling, and remodeling.

Porosity, Osteopenia, and Osteoporosis

Because of the channels and spaces within compact and cancellous bone, any particular region of a bone consists of certain amounts of bone tissue and nonbone tissue. The term **porosity** describes the proportion of nonbone tissue. At skeletal maturity the porosity of compact and cancellous bone is approxi-

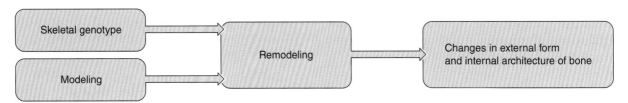

FIGURE 3.31 The relationships among skeletal genotype, modeling, and remodeling in the growth, development, and maintenance of bone.

mately 2% and 50%, respectively; the density (amount of bone tissue per unit volume) of compact bone is approximately double that of cancellous bone (Radin 1984, Tortora 2004). The density of bone tissue depends on the degree of mineralization. During ossification, the degree of mineralization of bone tissue gradually increases and reaches a maximum level at skeletal maturity (Bailey et al 1986). However, the amount of bone within the skeleton may continue to increase for 5 to 10 years after skeletal maturity, especially in people who are physically active (Stillman et al 1986, Talmage and Anderson 1984, Frost 2003). Consequently, bone mass peaks in males and females between 25 and 30 years of age. In terms of turnover, this means that from skeletal maturity to the age at which peak bone mass occurs, more new bone is formed than old and damaged bone is absorbed.

Following peak bone mass there is usually a stable period in which the amount of bone in the skeleton remains about the same; there is a balance between bone absorption and bone formation. This stable period is followed by a gradual decrease in bone mass for the rest of the person's life; the rate of bone absorption exceeds the rate of bone formation. Bone mass is the product of bone volume and bone density. The loss in bone mass that occurs with age following peak bone mass is the result of decreases in bone volume and bone density. **Osteopenia** refers to a level of bone density below the normal level for a person's age and sex (Bailey 1995, Frost 1997).

Bone mass starts to decrease earlier and at a greater rate in females than in males. In males, bone loss normally starts between 45 and 50 years of age and proceeds at a rate of 0.4% to 0.75% per year (Bailey et al 1986, Smith 1982). In females, bone loss has three phases. The first phase starts around 30 to 35 years of age and proceeds at a rate of 0.75% to 1% per year until menopause. From menopause until about 5 years after menopause, the rate of bone loss increases to between 2% and 3% per year. During the final phase, the rate of bone loss is approximately 1% per year. Thus, women may lose, on average, about 53% of their peak bone mass by the age of 80 years. In contrast, males may lose, on average, about 18% of their peak bone mass by the age of 80 years (figure 3.32).

Even though body weight tends to decrease with age, the rate of bone loss is usually much greater than the rate at which body weight decreases. Consequently, the effect of bone loss is that the bones, especially weight-bearing bones, become progressively weaker relative to the weight of the rest of the body. In addition to gradually losing strength, the bones also gradually lose their elasticity and become more

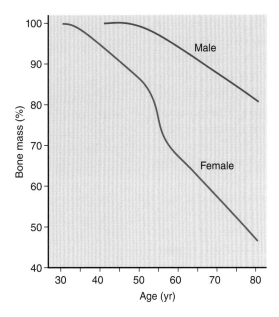

FIGURE 3.32 Effect of aging on bone mass.

brittle. In some people, especially women, a loss of bone mass and elasticity is eventually reached when some bones are no longer able to withstand the loads imposed by normal habitual activity. These bones become very susceptible to fracture. This condition, the most common bone disorder in elderly people, is called **osteoporosis** (Bailey et al 1986, Ferretti et al 2003). Osteoporosis may cause severe disfigurement, especially of the trunk, as a result of fractured or crushed vertebrae. Many deaths in elderly people are due to complications arising from bone fractures that occur as a result of osteoporosis (Kaplan 1983, Kado et al 1999, Cummings and Melton 2002).

Bone loss tends to occur earlier and to proceed at a faster rate in cancellous bone than in compact bone (Bailey et al 1986, Ferretti et al 2003). Consequently, regions of bones with a high proportion of cancellous bone, such as the bodies of the vertebrae, the head and neck of the femur, and the distal end of the radius, are particularly vulnerable to osteoporosis and fracture in elderly people. This vulnerability is reflected in studies that report a rapid increase in the incidence of bone fractures with age, especially in women (Bauer 1960, Chalmers and Ho 1970, Hagino et al 1991, Cummings et al 1993, Court-Brown and Caesar 2006). The results of one study showed that the incidence of fracture to the distal end of the radius was seven times higher in 54-year-old women than in 40-year-old women (Bauer 1960). In another study the incidence of fracture of the neck of the femur was found to be 50 times higher in 70-year-old women than in 40-year-old women (Chalmers and Ho 1970). Bone loss in compact bone occurs mainly on the endosteal surface so that bone width remains relatively unchanged into old age (Smith 1982, Ferretti et al 2003).

Although the cause of osteoporosis is not clear, there is general agreement that four variables are mainly responsible: genetic factors, endocrine status, nutritional factors, and physical activity (Bailey et al 1986, MacKinnon 1988, Ferretti et al 2003). The relative contribution of these variables has not been established, but physical activity seems to be the most important. In the absence of weight-bearing activity, no amount of endocrine or nutritional intervention will prevent rapid bone loss; there must be mechanical stress (Bailey et al 1986, Ferretti et al 2003). Research suggests that regular physical activity throughout life, within the moderate overload range (see chapter 11), can help to prevent osteoporosis in three ways (Bailey 1995, Greene and Naughton 2006, Baxter-Jones et al 2008):

1. Peak bone mass is directly related to the level of physical activity prior to peak bone mass; the higher the peak bone mass, the lower the risk of osteoporosis.
2. An above-average level of physical activity after peak bone mass will delay the onset of bone loss.
3. An above-average level of physical activity after peak bone mass will reduce the rate of bone loss.

Case study 2 discusses whether regular exercise can be used to prevent falls in osteopenic women.

Key Terms

porosity The proportion of nonbone tissue in a bone or region of a bone.

osteopenia Bone density below the normal level for the age and sex of the individual.

osteoporosis Loss of bone mass and elasticity to the extent that bones are no longer able to withstand the loads imposed by normal, habitual activity, resulting in a high susceptibility to fracture.

KEY POINT

From about 30 years of age in women and about 45 years of age in men, bone mass and bone elasticity gradually decrease. Many people, especially women, develop osteoporosis, which may cause severe disfigurement and in some cases death due to complications arising from osteoporotic bone fractures. Whereas the cause of osteoporosis is not clear, research suggests that one of the main causes is lack of mechanical stress. Regular physical activity throughout life appears to be the best way of preventing osteoporosis.

CASE STUDY 2 — USING EXERCISE TO PREVENT FALLS IN OSTEOPENIC WOMEN

Hourigan SR, Nitz JC, Brauer S, O'Neill S, Wong J, Richardson CA. 2008. Positive effects of exercise on falls and fracture risk in osteopenic women. *Osteoporosis International* 19:1077-1086.

The incidence of bone fractures increases with age. As the number of aged people worldwide is increasing, so is the total number of fractures in the aged population (age 65 years and older). Fractures in the aged population are a major health care cost and risk factor for permanent disability and death. Consequently, reducing the risk of fractures in the aged population is a challenge in many countries.

Physical trauma (in particular, falls) and weak bones (due to osteopenia or osteoporosis) are the major causes of fractures in the aged population. Poor balance and weak muscles are risk factors for falls. Osteopenia is defined as bone mineral density (BMD) between 1 and 2.5 standard deviations below the average for young women. Osteoporosis is defined as BMD more than 2.5 standard deviations below the average for young women.

Recent research indicates that muscle strength, bone strength, and balance in the aged population can be improved through properly prescribed physical activity training programs based on resistance exercises involving strength-training machines. The purpose of the present study was to determine the effect of a physical activity program based on weight-bearing activities (rather than strength-training machines) on balance, muscle strength, and bone mineral density in osteopenic women.

Ninety-eight community-dwelling (living more or less independently in their own homes) osteopenic women (age 62.01 ± 8.9; range 41-78 years) were randomly assigned to either a control group (n = 48: no intervention) or an exercise group (n = 50: two 1 hr exercise sessions per week for 20 weeks directed by a trained physiotherapist). Assessments at baseline and post intervention included balance (five measures), strength (quadriceps; hip abductors, adductors, and external rotators; trunk extensors), and BMD (proximal femur and lumbar spine). Baseline assessment showed no significant differences between the exercise and control groups in terms of balance, strength, BMD, or demographics.

The average number of sessions attended by the 42 members of the exercise group who completed the exercise program was 28.2 of a possible 40 sessions (71.2%). Following the intervention, the exercise group showed significantly better performance than the control group in 9 of 11 balance tests (ranging from 10% to 71% better performance) and 7 of 9 strength tests (ranging from 9% to 23% better performance). BMD of the exercise group increased but was not significantly greater than in the control group.

Application

A specific, well-directed program of weight-bearing exercises in a workstation format that emphasizes interaction, discussion, and enjoyment can significantly improve balance and strength in osteopenic women, which in turn is likely to reduce the risk of falling. This type of training may also positively influence BMD, but further research is needed.

Summary

The dominant structural feature of all connective tissues is a large mass of noncellular matrix; the physical characteristics of the matrix of each connective tissue determine its function. Connective tissues provide mechanical support at all levels of cellular organization and facilitate intercellular exchange. The mechanical function of connective tissues is clearly evident in the next chapter, which describes the articular system.

Review Questions

1. Describe the two main functions of connective tissues.
2. Describe the difference between the following:
   - Regular and irregular ordinary connective tissues
   - Loose connective tissue and adipose connective tissue
   - Capsular and noncapsular ligaments
   - Intracapsular and extracapsular ligaments
   - Ligaments and retinacula
   - Fibrous tissue and elastic tissue
3. Describe the types of cells found in ordinary connective tissues.
4. Describe the two main functions of cartilage.
5. Explain the difference between elasticity and viscoelasticity.
6. Describe the difference in structure and function between regular fibrous tissue and fibrocartilage.
7. Explain why repair in cartilage is usually slow and may not take place at all.
8. Differentiate between or among the following:
   - Primary and secondary centers of ossification
   - Appositional and interstitial growth
   - Epiphyseal and apophyseal plates
   - Osteoblasts, osteoclasts, and osteocytes
   - Compact and cancellous bone
   - Osteopenia and osteoporosis

The Articular System

The human body is capable of a broad range of movements facilitated by the combined effects of the open-chain arrangement of the bones, the number of joints linking the bones, the different types of joints, and the range of movement in the joints. Most joints allow a certain amount of movement, and all joints transmit forces. Joints differ in terms of the type and range of movement and the mechanism of force transmission; these differences are reflected in the structure of the joints. This chapter describes the structure and function of the various types of joints.

4

OBJECTIVES

After reading this chapter, you should be able to do the following:

1. Distinguish between temporary and permanent joints.
2. Describe the structural classification of joints.
3. Describe the structure and specific functions of the various forms of fibrous and cartilaginous joints.
4. Describe the structure of a synovial joint.
5. Describe the stability–flexibility classification of joints.
6. Explain the relationship of the structural classification to the stability–flexibility classification of joints.
7. Describe the various forms of synovial joints.
8. Describe the functions of joint capsules and ligaments.
9. Differentiate between flexibility, laxity, stability, and congruence.

A **joint,** also referred to as an articulation or an arthrosis, is defined as a region where two or more bones are connected. The adult skeleton normally has 206 bones linked by approximately 320 joints. **Articular system** refers to all of the joints of the body. The function of joints is largely mechanical—to facilitate relative motion between bones and transmission of force from one bone to another.

The joints of the adult articular system can be regarded as permanent joints because they are present throughout a person's life, although the structure and function of some of these joints may change with increasing age. In addition to having permanent joints, the immature articular system also has a large number of temporary joints, which are concerned with bone growth, that gradually become less distinct and are virtually obliterated at maturity. Consequently, in relation to the immature articular system, the term *joint* refers not only to the regions where the separate bones are connected but also to the regions that unite the bony parts of each immature bone.

Key Terms

joint A region where two or more bones are connected.

articular system All of the joints of the body.

All of the joints in the adult articular system—permanent joints—are present in the immature skeletal system. However, the immature articular system also has a number of temporary joints, concerned with bone growth, which gradually become less distinct and are virtually obliterated at maturity. The joints of the adult articular system facilitate relative motion between bones and the transmission of force from one bone to another.

Structural Classification of Joints

In terms of structure, there are two types of joints:

1. Joints in which the articular (opposed) surfaces of the bones are united by either fibrous tissue or cartilage are called fibrous joints and cartilaginous joints, respectively (figure 4.1a).

2. Joints in which the articular surfaces are not attached to each other but are held

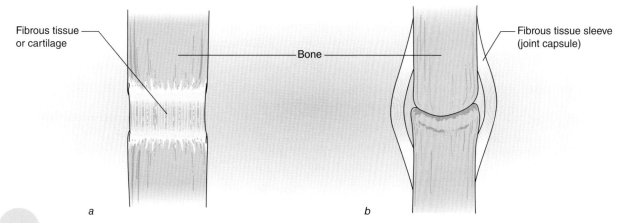

Fibrous tissue or cartilage

Bone

Fibrous tissue sleeve (joint capsule)

a

b

FIGURE 4.1 Two types of structure of joints. *(a)* Articular surfaces united by fibrous tissue or cartilage. *(b)* Articular surfaces not attached to each other but held in contact with each other by a fibrous tissue sleeve.

in contact with each other by a sleeve of fibrous tissue supported by ligaments (figure 4.1*b*) are referred to as synovial joints, and the fibrous sleeve is the joint capsule (see chapter 3).

Fibrous Joints

Fibrous joints are often referred to as syndesmoses (*syn* = with, *desmo* = ligament). The degree of movement in a **syndesmosis** is largely determined by the amount of fibrous tissue between the articular surfaces. In general, the smaller the amount of fibrous tissue, the more limited the movement. There are two types of syndesmoses: membranous and sutural.

Membranous Syndesmoses

In a membranous syndesmosis, the articular surfaces are united by a sheet of fibrous tissue called an interosseous membrane (*inter* = between, *osseous* = bone). The interosseous membrane acts rather like webbing. The interosseous borders of the radius and ulna are connected by an interosseous membrane (figure 4.2*a*). The majority of the fibers in the membrane run obliquely downward and medially from the interosseous border of the radius to that of the ulna. The remaining fibers run obliquely downward and laterally from the interosseous border of the ulna to that of the radius. The interosseous membrane stabilizes the radius and ulna throughout the range from

full supination to full pronation and provides areas of attachment for muscles on the anterior and posterior aspects of the lower arm (figure 4.2*c*). An interosseous membrane connects the medial border of the shaft of the fibula and the lateral border of the shaft of the tibia (figure 4.2*b*). This interosseous membrane, which is similar in structure to the interosseous membrane in the lower arm, stabilizes the tibia and fibula in all positions of the ankle joint and provides areas of attachment for muscles on the anterior and posterior aspects of the lower leg (figure 4.2*d*).

Whereas the interosseous membranes in the lower arms and lower legs are permanent features, a particular group of syndesmoses, the sutures and fontanels of the skull, originate as membranous syndesmoses and then change to sutural syndesmoses (see figure 2.6).

Sutural Syndesmoses

In a sutural syndesmosis the articular surfaces are united by a thin layer of fibrous tissue, rather like the thin layer of concrete between bricks in a wall (figure 4.3). By late childhood all of the sutures and fontanels of the skull are converted from membranous to sutural syndesmoses. The thin layer of fibrous tissue, together with the interlocking of the articular surfaces, tends to severely limit movement in these joints. With increasing age the fibrous tissue in sutures is gradually replaced by bone so that each suture is transformed into a **synostosis** (*syn* = with, *osteo* = bone); that is, the

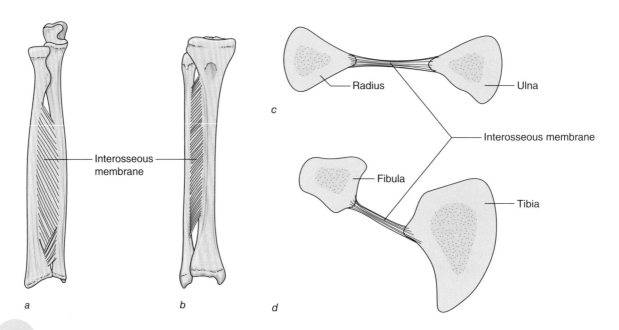

FIGURE 4.2 Anterior aspects and cross-sectional views of the membranous syndesmoses between *(a, c)* the right radius and ulna and *(b, d)* the right tibia and fibula .

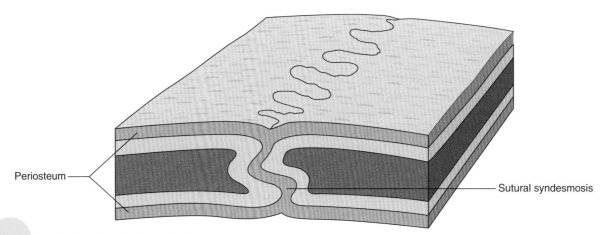

FIGURE 4.3 A typical sutural joint.

bones that were previously joined by fibrous tissue become a single bone.

Cartilaginous Joints

The degree of movement in cartilaginous joints is determined by the type and thickness of the cartilage. There are two kinds of **cartilaginous joints:** synchondroses and symphyses.

Synchondroses

In a **synchondrosis** (*syn* = with, *chondro* = cartilage), the articular surfaces are united by hyaline cartilage. There are two types of syn-

chondroses: temporary and permanent. The temporary synchondroses, sometimes referred to as physeal joints, include the following:

- Joints between bones that eventually fuse together to form larger bones in the adult skeleton. For example, each innominate bone is formed by the fusion of the corresponding ilium, ischium, and pubis.
- Joints formed by epiphyseal plates.
- Joints formed by apophyseal plates.
- The joints between the first ribs and the manubrium.

All temporary synchondroses are converted to synostoses at some stage of a person's life. A few synchondroses remain moderately flexible throughout life. These joints are the permanent synchondroses, and they include the joints between the anterior ends of the second to the tenth ribs and the sternum (see figure 2.19).

Symphyses

In a **symphysis** (*sym* = with, *physeal* = growth plate), the articular surfaces are united by a combination of hyaline cartilage and fibrocartilage. A layer of hyaline cartilage covers each articular surface, and sandwiched between the layers of hyaline cartilage is a relatively thick pad of fibrocartilage (figure 4.4). The fibrocartilage pad is often referred to as a disc even though it is usually kidney shaped or oval. The joint is normally supported by a number of ligaments that cross the outside of the joint and attach onto the periphery of the fibrocartilage pad. The bone, hyaline cartilage, and fibrocartilage in a symphysis joint are intimately connected; there are no sharp divisions between the three tissues but rather a gradual change from one tissue type to another. In effect, the joint consists of a single piece of material whose consistency varies across the joint.

Fibrocartilage readily deforms in response to bending and twisting loads, and the degree of movement in a symphysis joint is largely determined by the thickness of the fibrocartilage pad; the thicker the fibrocartilage, the greater the capacity for movement. The joints between the bodies of the vertebrae, the pubic symphysis, and the manubriosternal joint are all symphysis joints. Whereas most of the symphysis joints remain moderately flexible throughout life, some of them, such as the manubriosternal joint, may be converted to synostoses (Tortora and Anagnostakos 1984).

Synovial Joints

In the adult skeleton, approximately 80% of the joints are synovial (*syn* = with, *vial* = cavity). **Synovial joints** have greater range of movement than fibrous or cartilaginous joints. The body's capacity to adopt a broad range of postures is attributable largely to the range of movement in synovial joints. Virtually all of the joints in the upper and lower limbs are synovial.

Structure of Synovial Joints

As you recall from chapter 3, in a synovial joint each articular surface is covered with a layer of articular (hyaline) cartilage. The surfaces are not attached to each other but under normal circumstances are held in contact with each other, in all positions of the joint, by a joint capsule and various ligaments (figure 4.5). During movement of a synovial joint, the articular surfaces slide and roll on each other. The capsule encloses a joint cavity that is normally very small, given the close contact between the articular surfaces. In figure 4.5, the joint cavity is shown much larger than normal to differentiate the various features of the joint. The inner wall of the capsule and the nonarticular bony surfaces inside the joint are covered with synovial membrane. The synovial membrane consists of areolar tissue (see chapter 3) with specialized cells that secrete synovial fluid into the joint cavity. Synovial fluid is viscous and resembles the consistency of raw egg whites. It has two important functions.

1. A mechanical function: The fluid lubricates the articular surfaces so they slide over each other easily, preventing excessive wear.

2. A physiological function: The fluid seeps into the articular cartilage and nourishes the cartilage cells.

Figure 4.6 summarizes the structural classification of joints.

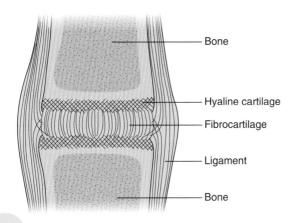

FIGURE 4.4 A typical symphysis joint.

Bone

Hyaline cartilage

Fibrocartilage

Ligament

Bone

Key Terms

fibrous joint A joint in which the opposed surfaces of the bones are united by fibrous tissue.

syndesmosis Another name for a fibrous joint.

synostosis A structure formed when the fibrous tissue in a fibrous joint becomes transformed into bone so that the bones that were previously joined by fibrous tissue become a single bone.

cartilaginous joint A joint in which the opposed surfaces of the bones are united by cartilage.

synchondrosis A cartilaginous joint in which the opposed surfaces are united by hyaline cartilage.

symphysis A cartilaginous joint in which the opposed surfaces are united by a combination of hyaline cartilage and fibrocartilage; a layer of hyaline cartilage covers each of the articular surfaces, and sandwiched between the layers of hyaline cartilage is a relatively thick pad of fibrocartilage.

synovial joint A joint in which the opposed surfaces are not attached to each other, but are held in contact with each other by ligaments and a joint capsule.

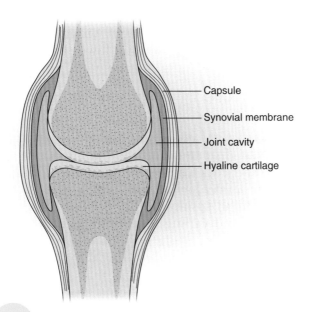

- Capsule
- Synovial membrane
- Joint cavity
- Hyaline cartilage

FIGURE 4.5 A typical synovial joint.

KEY POINT

In terms of structure, there are two types of joints:

- Fibrous or cartilaginous joints, where the articular surfaces are united by either fibrous tissue or cartilage
- Synovial joints, in which the articular surfaces are not attached to each other but are held in contact with each other by a joint capsule and ligaments

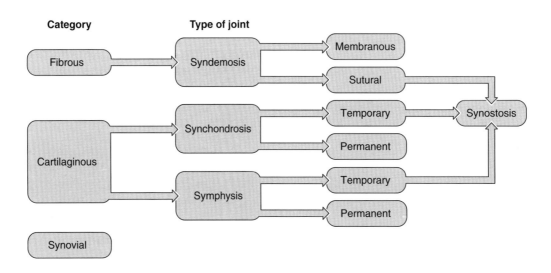

FIGURE 4.6 The structural classification of joints.

Congruence, Articular Discs, and Menisci

The articular surfaces in most synovial joints are reciprocally shaped, which normally results in a large contact area (relative to the area of the articular surfaces) between the opposed articular surfaces in all positions of the joint. For any particular joint position and joint reaction force, the larger the contact area between the articular surfaces—the larger the area over which the joint reaction force is transmitted—the lower the compressive stress on the articular surfaces and vice versa. Some synovial joints, such as the tibiofemoral joint, do not have reciprocally shaped articular surfaces so that the actual area of contact between the articular surfaces in any particular joint position is likely to be relatively small. However, in such joints the effective area of contact between the articular surfaces is normally as large as in joints with reciprocally shaped articular surfaces because of one or more pieces of fibrocartilage that form wedges between the unopposed parts of the articular surfaces and distribute the joint reaction force over a large area of the articular surfaces. The fibrocartilaginous wedges are not attached to the articular surfaces but are normally in contact with the articular surfaces. They are held in position by attachment to the inner wall of the joint capsule or by attachment to bone adjacent to the articular surface.

In the acromioclavicular joint, and sometimes the ulnar-carpal joint, a single fibrocartilage wedge in the form of a ring tapers from the outside toward the center (figure 4.7a). In the tibiofemoral joint there are normally two C-shaped fibrocartilage wedges; each wedge is called a **meniscus** (half-moon, crescent shape; figure 4.7b). Some joints, such as the sternoclavicular and ulnar-carpal joints, usually have a complete disc of fibrocartilage that effectively divides the joint into two joints (figure 4.7, c and d). A complete disc of fibrocartilage is referred to as an **articular disc.** The **congruence** of a joint refers to the area over which the joint reaction force is transmitted; in any particular joint position, the greater the area, the greater the congruence of the joint and vice versa. Normally, congruence is maximized to

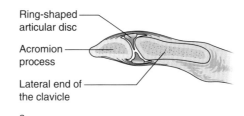

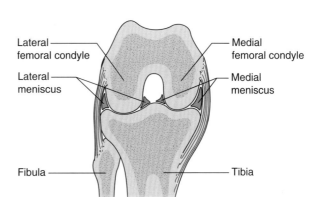

FIGURE 4.7 Articular discs and menisci. *(a)* Coronal section through the right acromioclavicular joint. *(b)* Coronal section through the right tibiofemoral joint with the joint flexed at approximately 90°. *(c)* Coronal section through the right sternoclavicular joint. *(d)* Coronal section through the left wrist joint.

minimize compressive stress on the articular surfaces. One of the main functions of articular discs and menisci is to improve congruence in joints. Improved congruence will tend to

- reduce compressive stress on the opposed articular surfaces,
- help maintain normal joint movements and effective distribution of synovial fluid over the articular surfaces, and
- improve shock absorption.

Key Terms

meniscus   A C-shaped wedge of fibrocartilage that helps to increase congruence in the tibiofemoral joint; there are normally two menisci in each tibiofemoral joint.

articular disc   A piece of fibrocartilage that helps to increase congruence in some synovial joints; the fibrocartilage may be in the form of a ring that tapers from the outside toward the center.

congruence   The area over which the joint reaction force is transmitted in a synovial joint; in any particular joint position, the greater the contact area between the articular surfaces, the greater the congruence and vice versa.

KEY POINT

The articular surfaces in most synovial joints are reciprocally shaped, which allows a high level of congruence in all positions of the joint. Some synovial joints do not have reciprocally shaped articular surfaces, and congruence in these joints is improved by articular discs.

Joint Movements

The type of movement that occurs in a joint, that is, the range and direction of movement of one articular surface with respect to the opposed articular surface (and therefore the change in the orientation of the bones to each other), depends on the type of joint and the shape of the articular surfaces.

Reference Axes and Degrees of Freedom

In describing a particular joint's movement, it is useful to refer to three mutually perpendicular reference axes. The three reference axes denote anteroposterior (front to back), mediolateral (side to side), and vertical directions with respect to the anatomical position. Figure 4.8 shows the position of the reference axes in relation to the shoulder joint. With respect to the reference axes, there are six possible directions, called **degrees of freedom,** in which the shoulder joint, or any other joint, might be able to move, depending on its structure. The six directions consist of three linear directions (along the axes) and three angular directions (around the axes). A joint with six degrees of freedom could move in any direc-

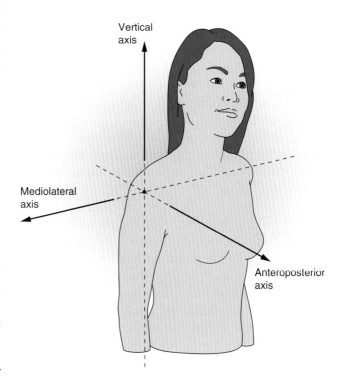

FIGURE 4.8   Reference axes with respect to the shoulder joint.

tion by a combination of linear and angular movements. Some cartilaginous joints have six degrees of freedom, albeit within a small range of movement. In contrast, the larger synovial joints tend to have no linear degrees of freedom, but they usually have one to three angular degrees of freedom with a relatively large range of movement.

Most movements in everyday life, such as walking, bending, reaching, and writing, involve simultaneous or sequential movement in two or more joints. In such multijoint movements, the degrees of freedom in the whole of the segmental chain responsible for the movement is the sum of the degrees of freedom of the individual joints in the chain. Consequently, an almost infinite number of combinations of joint movements can be used in all multijoint movements. Furthermore, temporary or permanent impairment in one joint in a segmental chain can usually be compensated by a change in the movement of other joints in the chain (see chapter 12).

Key Term

degrees of freedom The linear and angular directions of movement considered normal for a joint with respect to reference anteroposterior, mediolateral, and vertical axes providing six possible degrees of freedom—three linear directions (along the axes) and three angular directions (around the axes).

Angular Movements

Angular movements in joints refer to rotations around the three reference axes. With the anatomical position as a reference position, special terms describe the various angular movements.

In most joints the terms *abduction* and *adduction* refer to rotations around the anteroposterior axis. In the shoulder, wrist, and hip joints, abduction and adduction refer to movement of the arm, hand, and leg away from and toward the median plane, respectively (figures 4.9 and 4.10*a*). In the hand and foot, abduction of the metacarpophalangeal joints (hand) and metatarsophalangeal joints (foot) occurs when the fingers and toes are spread, and adduction occurs when the fingers and toes are returned to the reference position (figure 4.10*b*).

In most joints the terms *flexion* and *extension* refer to rotation around the mediolateral axis. In the shoulder, wrist, and hip joints, flexion refers to movement of the arm, hand, and leg forward, and extension refers to movement of the arm, hand, and leg backward (figure 4.10*c*; figure 4.11, *a* and *b*). In the elbow, knee,

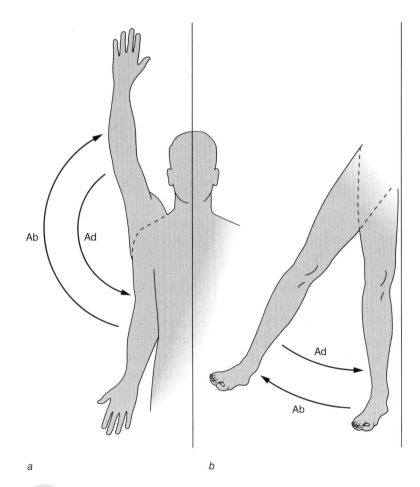

a

b

FIGURE 4.9 Abduction and adduction of *(a)* the shoulder joint and *(b)* the hip joint.

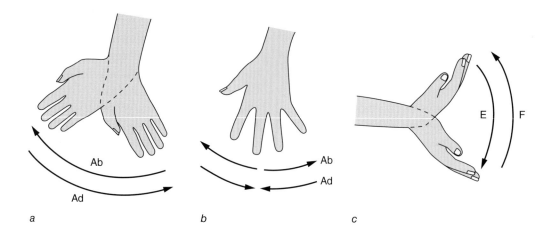

FIGURE 4.10 Angular movements of the wrist and fingers. *(a)* Abduction (Ab) and adduction (Ad) of the wrist. *(b)* Abduction and adduction of the fingers. *(c)* Flexion (F) and extension (E) of the wrist.

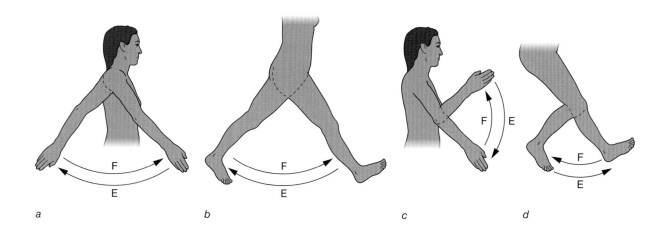

FIGURE 4.11 Flexion (F) and extension (E) of *(a)* the shoulder joint, *(b)* the hip joint, *(c)* the elbow joint, and *(d)* the knee joint.

metacarpophalangeal, metatarsophalangeal, and interphalangeal (hand and foot) joints, flexion occurs when the joints bend and extension occurs when the joints straighten (figure 4.11, *c* and *d*). In the trunk—the vertebral column as a whole—flexion refers to bending the trunk forward and extension refers to the reverse movement (figure 4.12, *a* and *b*). Lateral flexion of the trunk occurs when the trunk bends to the side about an anteroposterior axis (figure 4.12*c*).

Circumduction describes a movement in which the distal end of a bone or limb moves in a circle while the proximal end (at which joint movement takes place) stays in the same place.

Consequently, the movement of the bone or limb describes a cone shape. All joints capable of flexion and extension and abduction and adduction, such as the shoulder, wrist, metacarpophalangeal, and hip joints, are capable of circumduction.

Some joints, such as the shoulder, hip, and the vertebral column as a whole, can rotate about a vertical axis with respect to the anatomical position. This form of rotation is usually described as axial rotation, which is rotation around an axis passing along, parallel to, or close to parallel to the shaft of the moving bone. Axial rotation of the shoulder occurs when the humerus rotates about an axis

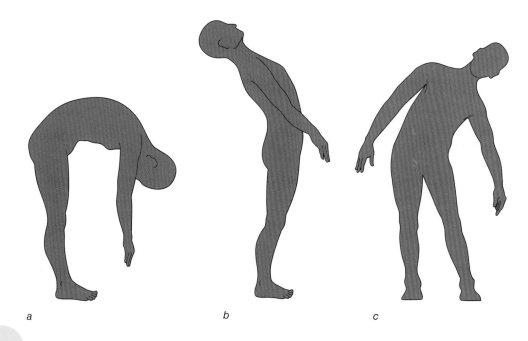

a   b   c

FIGURE 4.12 *(a)* Flexion, *(b)* extension, and *(c)* lateral flexion of the trunk.

parallel to its long axis. For example, internal rotation (medial rotation) and external rotation (lateral rotation) of the shoulder occur when one is cleaning a tabletop with a duster; in this situation the positions of the shoulder and elbow joints tend to move slightly, but the side-to-side sweeping movement of the hand is largely brought about by internal and external rotation in the shoulder. Axial rotation of the humerus should not be confused with pronation of the forearm, which occurs as a result of movement in the proximal and distal radioulnar joints. Rotation of the trunk about a vertical axis is usually described as axial rotation (or twist) to the left or right.

In addition to abduction, adduction, flexion, extension, and rotation, a number of other terms are used to describe specific movements in certain joints. These movements include supination and pronation (see figure 2.27), which occur, for example, in the forearm when one is using a screwdriver to put in or take out a screw.

> **KEY POINT**
>
> With respect to the anatomical position, abduction and adduction describe angular movement of a joint about the anteroposterior axis. Flexion and extension describe angular movement of a joint about the mediolateral axis, and axial rotation describes joint rotation about an axis passing along, parallel to, or close to parallel to the shaft of the moving bone.

Stability–Flexibility Classification of Joints

Joints are most frequently classified on the basis of structure, as described earlier. However, another fairly widely used classification is based on the degree of stability and flexibility of the joints. **Joint stability** refers to the strength of the bond between the bones; the stronger the bond, the more stable the joint. **Joint flexibility** refers to the degree of movement in the joint. In general, the greater the stability of a joint, the lower the flexibility and vice versa. There are three categories in the stability–flexibility classification of joints: synarthroses, diarthroses, and amphiarthroses (Tortora and Anagnostakos 1984).

A **synarthrosis** is a stable joint in which the degree of flexibility is nil or virtually nil. The sutures of the adult skull, the temporary synchondroses, and all synostoses fit into this category. In the structural classification the prefix *syn* means "with." In the stability–flexibility classification the prefix *syn* means "together," which refers to the direct attachment between the articular surfaces.

A **diarthrosis** is a relatively unstable joint with a relatively high degree of flexibility. All synovial joints fit into this category. The prefix *dia* means "apart;" this refers to the fact that the articular surfaces are not attached to each other, even though they are in contact.

An **amphiarthrosis** is a joint whose stability and flexibility characteristics are between the extremes represented by synarthroses and diarthroses. Amphiarthroses (*amphi* = both) are sometimes referred to as slightly moveable joints, whereas synarthroses and diarthroses are referred to as immovable and freely moveable joints, respectively. Membranous syndesmoses and symphyses fit into the category of amphiarthroses.

The stability–flexibility classification can be thought of as a continuum, with synarthroses and diarthroses at opposite ends of the continuum (figure 4.13). Each joint occupies a certain point on the continuum. The positions of these points may change as the structure and, therefore, the stability and flexibility of the joints change. Figure 4.14 summarizes the stability–flexibility classification of joints.

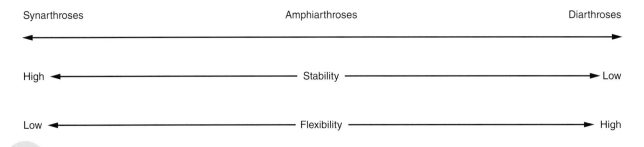

FIGURE 4.13 The stability–flexibility continuum.

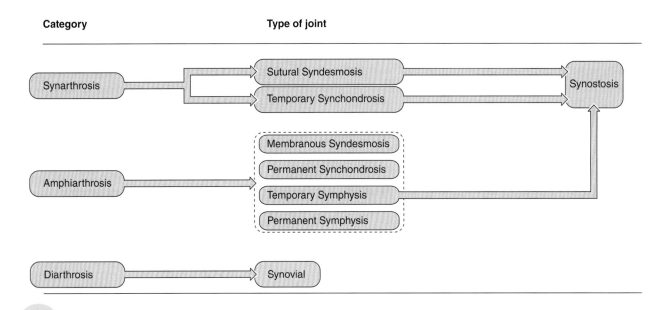

FIGURE 4.14 The stability–flexibility classification of joints.

Whereas most joints fit exclusively into one of the main categories of both the structural and the stability–flexibility classifications, some joints have characteristics of more than one category (Standring 2004). The following are examples of such joints:

- A joint cavity is one of the main characteristics of synovial joints. However, partial cavities may develop in some nonsynovial joints, such as the manubriosternal and pubic symphysis joints, and the joints between the ribs and the sternum.

- In some synovial joints the articular surfaces are separated or partially separated by articular discs or menisci.

- The anterior portion of each sacroiliac joint is basically synovial, whereas the posterior portion is a membranous syndesmosis (see chapter 5).

Consequently, the range of structure and function of joints is very broad and is subject to change over time; this reflects the capacity of the skeletal system for structural adaptation—the capacity to modify the structure of joints in relation to changes in functional requirements.

Key Terms

joint stability   The strength of the bond between the bones in a joint; the stronger the bond, the more stable the joint.

joint flexibility   The degree of movement in a joint.

synarthroses   Sutural syndesmoses and temporary synchondroses; also referred to as immoveable or fixed joints.

diarthroses   Synovial joints; also referred to as freely moveable joints.

amphiarthroses   Symphyses, membranous syndesmoses, and permanent synchondroses; also referred to as slightly moveable joints.

KEY POINT

Joints are classified on the basis of stability and flexibility into synarthroses, amphiarthroses, and diarthroses. Synarthroses such as sutures are stable with little or no flexibility. Diarthroses (including all synovial joints) are relatively unstable with a relatively high degree of flexibility. Amphiarthroses are joints whose stability and flexibility characteristics are between the extremes represented by synarthroses and diarthroses.

Synovial Joint Classification

Synovial joints are classified according to the type of movement that occurs in the joints. There are two kinds of synovial joints:

1. Joints with basically linear, limited movement. In these joints the articular surfaces, which are fairly flat, slide on each other. Consequently, these joints are called gliding or plane joints. Sliding occurs in all synovial joints to a certain extent, but in gliding joints it is the main type of movement. Gliding joints include the intercarpal and intertarsal joints and the joints between the superior and inferior facets of the vertebrae.

2. Joints with basically angular movements. Movement in these joints is normally a combination of rolling and sliding between the articular surfaces. There are three groups: uniaxial, biaxial, and multiaxial. Each group is divided according to the shape of the articular surfaces.

Uniaxial

In uniaxial joints, movement takes place mainly about a single axis. There are two types of uniaxial joints—hinge joints and pivot joints. In a hinge joint, a convex, pulley-shaped (bicondylar) articular surface articulates with a reciprocally shaped concave surface. The elbow (humeroulnar), interphalangeal, and ankle joints are the best examples of hinge joints (figure 4.15). The notch of the pulley prevents (or severely limits) side-to-side movement. The tibiofemoral joint is usually classified as a hinge joint even though the articular surfaces of the femoral and tibial condyles are not very congruent. However, in the normal tibiofemoral joint the congruence between the articular surfaces is considerably increased by the presence of menisci. If, for the purpose of comparison, the menisci are considered to be part of the tibia, the shapes of the articular surfaces of the knee joint are similar to those in the interphalangeal joints.

In a pivot joint, a cylindrical articular surface rotates about its long axis within a ring formed by bone and fibrous tissue. The joint between the dens of the axis and the fibro-osseous ring formed by the anterior arch and

transverse ligament of the atlas is a pivot joint (see figure 2.15). The proximal radioulnar joint is also a pivot joint (figure 4.16). The head of the radius is held against the radial notch by a ligament called the annular ligament (*annulus* = ring). During supination and pronation of the forearm, the head of the radius rotates within the ring formed by the annular ligament and radial notch.

Biaxial

In biaxial joints, movement takes place mainly about two axes at right angles to each other, usually the anteroposterior (abduction–adduction) and mediolateral (flexion–extension) axes. There are three types of biaxial joints: condyloid, ellipsoid, and saddle. In a condyloid joint an almost hemispherical convex condylar surface articulates with a smaller concave condylar surface. The metacarpophalangeal joints are condyloid joints (see figure 4.15, *c* and *d*). In an ellipsoid joint such as the radiocarpal joint, an elliptical convex surface articulates with an elliptical concave surface. The articular surface on the distal end of the radius is elliptical, concave, and shallow. This surface articulates with the proximal articular surfaces of the scaphoid and lunate, which together form a convex

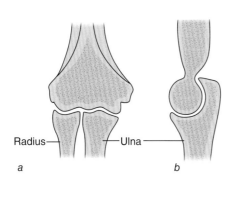

Radius — — Ulna —

a   *b*

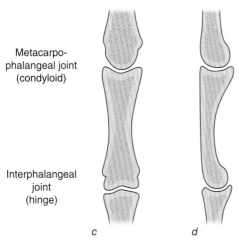

Metacarpo-
phalangeal joint
(condyloid)

Interphalangeal
joint
(hinge)

c   *d*

FIGURE 4.15 Hinge and condyloid joints. *(a)* Coronal section through the right elbow joint in extension. *(b)* Sagittal section through the right humeroulnar joint in extension. *(c, d)* Coronal and sagittal sections, respectively, through metacarpophalangeal and interphalangeal joints in extension.

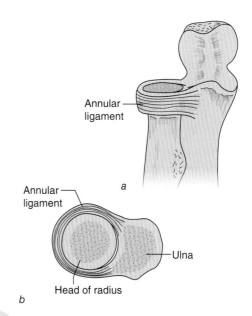

Annular —
ligament

a

Annular —
ligament

— Ulna

Head of radius

b

FIGURE 4.16 A typical pivot joint. *(a)* Anterior aspect of the right proximal radioulnar joint. *(b)* Transverse section through the right proximal radioulnar joint.

elliptical articular surface. Movements at the metacarpophalangeal joints and radiocarpal joints are normally combinations of flexion, extension, abduction, and adduction. In a saddle (or sellar) joint, the articular surfaces are concavo-convex, that is, saddle shaped (figure 4.17*a*). Each articular surface is convex in one direction and concave in a direction at right angles to the convex direction. Movement takes place mainly in two planes at right angles to each other. The carpometacarpal joint of the thumb (figure 4.17*b*) and the calcaneocuboid

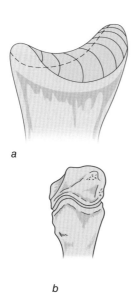

a

b

FIGURE 4.17 A typical saddle joint. *(a)* Saddle-shaped articular surface. *(b)* Anterior aspect of the carpometacarpal joint of the left thumb.

joints are saddle joints. The different sizes and shapes of the articular surfaces of condyloid, ellipsoid, and saddle joints facilitate different ranges of movement (Standring 2004).

Multiaxial

Some joints, such as the shoulder and hip, can rotate about all three reference axes. By combining rotations about the three reference axes, these joints can rotate about any axis in between the three. Consequently, these joints are referred to as multiaxial joints. In this type of joint, a very rounded articular surface, like part of a ball, articulates with a cuplike concavity. Because of the shapes of the articular surfaces, these joints are usually referred to as ball-and-socket joints. The best examples of ball-and-socket joints are the shoulder and hip (see figures 3.9*a* and 3.10*a*). Figure 4.18 summarizes the classification of synovial joints.

> **KEY POINT**
>
> Synovial joints are classified according to the type of movements that occur in the joints. Joints in which the main type of movement is linear are called gliding or plane joints. Joints in which the main type of movement is angular are classified into uniaxial (hinge and pivot), biaxial (condyloid, ellipsoid, and saddle), and multiaxial (ball-and-socket) joints.

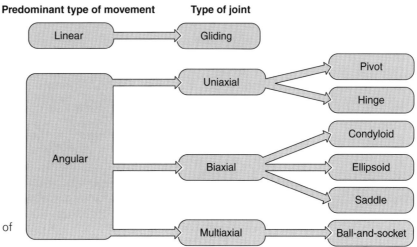

FIGURE 4.18 Classification of synovial joints.

Flexibility, Stability, and Laxity in Synovial Joints

In previous sections of this chapter the terms *flexibility* and *stability* have been used in a general sense. At this point it is necessary to define these terms more specifically in relation to synovial joints.

Flexibility

In a synovial joint, flexibility refers to the range of movement in those directions (degrees of freedom) considered normal for the joint. For example, the knee joint is designed primarily to rotate about a mediolateral axis—to flex and extend. Consequently, flexion and extension are considered normal movements at the knee. However, rotations about an antero-posterior axis—abduction and adduction—are considered abnormal movements at the knee.

The flexibility of a joint is determined by four factors:

1. Shape of the articular surfaces
2. Tension in the joint capsule and ligaments at the ends of the various ranges of motion
3. Soft tissue bulk, mainly skeletal muscle, surrounding the bones forming the joint
4. Extensibility of the skeletal muscles controlling the movement of the joint

Shape of Articular Surfaces

In some joints, flexibility is limited by the impingement or interlocking of the nonarticular surfaces of the bones at the end of certain ranges of motion. For example, elbow extension is restricted by the interlocking of the olec-

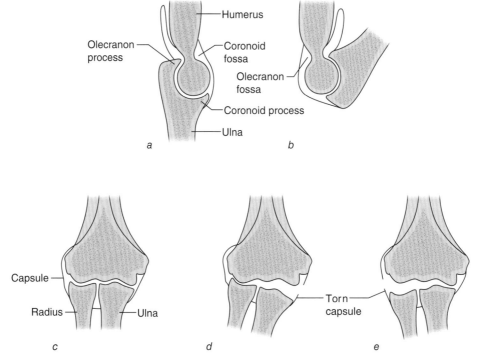

FIGURE 4.19 *(a)* Sagittal section through the right humeroulnar joint in extension. *(b)* Sagittal section through the right humeroulnar joint in flexion. *(c)* Coronal section through the right elbow joint in extension showing the normal orientation of the humerus, radius, ulna, and capsule. *(d)* Abduction of the elbow showing torn capsule on medial aspect of the joint. *(e)* Adduction of the elbow showing torn capsule on the lateral aspect of the joint.

ranon process of the ulna with the olecranon fossa of the humerus (figure 4.19*a*). Similarly, elbow flexion is restricted by the interlocking of the coronoid process of the ulna with the coronoid fossa of the humerus (figure 4.19*b*). Lateral and medial displacement of the radius and ulna relative to the humerus in the elbow joint is severely restricted by the interlocking of the articular surfaces (figure 4.19*c*).

Tension in Joint Capsule and Ligaments at Limit of Range of Movement

The function of ligaments is described in detail later in the chapter. At this point it is sufficient to appreciate that in a normal joint, the joint capsule and some of the ligaments that support the joint become taut at the end of each range of movement and thereby restrict further movement (see figure 4.19, *a* and *b*).

Soft Tissue Bulk

Soft tissue bulk restricts flexibility in some joints. For example, elbow and knee flexion may be restricted in a heavily muscled person because of the impingement of the adjacent body segments.

Extensibility of Muscles

The extensibility of a muscle is the maximum length that the musculotendinous unit can attain without injury. Muscle extensibility largely depends on the length range in which the muscle normally functions—the difference between its length when shortened and its length when extended: the shorter the range of length, the lower the extensibility. In the absence of regular flexibility training, games and sports involving highly repetitive

and exclusive movement patterns are likely to reduce extensibility in some muscles. In most people, muscle extensibility is probably the main factor limiting joint flexibility (Bach et al 1985, Steiner 1987, Nicholas and Marino 1987, Herbert 1988, Lenehan et al 2003, Smith and Fryer 2008).

> **KEY POINT**
>
> The flexibility of a synovial joint is determined by the shape of the articular surfaces and the passive (ligaments and joint capsule) and active (extensibility of musculotendinous units) joint support structures.

Stability and Laxity

In a synovial joint, stability refers to the degree of congruence between the articular surfaces. During joint movement, various parts of the articular surfaces come into contact with each other. However, in all positions of a joint, the better the congruence, the more stable the joint and the lower the risk of abnormal joint movements.

The term **laxity** refers to the degree of instability in a joint—the range of movement in those directions considered abnormal for the joint. The relationships among congruence, stability, and laxity are shown in figure 4.20.

Joint congruence is determined by three factors:

1. Shape of the articular surfaces
2. Suction between the articular surfaces
3. Supporting structures: skeletal muscles, ligaments, and the joint capsule

FIGURE 4.20 The relationships of congruence to stability and laxity in synovial joints.

Shape of the Articular Surfaces

Most synovial joints have fairly congruent articular surfaces in all joint positions. In those joints where the articular surfaces are not very congruent, articular discs or menisci are usually present to improve congruence. Perfect congruence between articular surfaces would maximize joint stability and minimize joint laxity. However, perfect congruence would impair nutrition of the articular cartilage. Consequently, most joints are slightly incongruent and usually have a slight degree of laxity.

Suction Between Articular Surfaces

In a normal joint the articular surfaces are usually in close contact, with a thin film of synovial fluid between the surfaces. The film of fluid allows sliding between the surfaces but keeps the surfaces in contact with each other via suction. In a similar manner, it is often difficult to separate two sheets of glass held in contact with each other by a thin film of water, even though the two sheets will slide on each other. When two surfaces are in contact with each other in closed space, like the articular surfaces in a normal synovial joint, any attempt to separate the surfaces will tend to increase the volume inside the closed space and, in turn, reduce the pressure inside the closed space (internal pressure) relative to the pressure outside the space (external pressure). Consequently, the external pressure will tend to restore the volume and pressure inside the closed space; the force experienced by the surfaces drawing them back together is called suction.

Supporting Structures

The supporting structures largely responsible for maintaining close contact between the articular surfaces in all positions of a joint are the skeletal muscles that control the movement of the joint (chapters 1 and 7) and the ligaments of the joint. As described in the next section, the lengths of the ligaments are such that they normally help to maintain maximum congruence in all positions of the joint and facilitate the normal sliding–rolling contact between the articular surfaces; that is, the ligaments normally guide the movements of the joint. In contrast to ligaments that provide passive (noncontractile) support, the muscles provide active (contractile) support by controlling the movement of the joint while helping to maintain maximum joint congruence.

Functions of Joint Capsule and Ligaments

The joint capsule has two main functions:

1. To assist joint stabilization by helping prevent (a) movement beyond normal ranges and (b) excessive laxity
2. To provide a base for the synovial membrane

At the end of each normal range of movement, part of the joint capsule will usually become taut to prevent movement beyond the normal range. For example, in full extension of the elbow, the anterior aspect of the capsule will be taut and the posterior aspect will be slack (see figure 4.19a). Similarly, in full flexion of the elbow, the posterior aspect of the capsule will be taut and the anterior will be slack (see figure 4.19b). With regard to abnormal ranges of movement, the regions of the joint capsule in the planes of abnormal ranges of movement are usually at their natural length—neither taut nor slack. However, these regions quickly become taut in response to abnormal movements, thereby helping to restrict the ranges of abnormal movements. For example, abduction and adduction are abnormal movements at the elbow, and the joint capsule normally helps to prevent these movements (see figure 4.19, c-e).

During normal movements the tension in ligaments is low to moderate—ligaments only become taut to prevent or restrict abnormal movements. An abnormal movement can be defined as any movement resulting in a decrease in normal congruence, that is, any movement that results in complete distraction of the articular surfaces, referred to as **luxation,** or partial distraction, referred to as **subluxation.**

Subluxations are usually transient; normal congruence is usually restored as soon as the load causing the subluxation is removed. Subluxations often result in joint sprains—partial tearing of ligaments and the joint capsule, together with effusion (swelling) in the joint. Like subluxations, luxations are usually transient. A luxation that persists after the load causing it is removed is a **dislocation** (figure 4.21). A dislocation usually requires manipulation by a physician or paramedic to restore the normal relationship between the articular surfaces (Grana et al 1987). Dislocation usually results in considerable damage to ligaments and the joint capsule.

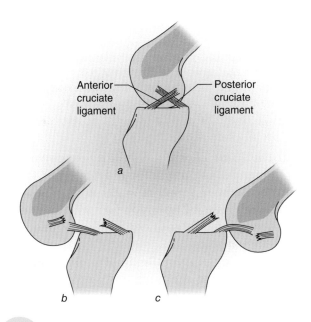

FIGURE 4.21 Dislocation of the tibiofemoral joint. *(a)* Normal orientation of the cruciate ligaments in the partially flexed tibiofemoral joint. *(b)* Anterior dislocation of the femur on the tibia showing torn posterior cruciate ligament. *(c)* Posterior dislocation of the femur on the tibia showing torn anterior cruciate ligament.

Considerable force is required to cause severe subluxations and luxations. Ligaments and joint capsules are likely to be subjected to very high forces in two particular situations:

1. Unexpected situations in which the degree of muscular control of joint movement is less than adequate, for example, twisting an ankle by stepping on an uneven surface. In this situation, with the body moving forward over the foot, the ankle is likely to be rapidly and forcibly twisted.

2. High-speed collisions, such as a tackle in American football or soccer.

There are two kinds of abnormal movements in joints: hyperflexibility and excessive laxity. These are described next with reference to the knee joint.

Key Terms

laxity The range of movement in those directions considered abnormal for the joint.

luxation A transient separation of the articular surfaces in a synovial joint.

subluxation A transient decrease in the normal area of contact between the articular surfaces in a synovial joint.

dislocation A permanent (in the absence of treatment) separation of the articular surfaces in a synovial joint.

Hyperflexibility

Hyperflexibility can be defined as movement beyond the normal range of movement in a direction considered normal for the joint. Extension is a normal movement in the knee joint. At full knee extension, the four main ligaments that support the knee will become taut to prevent further movement. If the knee is forcibly extended beyond this position, that is, hyperextended, the ligaments and joint capsule will be damaged (figure 4.22). Hyperextension of the knee is an example of hyperflexibility. In most people full knee extension normally corresponds to a position in which the leg is straight—the upper leg and lower leg are in line. Consequently, in most people hyperextension of the knee occurs when the knee is extended beyond the straight position. However, some individuals are able to voluntarily extend their knees slightly beyond the straight position. In these cases, extension of the knee beyond the straight position to the

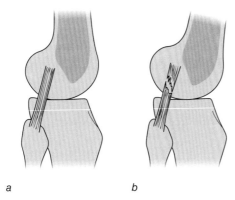

FIGURE 4.22 Hyperextension of the tibiofemoral joint. *(a)* Lateral aspect of the extended right tibiofemoral joint showing the normal orientation of the lateral ligament. *(b)* Hyperextension of the tibiofemoral joint showing torn lateral ligament.

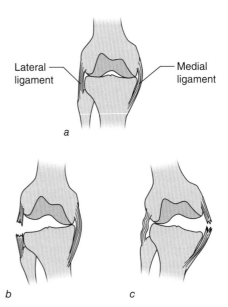

FIGURE 4.23 *(a)* Anterior aspect of the right tibiofemoral joint showing the normal orientation of the lateral and medial ligaments. *(b)* Adduction of the tibiofemoral joint showing torn lateral ligament. *(c)* Abduction of the tibiofemoral joint showing torn medial ligament.

point of full extension would be considered normal for the person provided that normal congruence is maintained.

Excessive Laxity

Excessive laxity is defined as movement beyond the normal degree of laxity in directions considered abnormal for the joint. For example, abduction and adduction are abnormal movements in the knee. In a normal knee joint, these movements are restricted to a minimal level by the medial and lateral ligaments, respectively (figure 4.23). Adduction of the knee beyond a minimum level will damage the lateral ligament and joint capsule. Similarly, abduction of the knee beyond a minimal level will damage the medial ligament and joint capsule.

Movements of the femoral condyles backward and forward (along the anteroposterior axis) relative to the tibial condyles are abnormal movements. In a normal knee joint, these movements are restricted to a minimal level by the anterior and posterior cruciate ligaments, respectively (see figure 4.21). Forward movement of the femoral condyles (from the position shown in figure 4.21*a* toward that shown in figure 4.21*b*) beyond a minimal level will damage the posterior cruciate ligament

and the joint capsule. Similarly, backward movement of the femoral condyles (from the position shown in figure 4.21*a* toward that shown in figure 4.21*c*) beyond a minimal level will damage the anterior cruciate ligament and the joint capsule.

Any degree of hyperflexibility or excessive laxity in a joint tends to damage not only ligaments and joint capsules but also articular cartilage. The latter is due to localized overloading of articular cartilage resulting from reduced congruence caused by the abnormal movements.

> **KEY POINT**
>
> A joint capsule provides a base for the synovial membrane and assists joint stabilization. In association with muscles, ligaments bring about normal joint movements and help maintain joint stability. During normal movements ligaments only become taut to prevent or restrict abnormal movements—hyperflexibility and excessive laxity.

Anterior Cruciate Ligament Injuries

Anterior cruciate ligament (ACL) injuries are common, and approximately 70% of ACL injuries occur in sport (Renström et al 2008). Up to two thirds of patients who have complete ACL tears develop knee instability and subsequently experience damage to the menisci and articular surfaces, which in turn significantly affects knee function and decreases one's level of physical activity. ACL reconstruction was the sixth most common orthopedic procedure performed in the United States in 1999-2000, with approximately 100,000 ACL reconstructions being performed annually at a cost of more than US$2 billion for surgery alone.

From 70% to 90% of ACL injuries occur in noncontact situations (no direct contact with the knee) and are associated with actions that involve one or more of the following maneuvers: foot strike with the knee close to full extension, landing, deceleration, rapid change in direction (cutting, sidestepping). These actions are common in sports such as basketball, volleyball, netball, and soccer, and the incidence of noncontact ACL injury in these sports is particularly high.

ACL injury occurs as a result of insufficient stability of the tibiofemoral joint (Hughes and Watkins 2006). The stability of the tibiofemoral joint is maintained by passive mechanisms (including the shape of the articulate surfaces, the menisci, the ACL and other ligaments, and the joint capsule) and active mechanisms (the musculotendinous units that cross the tibiofemoral joint and, in particular, the stretch reflexes between ligaments and the associated muscles). The incidence of noncontact ACL injuries in females is reported to be six to eight times greater than in males competing in similar activities at the same level of competition. A number of risk factors, including the Q angle (see figure 6.18b), intercondylar notch width, joint laxity, and muscle stiffness, have been proposed to account for the sex difference in the incidence of ACL injuries, but the evidence is weak. However, there is mounting evidence that females have a significantly greater risk of ACL injury during the preovulatory phase of the menstrual cycle than during the postovulatory stage. The hormonal balance during the preovulatory phase may temporarily reduce the stiffness of ligaments and tendons, which in turn may adversely affect joint stability directly (passive support) and indirectly (active support in terms of the speed and coordination of associated muscles). These temporary changes in the passive and active joint support mechanisms of the tibiofemoral joint would increase the risk of ACL injury.

The general consensus is that the risk of ACL injury in sport can be reduced by

- proprioception training (see chapter 7) of the leg extensor muscles to reduce muscle reaction time and improve neuromuscular coordination, in particular, coordination between the quadriceps and hamstrings;

- technique training to improve foot position (in line with intended direction) and avoid straight-leg landing in the deceleration phase of whole-body direction changes; and

- avoiding training and playing when fatigued, because fatigue will reduce neuromuscular coordination (Fagenbaum and Darling 2003, Chappell et al 2005).

Movement Between Articular Surfaces

Three types of movement occur between articular surfaces: spinning, sliding, and rolling (figure 4.24). In most joints, normal movement involves a combination of these three types of movement, either simultaneously or sequentially. Sliding and, to a lesser extent, spinning subject the articular cartilage to a combination of compression and shear loading.

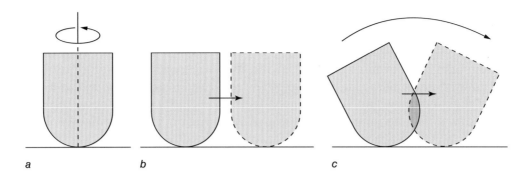

FIGURE 4.24 Types of movement between articular surfaces in synovial joints: *(a)* spinning, *(b)* sliding, and *(c)* rolling.

Rolling subjects the articular cartilage to compression loading. The compression and shear loads tend to cause mechanical wear of the articular cartilage, breaking up the cartilage internally and wearing away the cartilage at the surface. Under normal circumstances, when the articular surfaces are congruent and properly lubricated, there is a balance between the amount of wear and the production of new cartilage. However, during subluxations, those parts of the articular cartilage that remain in contact are subjected to abnormally high loads. Over a period of time, the wear caused by the abnormal loads may outpace the production of new cartilage and permanently damage the cartilage (see chapter 10).

> **KEY POINT**
>
> In most synovial joints, movement between the articular surfaces involves spinning, rolling, and sliding. In a normal joint, the combination of spinning, sliding, and rolling minimizes wear of the articular cartilage. Excessive laxity in a joint tends to overload certain parts of the articular cartilage, resulting in excessive wear.

Flexibility Training

In normal, healthy individuals, ligaments and joint capsules, like musculotendinous units, adapt to the length range in which they normally function. Consequently, if a joint is not regularly moved through the full range of movement, the ligaments and joint capsule shorten, preventing full range of normal movement. In this context, full range of normal movement is the range of movement between positions of the joint where subluxation of the articular surfaces occurs.

Flexibility training, in the form of a carefully prescribed program of exercises designed to lengthen ligaments and joint capsules and to increase the extensibility of musculotendinous units, may restore full ranges of movement. As previously mentioned, lack of muscle extensibility rather than shortened ligaments and joint capsules is probably the main factor limiting flexibility in most people. During flexibility training, ligaments and joint capsules must not be stretched to the point where joints become hyperflexible. If this happens, the overstretched ligaments and joint capsules will not be able to function properly; the joints will be less stable in all joint positions, and abnormal movements will become more likely.

Summary

Joints facilitate relative motion and transmission of force between bones. There is considerable variation in the type and range of movement and the mechanism of force transmission in different joints. The next two chapters consider the joints of the axial skeleton (chapter 5) and the appendicular skeleton (chapter 6).

Review Questions

1. Differentiate between permanent and temporary joints.
2. Describe the two main approaches to classifying joints.
3. Differentiate between the following:
   - A syndesmosis and a symphysis
   - A synarthrosis and a diarthrosis
4. Describe the basic structure of a synovial joint.
5. With respect to synovial joints, differentiate between the following:
   - Flexibility and laxity
   - Stability and congruence
   - Subluxation and dislocation
6. Describe the factors affecting the flexibility of a synovial joint.
7. Describe the function of ligaments in relation to synovial joints.
8. Describe the types of movement that occur between the articular surfaces in a synovial joint.

Joints of the Axial Skeleton

A number of small uniaxial synovial joints are found in the adult axial skeleton, but the larger joints of the axial skeleton and the joints of the pelvis are mainly cartilaginous. In contrast, the large majority of joints in the adult appendicular skeleton are synovial. The differences in the types of joints that make up the axial and appendicular parts of the skeleton reflect differences in function. The synovial joints of the appendicular skeleton facilitate a large range of movement with limited shock absorption, whereas the cartilaginous joints of the axial skeleton and pelvis provide a high level of shock absorption with limited range of movement. This chapter describes the structure and function of the joints of the axial skeleton and pelvis.

OBJECTIVES

After reading this chapter, you should be able to do the following:

1. Describe the structure and function of intervertebral joints and facet joints.
2. Explain the relationship between intervertebral joints and facet joints in terms of transmission of load between vertebrae.
3. Describe the location and function of the ligaments that support the intervertebral and facet joints.
4. Describe the range of movement in different regions of the vertebral column.
5. Describe the effects of aging and excessive overload on degeneration and damage in intervertebral discs and vertebrae.
6. Describe normal and abnormal curvatures of the vertebral column in the median and frontal planes.
7. Describe the structure and function of the sacroiliac joints and pubic symphysis joint.
8. Explain the relationship between the sacroiliac and pubic symphysis joints in terms of transmission of load across the pelvis.

Joints Between the Vertebrae

Adjacent pairs of vertebrae, apart from the atlas and axis, articulate with each other by the following means:

1. A symphysis joint between the vertebral bodies. These joints are usually referred to as **intervertebral joints** (figure 5.1).
2. Two synovial gliding joints between the inferior articular processes of the upper vertebra and the superior articular processes of the lower vertebra. These joints are usually referred to as **facet joints** or apophyseal joints (figure 5.1).
3. A number of longitudinal and intersegmental ligaments. These ligaments are described later in the chapter. As discussed in chapter 4, some of the intersegmental ligaments constitute membranous syndesmoses.

There is no intervertebral joint (symphysis joint) between the atlas and axis, but there is a synovial pivot joint between the dens of the axis and the anterior arch of the atlas (see figure 2.15). In other respects, the links

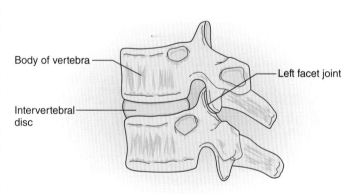

Body of vertebra
Left facet joint
Intervertebral disc

FIGURE 5.1 Left lateral aspect of the tenth and eleventh thoracic vertebrae showing the intervertebral disc and left facet joint.

between the atlas and axis are similar to those between other vertebrae.

Movement between each pair of adjacent vertebrae involves simultaneous movement in all of the joints (cartilaginous, synovial, and fibrous) that link them together. It is not surprising, then, that the various joints function as a single unit to transmit loads between the vertebrae in the most mechanically efficient manner (Dunlop et al 1984). The functional unit consisting of two adjacent vertebrae and the joints and supporting soft tissues linking them together is often referred to as a **motion segment** (Brinckmann 1985). The vertebral column consists of a chain of motion segments.

Just as the various joints within a single motion segment function as one unit, so the chain of motion segments function as a unit to transmit loads along the vertebral column. It follows that damage to any particular joint in a motion segment will likely affect the function of adjacent motion segments. For example, damage to an intervertebral joint may lead a person to adopt compensatory movements in adjacent motion segments to reduce the load on the injured joint. Even a slight amount of compensatory movement in a joint can alter the normal pattern of loading in the joint (Riegger-Krugh and Keysor 1996).

KEY POINT

There is no intervertebral joint between the atlas and axis, but otherwise each adjacent pair of vertebrae articulates by means of the following:

- An intervertebral joint
- Two facet joints (apophyseal joints)
- Longitudinal and intersegmental ligaments

Key Terms

intervertebral joint Symphysis joint linking the bodies of adjacent vertebrae in a motion segment.

facet joint Synovial gliding joint between superior and inferior articular processes in a motion segment.

motion segment Functional unit consisting of two adjacent vertebrae and the joints and supporting soft tissues linking them together.

Intervertebral Joints

The intervertebral joints and vertebral bodies form the main weight-bearing part of the ver-

tebral column. The intervertebral joints are specialized symphysis joints. A thin layer of hyaline cartilage covers each articular surface of the vertebral bodies. Sandwiched between the layers of hyaline cartilage is an **intervertebral disc,** a fairly thick pad of highly specialized fibrocartilage. The vertebral bodies, hyaline cartilage, and intervertebral disc are intimately connected to each other.

In the transverse plane, the shape of the discs is the same as that of the vertebral bodies—kidney shaped. In the frontal plane, the discs are rectangular. In the median plane, the discs in the cervical and lumbar region tend to be thicker in front than behind, so that the discs contribute to the anterior convexity of these regions (Standring 2004). The discs in the thoracic region tend to be of uniform thickness in the median plane. The anterior concavity of the thoracic region is largely due to the shape of the vertebral bodies, which are slightly shorter at the front compared with the back (Standring 2004) (see figure 2.10*a*). The thickness of the discs, like the size of the vertebral bodies, increases from the superior end of the column downward. On average, the thickness of the discs in the cervical, thoracic, and lumbar regions in an adult is approximately 3 mm, 5 mm, and 9 mm, respectively (Kapandji 1974). In a normal adult the discs contribute approximately 20% to the total length of the vertebral column (Standring 2004).

KEY POINT

The intervertebral joints and vertebral bodies form the main weight-bearing part of the vertebral column. The size of the vertebral bodies and thickness of the intervertebral discs increase downward.

Each intervertebral disc consists of an outer ring of fibrous material called the annulus fibrosus (fibrous ring), which surrounds an inner gelatinous mass called the nucleus pulposus (pulpy center) (figure 5.2). The annulus fibrosus consists of a relatively thin outer ring of regular fibrous tissue and a broader inner ring of fibrocartilage. The fibrocartilage portion is in the form of a number of concentric layers. The collagen fibers in each layer run in the

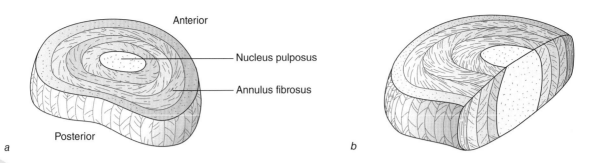

FIGURE 5.2 Intervertebral disc. *(a)* Complete disc. *(b)* Vertical section through disc to show relationship between nucleus pulposus and annulus fibrosus.

same direction, but the fibers in adjacent layers run in slightly different directions. In association with the nucleus pulposus, this layered arrangement of the fibrocartilage ring enables the disc to strongly resist all types of loading.

The nucleus pulposus is structurally similar to proteoglycan gel (see chapter 3) and thus has a high affinity for water; the nucleus pulposus continually exerts an **osmotic swelling pressure (OSP)** to absorb water (Lindh 1989). The annulus fibrosus also exerts a certain amount of OSP but less than that of the nucleus pulposus. Water is absorbed into the disc from the vertebral bodies via the hyaline cartilage. Under normal loading conditions a healthy nucleus pulposus is approximately 80% to 90% water (Kapandji 1974, Lindh 1989). The nucleus pulposus is enclosed within a virtually inextensible casing formed by the vertebral bodies and the annulus fibrosus. Consequently, as the volume of water absorbed into the nucleus pulposus increases, there is a simultaneous increase in the pressure within the nucleus, because its capacity for expansion is limited. The pressure in the nucleus pulposus is transmitted to the vertebral bodies and annulus fibrosus such that the thickness of the disc increases as more water is absorbed into the nucleus pulposus (figure 5.3). The flow of water into the nucleus pulposus ceases when the pressure exerted on the nucleus pulposus via the bony-fibrous casing is equal to the OSP.

When a person is recumbent (lying down), the vertebral column does not have to support the weight of the upper body. In this situation the amount of water that the nucleus pulposus of each disc can absorb is limited only by the extensibility of the corresponding annulus

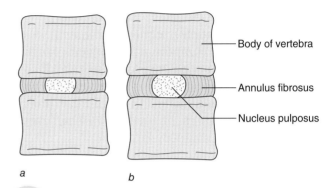

FIGURE 5.3 Effect of osmotic swelling pressure (OSP) on thickness of intervertebral disc. *(a)* Disc thickness as a result of equilibrium between OSP and compression load on motion segment. *(b)* Increased disc thickness following establishment of a new state of equilibrium between OSP and reduced compression load on the motion segment.

fibrosus. Consequently, during rest or sleep in a horizontal position, and any other condition where the compression load on the vertebral column is considerably reduced (such as during space flight), the nucleus pulposi absorb the maximum possible amount of water and the discs achieve maximum thickness. For this reason, most people are at their tallest just after getting out of bed in the morning.

During normal daily activities, with the trunk in a fairly upright position, the vertebral column has to support the weight of the upper body. Consequently, the discs are subjected to almost continuous compression loading. In this situation, the amount of water that the nucleus pulposus of each disc can absorb is limited by the size and duration of the compression loads. In most people the compression loads exerted on the nucleus pulposi as a result of maintaining an upright posture exceed the OSP, at least during the first few hours after getting out of bed in the morning. Consequently, water is gradually forced out of the nucleus pulposi and back into the vertebral bodies such that the discs become thinner. In healthy joints the loss of water is gradual because of the low permeability of the hyaline cartilage layers. Nevertheless, the cumulative reduction in the thickness of the discs during a normal day can be as much as 2 cm (Kapandji 1974). This represents a reduction of approximately 15% in the combined maximum thickness of all the discs.

Because of the OSP of the nucleus pulposus, the annulus fibrosus of each disc and the disc as a whole are constantly under load; that is, the annulus fibrosus is constantly under a certain degree of tension, which is analogous to the tension in the casing of an inflated ball: The greater the degree of inflation—comparable to the amount of water in the nucleus pulposus— the greater the tension in the casing. Because a healthy annulus fibrosus is under considerable load even when there is no external force acting on the disc, the disc is said to be permanently preloaded (Kapandji 1974). The preloaded condition, which can only be created when the nucleus pulposus and the annulus fibrosus function normally, is mechanically significant, as described subsequently.

Key Terms

intervertebral disc The specialized fibrocartilage component of an intervertebral joint.

osmotic swelling pressure (OSP) The water-absorbing force exerted by an intervertebral disc.

> **KEY POINT**
>
> Each intervertebral disc consists of a peripheral annulus fibrosus surrounding a central nucleus pulposus. In a healthy disc, the nucleus pulposus and annulus fibrosus continually exert an osmotic swelling pressure that simultaneously increases the thickness of the disc and preloads the disc. The preloaded condition is necessary for normal disc function.

Restoration of Normal Vertebral Column Alignment

When the vertebral column is flexed, extended, or laterally flexed with respect to its normal alignment, the vertebral bodies of each motion segment pivot about the corresponding nucleus pulposus, which, again, is comparable to an inflated ball. Consequently, as one side of the annulus fibrosus is compressed, the other side is stretched (subjected to greater than normal tension). In addition, the nucleus pulposus is displaced laterally in the direction of the stretched side of the joint; the greater the angular displacement in the motion segment, the greater the lateral displacement of the nucleus pulposus (figure 5.4). Consequently, the increased tension in the annulus fibrosus on the stretched side of the joint is due not only to the angular displacement of the vertebral bodies but also to the pressure exerted by the laterally displaced nucleus pulposus. The combined compression on one side of the disc and tension on the other produce considerable bending stress in the disc that tends to restore the normal orientation of the annulus fibrosus and nucleus pulposus.

When the nucleus pulposus does not function normally (reduced capacity to preload), the deformation of the annulus fibrosus tends to be asymmetric because of differences in the tensile and compressive strength of the annulus fibrosus. This reduces the bending strain on the disc and consequently reduces the capacity of the disc to restore the normal (symmetrically loaded) orientation of the annulus fibrosus. It

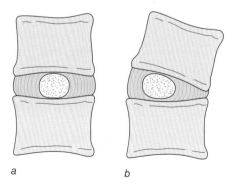

a   *b*

FIGURE 5.4 *(a)* Orientation of annulus fibrosus and nucleus pulposus in normal upright posture. *(b)* Lateral displacement of the nucleus pulposus in response to angular displacement of the motion segment.

follows that the normal response of a disc to flexion, extension, and lateral flexion of the vertebral column is impaired whenever the disc's normal capacity to preload is impaired.

When the vertebral column is flexed, extended, or laterally flexed with respect to its normal alignment, the intervertebral discs are stretched on one side and compressed on the other. This pattern of loading results in considerable bending stress on the discs, which helps to restore the normal alignment of the column.

Protection of Vertebrae From Excessive Loading

In response to external loading imposed on the vertebral column by normal everyday activities, the discs cushion the effect of the external loading and protect the vertebrae. The type of deformation the discs experience depends on the rate at which the external load is applied. In response to impact loads—loads with a high rate of loading and short duration—the discs behave elastically. For example, in situations such as heel strike in walking and running, kicking a football, and hitting a tennis ball, there is insufficient time for water to be squeezed out of the discs. Consequently, the discs behave elastically; they deform instantaneously on loading and instantaneously recover their original shapes on unloading. The degree of strain experienced by a disc in response to impact loading depends on the magnitude of the load and the degree of preloading in the disc. Generally speaking, the higher the state of preloading, the lower the strain and the better the cushioning effect of the disc. This cushioning effect of the discs in response to impact loading is often referred to as shock absorption.

In contrast to their elastic response to impact loading, discs behave viscoelastically in response to loads applied relatively slowly and for a prolonged period. For example, in response to prolonged compression loading, such as would occur when standing upright for a long period, especially if a person is holding a

weight such as a young child, water is gradually squeezed out of the discs so that they become progressively thinner. This response is similar to water being squeezed out of a sponge or to air being squeezed out of an inflated plastic bag that has a number of pinholes. However, when the load on the discs is removed or reduced to a level below the OSP of the discs, they will begin absorbing water to restore their original size and shape.

The reduction in disc thickness that results from supporting the weight of the upper body in upright postures, followed by the restoration of the thickness of the discs during sleep, exemplifies the viscoelastic behavior of the discs. The extent of the reduction in disc thickness in response to prolonged compression loading depends on the magnitude and duration of the load: The greater the magnitude and duration, the thinner the discs become. However, as mentioned earlier in this chapter, in a healthy vertebral column the reduction in the thickness of the discs that occurs as a result of normal everyday activity is unlikely to be greater than 15% of the combined maximum thickness of all the discs.

In addition to its mechanical importance in protecting the vertebrae from excessive loading, the viscoelastic behavior of the discs is important physiologically in terms of the nutrition of the discs. The alternate expulsion and absorption of water create a circulation of water through the discs similar to the circulation of synovial fluid through articular hyaline cartilage. The circulation of water enables the chondrocytes of the discs to receive nutrients and eliminate waste products (Adams and Hutton 1985).

In response to external loading by normal everyday activities, the intervertebral discs deform to cushion the load and thereby protect the vertebrae. In response to impact loads, the discs respond elastically; in response to nonimpact loads, the discs respond viscoelastically. The capacity of the discs to respond effectively to loading is reduced when the capacity of the discs to preload is impaired.

Facet Joints

The facet joints in each motion segment are synovial gliding joints formed by the inferior articular processes of the upper vertebra and the superior articular processes of the lower vertebra (figure 5.5). The joint capsule of each facet joint is supported by relatively thick capsular ligaments on the anterior and posterior aspects. The function of the facet joints is largely determined by the orientation of the articular surfaces, which gradually change throughout the length of the vertebral column (see figure 2.10). The facet joints assist the intervertebral joints to transmit loads between the vertebrae. Specifically, the facet joints contribute to load transmission in certain postures and restrict certain ranges of movement of the intervertebral joints to prevent injury to the intervertebral discs (Adams and Hutton 1983).

With respect to standing postures, the load on the facet joints depends on the thickness of the discs and the alignment of the vertebral column. In relaxed upright standing, the compression load on the facet joints of

a healthy vertebral column, with the discs at maximum thickness, is small (Adams and Hutton 1980). However, during prolonged standing the thickness of the discs gradually decreases such that the inferior articular processes slide progressively downward relative to the superior articular processes, and the compression stress on the articular surfaces gradually increases (figure 5.6). For example, it has been shown that after a person stands upright for 3 hr, the facet joints of the lumbar motion segments support approximately 16% of the total vertical compression load on the motion segments (Adams and Hutton 1980). If disc thickness continues to decrease, eventually the tips of the articular processes impinge on the vertebral arch; that is, the tips of the inferior articular processes impinge on the

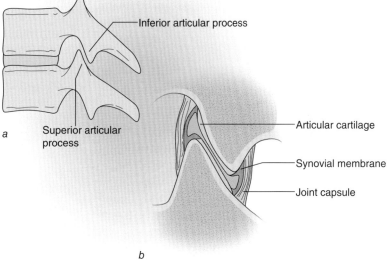

a

b

FIGURE 5.5 Facet joint. *(a)* Orientation of articular surfaces in a facet joint of a typical thoracic motion segment. *(b)* Sagittal section through a facet joint.

FIGURE 5.6 Effect of disc thickness on compression load on facet joints. *(a)* Normal disc thickness. *(b)* Reduced disc thickness resulting in increased compression load on facet joint. *(c)* Extra-articular impingement. A = impingement of the tip of the inferior articular process of the upper vertebra on the superior aspect of the lamina of the lower vertebra. B = tip of the superior articular process of the lower vertebra close to impingement on the inferior aspect of the pedicle of the upper vertebra.

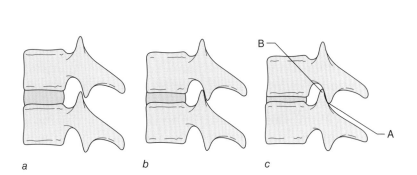

a   *b*   *c*

laminae of the lower vertebra or the tips of the superior articular processes impinge on the pedicles of the upper vertebra (figure 5.6c). This condition is referred to as **extra-articular impingement (EAI).**

EAI, which can involve impingement of the facet joint synovial membranes, is a common feature of motion segments in which the discs are badly degenerated. In such cases, the compression loads on the facet joints and areas of EAI can be very high. For example, in badly degenerated lumbar motion segments, the facet joints and areas of EAI support approximately 70% of the total compression load on the motion segments in the upright standing position (Adams and Hutton 1980).

Key Term

extra-articular impingement (EAI) Condition that occurs when the tips of the inferior articular processes impinge on the laminae of the lower vertebra or the tips of the superior articular processes impinge on the pedicles of the upper vertebra in a motion segment.

KEY POINT

The facet joints contribute to load transmission between vertebrae in certain postures and restrict certain ranges of motion to prevent injury to intervertebral discs. In relaxed upright standing, the compression load on the facet joints of a healthy vertebral column may be small. Prolonged standing decreases the thickness of the discs (as a result of expulsion of water), which increases the compression load on the facet joints and possibly results in EAI.

The loads on the facet joints between the fourth and fifth lumbar vertebrae, L4 and L5, and between L5 and the first sacral vertebra, S1, tend to be greater than the loads on facet joints in other regions of the column because of the orientation of the L4, L5, and S1 vertebrae. In relaxed upright standing, the plane of the inferior end plate of L4 makes an angle (L4 angle) of approximately 15° with the horizontal, and the plane of the inferior end plate of L5 makes an angle (L5 angle) of approximately 25° with the horizontal (figure 5.7, a and b). Consequently, there is a tendency for the vertebral bodies of L4 and L5 to slide forward and downward on L5 and S1, respectively, and thereby subject the L4-L5 and L5-S1 discs to considerable shear loading.

However, in a healthy vertebral column, the facet joints and supporting ligaments prevent L4 and L5 from sliding (see next section; figure 5.7, b and c). In each vertebra, the parts of the vertebral arch between the superior and inferior articular processes—at the junctions of the pedicles and laminae—are called the pars interarticularis (PI: interarticular parts) (figure 5.7d). In resisting the tendency of L4 and L5 to slip forward and downward, the **pars interarticularis** of the L4-L5 and L5-S1 motion segments are subjected to shear loading (figure 5.7c). As shown in figure 5.7c, F_1 is the component of the load exerted by L4 on L5 perpendicular to the end plate; F_2 is the component of the load exerted by L4 on L5 parallel to the end plate; and F_3 and F_4 are components of the load exerted by the sacrum on L5 perpendicular and parallel to the end plate, respectively. F_2 and F_4 constitute a shear load on the pars interarticularis of L5.

With reference to the shape of the vertebral column in relaxed upright standing (figure 5.7, a and b), extension of any part of the vertebral column (as in the position shown in figure 5.7, e and f) tends to increase the compression loading on the articular surfaces of the facet joints in the extended region. The increase in compression loading is offset to a certain extent by a simultaneous increase in the area of contact between the articular surfaces so that the increase in compressive stress (compression load divided by contact area) on the articular surfaces is much less than the increase in compression load (figure 5.7, e and f). However, the increase in compression load on the articular facets can increase the shear stress on the pars interarticularis (Silver et al 1986, Aspden 1987, Ranawat et al 2003). Consequently, repeated forceful extension of any part of the vertebral column is likely to result in partial or complete fracture of the PI in the affected region (see discussion later in the chapter).

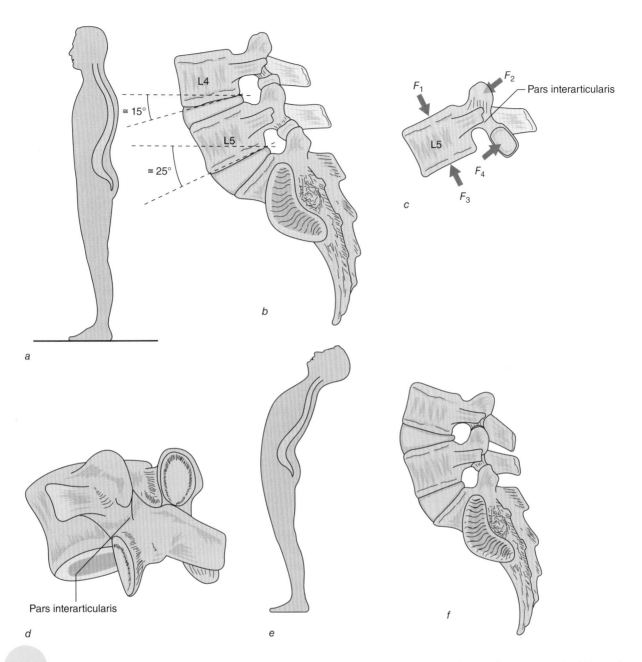

FIGURE 5.7 Orientation of fourth (L4) and fifth (L5) lumbar vertebrae and sacrum in upright postures. *(a-c)* Normal upright standing. *(d)* Oblique view of a lumbar vertebra showing the left pars interarticularis. *(e, f)* Trunk extension during upright standing.

Flexion of any part of the vertebral column with respect to the shape in relaxed upright standing decreases the compression load on the articular surfaces of the facet joints. For example, a very small amount of lumbar flexion has been shown to completely unload the lumbar facet joints (Adams and Hutton 1980). Because most relaxed sitting postures involve a certain amount of lumbar flexion, the lumbar facet joints are likely to be unloaded in these situations (figure 5.8). The only situation where the facet joints are likely to be loaded during flexion is when a person is lifting a heavy weight with the trunk flexed forward and the legs kept fairly straight. In this situation, any tendency of trunk flexion to unload the facet

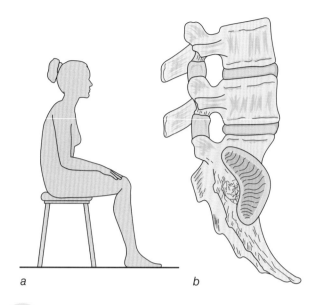

a *b*

FIGURE 5.8 Effect of lumbar flexion on loading on facet joints during relaxed sitting on a stool. *(a)* Relaxed sitting posture. *(b)* Separation and unloading of facet joints as a result of lumbar flexion.

joints is more than offset by the tendency of the vertebrae to slide forward and downward on each other, thereby increasing the compression loading on the facet joints.

Key Term

pars interarticularis (PI) The parts of the vertebral arch between the superior and inferior articular processes.

Flexion of any part of the vertebral column tends to decrease the compression load on the affected facet joints, whereas extension tends to increase the compression on the facet joints. The greater the compression load on the facet joints of a vertebra, the greater the shear load on the pars interarticularis of the vertebra.

Ligaments of the Vertebral Column

The vertebral column is supported by three longitudinal ligaments, which run almost the whole of the length of the column, and by three groups of intersegmental ligaments (Standring 2004). The longitudinal ligaments are the anterior longitudinal ligament, posterior longitudinal ligament, and supraspinous ligament. The anterior longitudinal ligament is a broad band running down the anterior aspect of the vertebral bodies from the anterior aspect of the occipital bone to the anterior aspect of the sacrum (figure 5.9). It is attached to the intervertebral discs and to the anterior aspects of the vertebral bodies except for the upper and lower edges. The posterior longitudinal ligament is a broad band running down the posterior aspect of the vertebral bodies from the axis to the posterior aspect of the sacrum. It is attached to the intervertebral discs and to the upper and lower edges of the vertebral bodies. In the lower thoracic and lumbar regions, the ligament fans out at the level of each intervertebral joint to gain a broad attachment to the intervertebral disc (figure 5.10).

The supraspinous ligament is a thick cord that links the external occipital protuberance and the tips of the vertebral spines between the seventh cervical vertebra and the posterior aspect of the sacrum (see figure 5.9). In the cervical region, the supraspinous ligament expands anteriorly in the median plane by means of a series of radiating fibers that attach onto the tips of the spines of the upper six cervical vertebrae. The series of radiating fibers constitute a membranous syndesmosis called the ligamentum nuchae (figure 5.11). The lateral aspects of the ligamentum nuchae provide areas of attachment for muscles of the neck. The ligamentum nuchae and that part of the supraspinous ligament from which it arises consist largely of regular elastic tissue. The elasticity of these ligaments helps to restore the head to its normal orientation following flexion of the neck and relieves the load on the extensor muscles of the neck. The ligamentum nuchae is extensive in animals, such as the horse and giraffe, that normally stand with the head suspended in front of the rest of the body.

The three groups of intersegmental ligaments are the ligamentum flava, interspinous ligaments, and intertransverse ligaments. The ligamentum flava link the laminae of adjacent

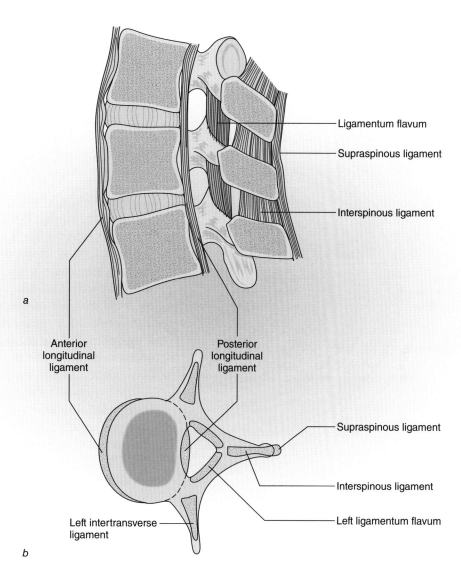

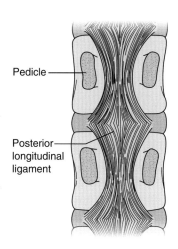

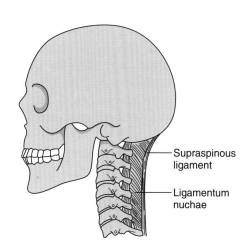

FIGURE 5.9 Ligaments of the vertebral column. (a) Median section through the lumbar region. (b) Superior aspect of a typical vertebra showing the location of ligaments.

Ligamentum flavum

Supraspinous ligament

Interspinous ligament

a

Anterior longitudinal ligament

Posterior longitudinal ligament

Supraspinous ligament

Interspinous ligament

Left ligamentum flavum

Left intertransverse ligament

b

Pedicle

Posterior longitudinal ligament

FIGURE 5.10 Vertical section through the pedicles in the lumbar region. Posterior aspect of the bodies of the vertebrae showing the location of the posterior longitudinal ligament.

Supraspinous ligament

Ligamentum nuchae

FIGURE 5.11 Ligamentum nuchae.

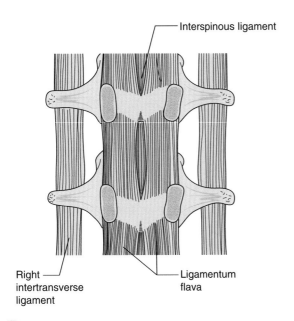

— Interspinous ligament

Right intertransverse ligament

Ligamentum flava

FIGURE 5.12 Anterior aspect of a vertical section through the pedicles in the lumbar region showing the location of the intertransverse ligaments and ligamentum flava.

The vertebral column is supported by three longitudinal ligaments (anterior longitudinal ligament, posterior longitudinal ligament, supraspinous ligament) and three groups of intersegmental ligaments (ligamentum flava, interspinous ligaments, and intertransverse ligaments).

vertebrae (see figures 5.9 and 5.12). Each motion segment contains a left and a right ligamentum flavum with a space between them in the median plane for the passage of blood vessels (figure 5.12). Each ligamentum flavum consists of a sheet of regular elastic tissue. The ligamentum flava, like the ligamentum nuchae, are stretched during flexion of the vertebral column and therefore help to restore the normal orientation of the column after flexion. In doing so, the ligamentum flava relieve the load on the back extensor muscles.

The interspinous ligaments are membranous syndesmoses that link together the spines of adjacent vertebrae (see figure 5.9). The interspinous ligaments are continuous with the supraspinous ligament and provide areas of attachment for muscles on both lateral aspects. The intertransverse ligaments link together the transverse processes of adjacent vertebrae. They tend to become progressively broader from the top to the bottom of the column.

Range of Movement in the Vertebral Column

Movement in the individual motion segments and in the whole vertebral column is limited

by the combined effects of the longitudinal and intersegmental ligaments, the thickness of the intervertebral discs, the orientation of the facet joints, the shape of the vertebral spines, and, in the case of the thoracic region, the splinting effect of the ribs. The extensibility of the supraspinous ligaments, interspinous ligaments, intertransverse ligaments, posterior longitudinal ligament, capsules of facet joints, and posterior aspects of the intervertebral discs limits flexion of the column. Extension is limited by the extensibility of the anterior longitudinal ligament and anterior aspects of the intervertebral discs and by impingement of the vertebral spines on each other. Lateral flexion, always associated with a certain amount of axial rotation, is limited by the extensibility of the supraspinous and interspinous ligaments and, on the convex side, by the intertransverse ligaments and lateral aspects of the intervertebral discs (Kapandji 1974). Axial rotation is limited by the extensibility of the supraspinous, interspinous, and intertransverse ligaments; by the extensibility of the intervertebral discs (in response to torsion); and, in the lumbar region, by the orientation of the facet joints.

In addition to being limited by the shapes of the vertebrae and the various fibrous supporting structures, the ranges of movement of the vertebral column about the three principal axes are limited by the extensibility of the surrounding muscles. Whereas the range of movement in each motion segment is quite small, the cumulative ranges of movement are quite large (table 5.1). Similarly, the shock-absorbing capacity of the vertebral column is determined to a certain extent by the cumulative shock-absorbing capacity of the individual motion segments. However, the shock-absorbing capacity of the vertebral column also depends on its curved shape in the median plane.

Table 5.1 Normal Ranges of Movement (in Degrees) in the Vertebral Column Relative to Relaxed Upright Standing

	Flexion		Extension		Lateral flexion		Axial rotation	
	A	B	A	B	A	B	A	B
Cervical	6.7	40	12.5	75	5.8	35	8.3	50
Thoracic	3.7	45	2.1	25	1.7	20	2.9	35
Lumbar	12.0	60	7.0	35	4.0	20	1.0	5
Total		145		135		75		90

A = approximate range of movement per intervertebral joint; B = approximate range of movement in each region.

Based on I.A. Kapandji, 1974, *The physiology of the joints*, Vol. 3 (Edinburgh: Churchill Livingstone).

Whereas a vertical metal rod will provide little or no shock absorption in response to a vertical impact load, the same piece of metal can be converted into an excellent shock absorber by making it into a helical spring (Adams and Hutton 1985). In the same way, the cervical, thoracic, and lumbar curves of the vertebral column considerably increase its shock-absorbing capacity in activities involving axial compression of the vertebral column, such as landing from a jump.

> **KEY POINT**
>
> Movement in the individual motion segments and in the vertebral column as a whole is limited by the combined effects of the discs, ligaments, orientation and size of the vertebral spines and, in the thoracic region, the splinting effect of the ribs. Although the range of movement in each motion segment is quite small, the cumulative ranges of movement are quite large.

Degeneration and Damage in the Vertebral Column

Back pain, particularly low back pain (pain in the lumbar region), is a common medical disorder in industrial countries and is a major cause of absence from work (Hagen and Thune 1998, Kaaria et al 2005, Nyman et al 2007). The source of back pain is often difficult to diagnose (Nachemson 1992), and it is suggested that back pain may occur with or without structural damage (Plum and Ofeldt 1985, Luoto et al 1995, Sparto et al 1997). In the absence of structural damage, the cause of back pain may be muscle fatigue arising from relatively low levels of muscle strength or endurance (Plum and Ofeldt 1985, Luoto et al 1995, Sparto et al 1997). However, persistent back pain is likely to be due to structural damage or degeneration in one or more motion segments (Nachemson 1992, Nyman et al 2007).

Muscle Fatigue

The maximum amount of force that a muscle (or muscle group) can exert—the strength of the muscle—largely depends on how much it is used in everyday physical activities: The greater the level of activity, the greater the strength of the muscle, and vice versa. When a person's level of habitual physical activity decreases, his or her muscles atrophy—decrease in size—and become weaker. As muscles weaken, the proportion of the strength needed to bring about or maintain a particular movement or posture progressively increases; that is, the intensity of the muscular effort progressively increases. The blood flow through a muscle supplies oxygen to and removes metabolic waste products from the muscle. This blood flow depends on the intensity of muscular effort; the greater the intensity, the more slowly the blood flows and the faster the muscle fatigues.

Muscle fatigue is characterized by the buildup of metabolic waste products, including toxic substances such as lactic acid. Because this buildup results in pain, the person is likely to restrict movement of the affected muscles and associated joints, which can result in more pain due to stiffness, cartilage degeneration, and increased muscle atrophy (Nelson et al 1995) or local overload as a result of compensatory movements in response to muscle fatigue (Plum and Ofeldt 1985, Risch et al 1993, Nelson et al 1995).

Structural Damage and Degeneration

Structural damage to a motion segment or its supporting structures is likely to result in pain. Minor muscle tears are a common cause of temporary low back pain (LBP) (Clarkson and Tremblay 1988). Minor muscle tears occur in a variety of everyday situations when the vertebral column is subjected to above-average loading, such as when a person tries to lift a heavy load. With rest, minor muscle tears tend to heal quickly such that the LBP gradually goes away within a few days. Major muscle tears are usually associated with other types of damage that prolong LBP.

Whereas minor muscle tears tend to result in temporary LBP, damage or degeneration in intervertebral discs can result in persistent LBP. The intervertebral discs have no nerve supply, so damaged or degenerated discs are not a direct source of pain. However, the mechanical integrity of the discs, especially the ability to preload to normal levels, profoundly influences the loads on other parts of the motion segments. Damage or degeneration in an intervertebral disc can cause overloading of other parts of the motion segment, resulting in pain from a variety of sources.

> **KEY POINT**
>
> Low back pain is one of the most common medical disorders in industrial countries and the main cause of absence from work. The source of back pain is often difficult to diagnose, and it can occur with or without structural damage.

Degeneration and Damage in Intervertebral Discs

Degeneration and damage in intervertebral discs occur as a result of aging and mechanical overload.

Aging

After the second decade of life, degenerative changes that impair the capacity of a disc to preload are likely to occur in both the nucleus pulposus and the annulus fibrosus (Brinckmann 1985, Martin and Buckwalter 2002, Roughley 2004). These degenerative changes bring about a gradual reduction in the OSP of the nucleus pulposus and a gradual softening and weakening of the annulus fibrosus, which include the occurrence of fissures between and across the layers of fibrocartilage, especially those close to the nucleus pulposus (Adams and Hutton 1982) (figure 5.13). The fissures are similar to the circumferential and radial tears that occur in the menisci of the knee. In the medium to long term, these degenerative changes are likely to cause pain, either directly or indirectly, in the following three ways.

1. Abnormal loading on end plates of vertebral bodies. In each motion segment, the central areas of the end plates of the vertebral bodies tend to be weaker than the surrounding circumferential areas. When the disc is healthy, most of the compression load on the motion segment is transmitted through the circumferential areas of the end plates. However, as the disc degenerates, the compression loads on the central areas tend to increase, leading to increased compression loading on the subchondral bone and the likelihood of pain.

2. Internal derangement of the nucleus pulposus. During flexion, extension, and lateral flexion of the vertebral column, the vertebrae in each motion segment pivot about the nucleus pulposus, which is itself laterally displaced toward the stretched side of the joint (see figure 5.4). As the normal orientation of the vertebrae is restored, the nucleus pulposus in a healthy disc moves back to its normal position. However, in a degenerated disc small portions of the nucleus pulposus can become trapped in the fissures of the degenerated annulus fibrosus—portions of the nucleus pulposus stick in an off-center position (figure 5.13*b*). This situation may result in unequal tension in the muscles on opposite sides of a motion segment, with the muscles under the greater tension going into spasm. The muscle

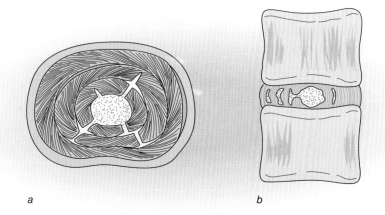

FIGURE 5.13 Degeneration and damage in an intervertebral disc. *(a)* Superior aspect of a disc showing fissures. *(b)* Vertical section through a disc showing portions of the nucleus pulposus trapped in fissures.

spasm can stretch the adjacent spinal nerve, resulting in pain (Nachemson 1992).

3. Reduced thickness of the disc. In degenerated discs the capacity to preload is impaired; the greater the degeneration, the greater the impairment. Consequently, under normal loading conditions degenerated discs are thinner than healthy discs. Reduction in the thickness of a disc has a number of potentially painful consequences (Adams and Hutton 1980, Letts et al 1986, Rydevik et al 1984):

- Impingement of the spinal cord, spinal nerves, and nerves of supporting ligaments and muscles caused by reduced size of intervertebral foramen and protrusion of discs, especially at posterior and posterolateral aspects (figure 5.14, *a* and *b*)
- Increased loading on articular surfaces of facet joints resulting in pain from subchondral bone and progressive damage to articular cartilage (see figure 5.6)
- Fractures of vertebral bodies and pars interarticularis due to reduced shock-absorbing capacity and increased shear load on the pars interarticularis
- Extra-articular impingement of the facet joints (see figure 5.6*c*)

Mechanical Overload

Repetitive or continuous high-level loading for long periods of time is referred to as **chronic**

loading. The intervertebral discs are subjected to chronic loading, especially in the form of bending and torsion, during heavy manual work—activities involving fairly strenuous lifting, carrying, shoveling, pulling, and pushing. Chronic loading accelerates the degeneration of the discs as they age.

Fissures in the annulus fibrosus (figure 5.13*a)* result from a combination of degeneration of the fibrocartilage and strain in response to loading. Given adequate rest, the fissures may heal somewhat. However, the supply of nutrients to the annulus fibrosus is relatively poor and the load on the discs is usually high, even when simply supporting the weight of the upper body; consequently, healing is usually slow. Under chronic loading conditions, the repair process may be outpaced by the rate at which fissuring occurs so that the damage to the annulus fibrosus gradually accumulates. This form of cumulative damage is referred to as **fatigue damage** (Adams and Hutton 1982). **Fatigue failure** occurs when the annulus fibrosus becomes so badly damaged that the capacity of the disc to preload is severely and permanently impaired. The effects of fatigue failure are the same as those that result from aging, except that the various sources of pain that result from disc damage tend to occur earlier.

As the annulus fibrosus of a disc degenerates, it becomes progressively weaker and more vulnerable to damage from all types of loads. In particular, the annulus fibrosus is less able

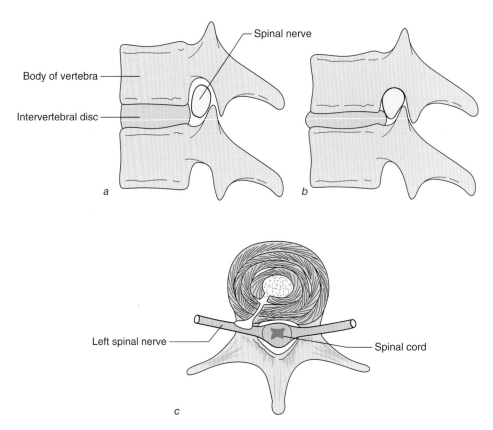

FIGURE 5.14 Impingement of a spinal nerve. *(a)* Normal orientation of vertebrae, disc, and spinal nerve. *(b)* Impingement of spinal nerve by reduced intervertebral foramen and posterior protrusion of the disc. *(c)* Superior aspect of an intervertebral joint showing a prolapsed disc and impingement of the left spinal nerve by part of the nucleus pulposus.

to withstand the tension loads imposed on it during bending movements of the vertebral column. In these circumstances the nucleus pulposus can burst through the annulus fibrosus, resulting in what is referred to as a **prolapsed,** herniated, or "slipped" disc, or an external derangement of the nucleus pulposus (Brinckmann 1985, Adams and Hutton 1982, see figure 5.14*c*). A prolapse usually occurs suddenly because of an **acute failure** of the annulus fibrosus—the annulus fibrosus suddenly bursts in a manner similar to a blowout of a tire.

A prolapse can occur in a previously healthy annulus fibrosus, but in the majority of cases it is likely that the disc was degenerated to a certain extent prior to the prolapse (Brinckmann 1985). In a slightly degenerated disc, the nucleus pulposus still has a high OSP and can still exert a high pressure on the annulus fibrosus. Consequently, a slightly degenerated disc may be more vulnerable to prolapse than a severely degenerated disc in which the nucleus

pulposus has lost its osmotic swelling property.

Many people between 30 and 50 years of age have slightly to moderately degenerated discs. Because many people in this age group are still active and impose large loads on the vertebral column, it follows that the discs are particularly vulnerable to prolapse during this period. If disc degeneration does not occur until later in life, when the person's level of activity is considerably lower than in the middle years, it is unlikely that a vulnerable disc will be subjected to the high loading necessary to cause prolapse (Brinckmann 1985).

Prolapses can occur in the cervical, thoracic, and lumbar regions of the vertebral column but occur most frequently in the L4-L5 and L5-S1 discs (Adams and Hutton 1982, Brinckmann 1985). Prolapses in the lumbar region are usually the result of postures in which the lumbar region is heavily loaded while fully flexed, such as when someone stoops to lift an object from the floor while keeping the legs straight (figure 5.15*b*). In this situation the combined weight of the upper body and the object imposes a considerable load on the lumbar region. The tension load on the stretched posterior aspect of a lumbar disc can cause the disc to prolapse, usually in a posterolateral direction, so that the nucleus pulposus shoots through the tear in the annulus fibrosus and rams into the adjacent spinal nerve, causing a sharp severe pain (see figure 5.14c). After the prolapse, the extruded part of the nucleus pulposus can continue to impinge on the spinal nerve, resulting in continuous pain. The posterior longitudinal ligament strengthens the posterior part of the disc (see figures 5.9 and 5.10) so that prolapses usually occur in the relatively weaker posterolateral aspects of the disc.

Key Terms

chronic loading   Repetitive or continuous high-level loading for long periods of time.

fatigue damage   Cumulative structural damage caused by chronic loading.

fatigue failure   Failure of a structure caused by fatigue damage.

prolapsed disc   Acute failure of the annulus fibrosus of an intervertebral disc resulting in external derangement of the nucleus pulposus; sometimes called a slipped disc or herniated disc.

acute failure   Sudden failure of a structure caused by overload.

With regard to lifting postures, the load on the lumbar region increases as the inclination of the trunk increases. Consequently, lifting an object with a bent back and straight legs—a stoop lift—puts maximum load on the lumbar region, whereas lifting the same load with bent legs and a fairly straight back—a squat lift—minimizes the load on the lumbar region (figure 5.15). In a squat lift, the back muscles

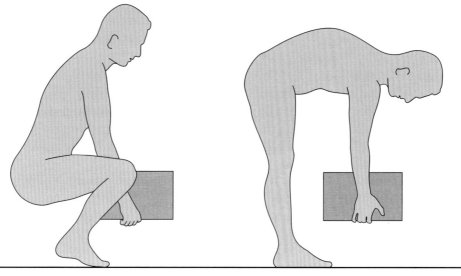

FIGURE 5.15 *(a)* Squat lifting posture. *(b)* Stoop lifting posture.

a

b

hold the trunk steady while the extensor muscles of the legs do most of the work in lifting the object. However, in a stoop lift, the leg extensor muscles contribute little to the work of lifting the object; the extensor muscles of the lumbar region do most of the work of lifting the object, which puts a considerable load on the lumbar structures.

The load on the lumbar region when one is lifting loads in the stooped posture depends not only on the overall inclination of the trunk but also on the shape of the lumbar region. With any particular trunk inclination, the greater the flexion of the lumbar region, the lower the mechanical advantage of the back extensor muscles (see chapter 9) adjacent to the lumbar region and, consequently, the greater the force exerted by the muscles and the greater the intervertebral joint reaction forces. The greater the muscle forces and joint reaction forces, the greater the risk of injury. Maintaining a **lumbar lordosis**—posterior concavity in the lumbar region—as shown in figure 5.16, reduces the muscle forces and joint reaction forces and the risk of injury (Aspden 1987). Consequently, maintaining a lumbar lordosis will reduce the risk of injury in any lifting posture, but the risk of injury will still be greater in the stoop posture compared with a squat posture because of the increased inclination of the trunk in the stoop posture.

The ability to maintain a lumbar lordosis in the stooped posture may depend to a certain extent on **intratruncal pressure**—the pressure developed in the thoracic and the abdominal cavities in response to all types of activity that impose large loads on the lumbar region (Troup 1970, Aspden 1987, McGill and Norman 1987). The thorax contains air and the abdomen contains liquid and semisolid material. When a person inhales and then holds her breath, the pressure inside the thorax—intrathoracic pressure—can significantly increase above normal (figure 5.16). Similarly, in association with intrathoracic pressure, coordinated activity in the back extensor muscles and abdominal muscles considerably increases the pressure inside the abdomen—intraabdominal pressure (IAP). The role of IAP

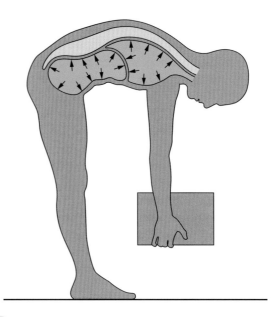

FIGURE 5.16 Intratruncal pressure in the thoracic and abdominal cavities.

in lifting mechanics is not clear (McGill and Norman 1987), but it is suggested that IAP, in conjunction with the force exerted by the back extensor muscles, stabilizes the lumbar lordosis and thereby reduces the likelihood of overloading the lumbar structures (figure 5.16) (Aspden 1987).

Figures 5.17 and 5.18 summarize the sources of back pain that can result from degeneration and damage in intervertebral discs.

Key Terms

intratruncal pressure   The pressure developed in the thoracic and abdominal cavities by coordinated activity of the muscles surrounding the thorax and abdomen.

lumbar lordosis   Posterior concavity in the lumbar region in the median plane.

KEY POINT

Intratruncal pressure can help to maintain a lumbar lordosis in lifting postures and thereby reduce the likelihood of overloading the lumbar structures.

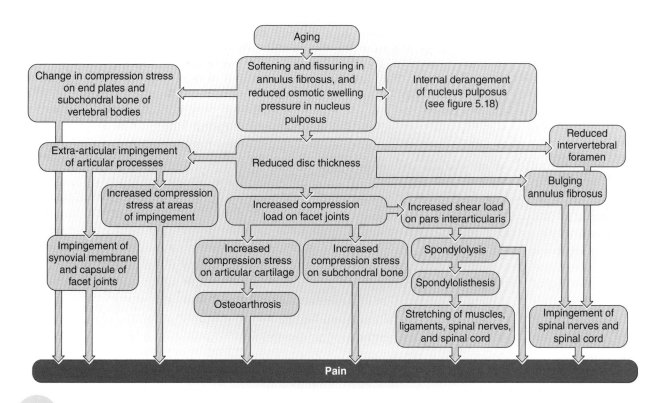

FIGURE 5.17 Sources of back pain that can result from degeneration and damage in intervertebral discs due to aging.

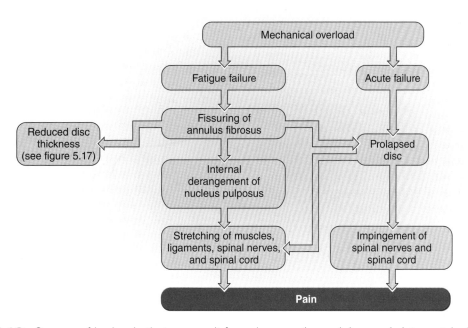

FIGURE 5.18 Sources of back pain that can result from degeneration and damage in intervertebral discs due to mechanical overload.

Degeneration and Damage in Vertebrae

Like intervertebral discs, the vertebrae can experience degeneration and damage due to aging and mechanical overload.

Aging

The main effect of aging on vertebrae is the gradual reduction in bone mass beginning at approximately 30 and 45 years of age in women and men, respectively (see chapter 3). The gradual loss in bone mass results in a gradual decrease in the strength of the vertebrae; for example, the compression strength of lumbar vertebral bodies in 60- to 70-year-olds is approximately half of that found in 20- to 30-year-olds (Brinckmann 1985). Fractures of the vertebral bodies, vertebral arches, spines, and transverse processes are fairly common in elderly people (Old and Calvert 2004). Approximately 25% of postmenopausal women in the United States are affected by vertebral fractures (Melton 1997).

Mechanical Overload

In young and middle-aged people, the most frequently reported form of damage to the vertebrae is fracture of the pars interarticularis (PI). Fractures of the PI occur most frequently in the cervical and lumbar regions and are usually associated with motor vehicle accidents or participation in sports that subject the vertebral column to high bending, torsion, and shear loads (Lowe et al 1986, Letts et al 1986, Silver et al 1986, Tall and Devault 1993, Motley et al 1998, Van der Wall et al 2006, Adams et al 2007, Annear et al 2008). These sports include gymnastics, diving, trampolining, judo, wrestling, weightlifting, soccer, rugby football, and American football. Acute failure—sudden, complete fracture of the PI—can occur in previously undamaged PI. When this occurs it is usually the result of a sudden, violent hyperextension or hyperflexion. The cervical region is particularly vulnerable to damage from this type of movement, and many neck injuries resulting from motor vehicle accidents (whiplash injuries) and trampolining (landing on the head) occur in this manner. Many of these injuries involve severe damage to the spinal cord and spinal nerves (Silver et al 1986, Gozna and Harrington 1982, Adams et al 2007).

Fatigue Damage   Although acute failure of the PI is fairly common, it is likely that most complete fractures of the PI, like prolapsed intervertebral discs, are the result of progressive fatigue damage that eventually results in fatigue failure—complete fracture. Fatigue failure of the PI occurs in response to chronic overloading, when the rate of loading is such that the occurrence of microfractures outpaces the capacity of the healing processes to repair them. Fatigue failure of the PI is often the result of chronic bending loads—fairly high frequency flexion and extension of the lumbar region while it is under fairly high loading. This type of movement is a common feature of gymnastics and trampolining. The effect of this type of movement on the lumbar region as a whole and on the PI in particular is similar to that of repeatedly bending a piece of wire alternately in one direction then the other until it breaks (Hutton and Cyron 1978).

Fatigue failure of the PI of a vertebra occurs in several stages. In the first stage, microfractures gradually accumulate until a stress fracture—partial fracture—is noticeable in one PI. The stress fracture progresses until the PI is completely fractured, but without separation of the fractured ends. This condition is referred to as **spondylolysis** (figure 5.19*a*). The same events occur in the other PI so that eventually both PI are completely fractured—bilateral spondylolysis. At this stage the fractured ends are held together by the supporting structures—ligaments, intervertebral discs, and muscles. However, if the chronic loading continues, the vertebral body may slide forward so the posterior part of the vertebral arch separates from the rest of the vertebra. This condition is called **spondylolisthesis** (figure 5.19). Prognosis of spondylolysis and spondylolisthesis depends largely on the severity of the injury and the age of the person (Adams et al 2007). In young people, spondylolysis and spondylolisthesis that do not involve damage to the spinal cord or spinal nerves tend to heal well with no long-term consequences if the person has adequate rest and refrains from activity that is likely to aggravate the injury (high-frequency, forceful flexion and extension of the lumbar region). However, as a person ages, complete recovery is less likely, especially when the injury involves damage to the spinal cord or spinal nerves (Muschik et al 1996, Standaertand and Herring 2000, Adams et al 2007).

Key Terms

spondylolysis Complete fracture of the pars interarticularis without separation of the fractured ends.

spondylolisthesis Separation of the vertebral arch from the vertebral body subsequent to bilateral spondylolysis.

KEY POINT

In young and middle-aged people, the most common form of damage to the vertebrae is fracture of the pars interarticularis. Acute failure of the pars interarticularis can occur without prior damage, but it is likely that most acute failures, like prolapsed discs, are the result of progressive fatigue damage. There are four stages in fatigue failure of the pars interarticularis: stress fracture, unilateral spondylolysis, bilateral spondylolysis, and spondylolisthesis.

Spondylolisthesis subjects the associated intervertebral disc to shear loading, stretches the supporting ligaments, and, depending on the degree of slippage, compresses or stretches the adjacent spinal cord and spinal nerves. In the general population, spondylolysis and spondylolisthesis are uncommon in children younger than 8 years (Sonne-Holm et al 2007). The incidence of spondylolysis and spondylolisthesis in older children and adults is between 4% and 7%, and the percentage increases with age (Rossi and Dragoni 1994, Sonne-Holm et al 2007, Kalichman et al 2009). Participation in high-level training and competition in a variety of sports tends to increase the incidence of spondylolysis and spondylolisthesis (Engelhardt et al 1997, Lawrence et al 2006).

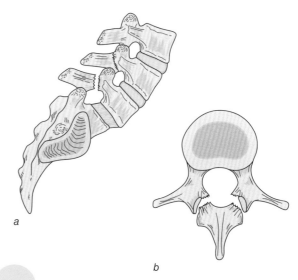

FIGURE 5.19 Spondylolysis and spondylolisthesis. *(a)* Right lateral aspect of lower lumbar and sacral region showing spondylolysis of the right pars interarticularis of L4 and spondylolisthesis of L5. *(b)* Superior aspect of a lumbar vertebra showing spondylolisthesis.

For example, in a group of female gymnasts, the incidence of spondylolysis and spondylolisthesis was found to be 11% and 6%, respectively (Jackson et al 1976). In a group of high school and university male sports participants from 20 sports ranging from rifle shooting to American football, the incidence of spondylolysis was nearly 21% (Hoshina 1980). Soler and Calderon (2000) reported an incidence of spondylolysis of 27% in elite athletes in throwing sports, and Gregory and colleagues (2009) reported incidences of spondylolysis of 67% in cricketers and 66% in soccer players.

Weightlifting The frequency of flexion and extension of the lumbar region in weightlifting is much lower compared with activities like gymnastics and trampolining, but the load on the lumbar region tends to be much higher so that the lumbar region and the PI in particular are subjected to chronic bending loads. In addition, some of the static and quasi-static postures that occur in weightlifting subject the PI to considerable shear loading. To illustrate the types of loading on the lumbar structures during weightlifting, let's consider three postures (Troup 1970):

1. Just after the start of a clean-and-press lift (figure 5.20). This posture is basically the same as the stoop lifting position (see figure 5.15*b*). The lumbar region is in a flexed position and subject to considerable loading.

   - All of the ligaments posterior to the vertebral bodies are likely to be stretched.
   - The posterior aspects of the discs are likely to be stretched, with the possibility of posterior prolapse.
   - The shear load on the PI of the lower lumbar vertebrae is likely to be high, with the possibility of fatigue failure or acute failure of the PI, leading to spondylolisthesis.

2. Holding a barbell at shoulder level or overhead with both feet in a frontal plane (figure 5.21). During the clean stage of a clean and press, the heavily loaded lumbar region changes from a flexed to an extended position (see fig-

ures 5.20*a* and 5.21*a)*. The lumbar region remains extended during the press stage and during the period when the weight is held overhead at the end of the lift. In the extended position, the loads on the lumbar structures are considerable (figure 5.21*b):*

- The shear load on the PI of L5 can result in fatigue failure or acute failure.
- The spines of the lower lumbar vertebrae can impinge on each other with the possibility of fracture of one or more spines.
- Extra-articular impingement (EAI) of the articular processes of the facet joints can occur.
- The anterior longitudinal ligament can be overstretched, resulting in excessive tension in the anterior aspects of the intervertebral discs with the possibility of anterior prolapse of one or more discs.

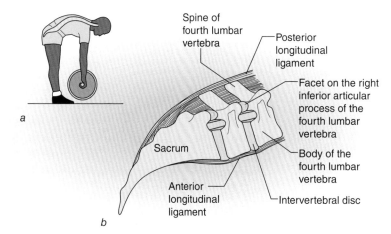

FIGURE 5.20 Stoop lifting posture. *(a)* Orientation of the vertebral column. *(b)* Orientation of L4, L5, and sacrum.

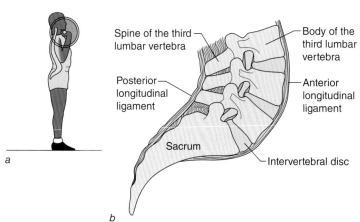

FIGURE 5.21 Upright posture when supporting a weighted barbell at shoulder height. *(a)* Orientation of the vertebral column. *(b)* Orientation of L3, L4, L5, and sacrum.

3. A shoulder press involving an exaggerated lean-back (figure 5.22). This posture involves full extension of the lumbar region so that the lower thoracic to upper lumbar region is more or less horizontal:

- L2 tends to be displaced downward relative to L3. The L2-L3 disc, subjected to shear loading, resists this. The downward displacement of L2 is also resisted by most of the ligaments supporting the L2-L3 motion segment, including the ligaments and capsules of the facet joints, which are luxated in this posture.

- The anterior longitudinal ligament can be overstretched, resulting in excessive tension in the anterior aspects of the intervertebral discs with the possibility of anterior prolapse.

- Whereas the L2-L3 facet joints are luxated, the L4-L5 and L5-S1 facet joints are heavily compressed, with the possibility of extra-articular impingement of the articular processes.

- The spines of the lumbar vertebrae can impinge on each other with the possibility of fracture of one or more spines.

The three postures described are usually symmetrical about the median plane. If the postures are accompanied by axial rotation, a certain degree of twisting, the loads on the lumbar structures can be greater than in the symmetrical position. Figure 5.23 summarizes the sources of back pain that can result from degeneration and damage in vertebrae.

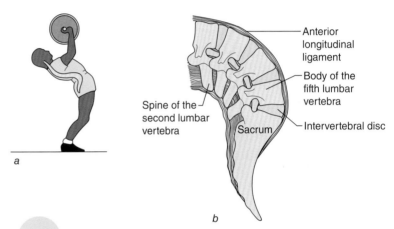

FIGURE 5.22 Posture involving an exaggerated lean-back for pressing a weighted barbell upward from the shoulders. *(a)* Orientation of the vertebral column. *(b)* Orientation of L2, L3, L4, L5, and sacrum.

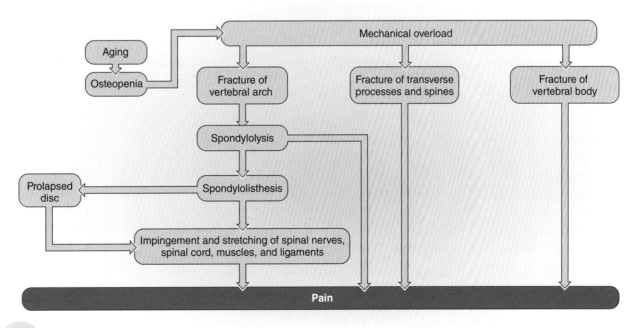

FIGURE 5.23 Sources of back pain that can result from degeneration and damage in vertebrae.

Normal Shape of the Vertebral Column

When the trunk is in an upright position, as in standing, the normal vertebral column has four moderate curves in the median plane but is straight in the frontal plane (figure 5.24, *a* and *b*). The curvature in the median plane represents a compromise between mechanical and morphological requirements. Mechanically, the vertebral column provides stability and flexibility to support the weight of the upper body, acts as a shock absorber in response to impact loads, and protects the spinal cord. Morphologically, the anterior concavities of the thoracic and sacral regions provide spaces for the thoracic and lower abdominal organs, respectively. In a rigid upright structure supported by a single pillar, the pillar is situated centrally to correspond to the line of action of the weight of the structure and thereby prevent any bending load on the supporting pillar (figure 5.24c). Because the vertebral column is curved, the weight of the upper body always exerts a bending load, which increases the curvature of the column. However, in normal upright posture, the line of action of upper-body weight passes through the cervical and

lumbar vertebrae, which minimizes the bending load on the column (Kapandji 1970).

> **KEY POINT**
>
> In the anatomical position, the normal vertebral column has four moderate curves in the median plane (cervical, thoracic, lumbar, and sacral) but is straight in the frontal plane. The curvature in the median plane represents a compromise between mechanical and morphological requirements.

The vertebral column has three flexible regions—cervical, thoracic, and lumbar—and one rigid region, the sacrum. Because the four regions are linked together in a chain, a change in the orientation or curvature of one region results in compensatory changes in the orientation or curvature of the other regions. In normal upright posture, the degree of curvature of the vertebral column is usually moderate—the alternating curves change direction gradually so that the anteroposterior depth

 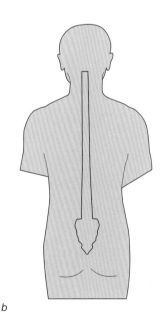 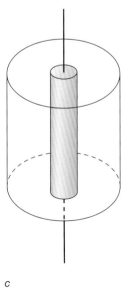

a   *b*   *c*

FIGURE 5.24 Shape of the vertebral column during normal upright standing posture. *(a)* Median plane, right lateral aspect showing the line of action of body weight. *(b)* Frontal plane, posterior aspect. *(c)* Circular structure supported by a central pillar.

of the column as a whole is relatively small. This distance increases as the curvature of the column increases, for example, as a result of lowering the shoulders in a slouched upright posture.

Descriptions of correct upright posture can be functional or anatomical. Functional descriptions refer to the degree of muscular effort required to maintain a balanced position. For example, upright posture may be considered to be correct "if it is effortless, nonfatiguing and painless when the individual remains upright for reasonable periods" (Bullock-Saxton 1988, p 94). Anatomical descriptions of correct upright posture refer to the orientation or curvature of the vertebral column. For example, in correct upright posture "the line of action of body weight lies in the median plane

and passes through the mastoid processes, just in front of the shoulder joints, through or just behind the hip joints, through the knee joints and just in front of the ankle joints" (Basmajian 1965, p 27) (figure 5.25a).

More detailed anatomical descriptions may refer to the degree of curvature of the regions of the column (figure 5.25b). The curvature of a particular region is assessed by measuring, with an electronic image such as an MRI (magnetic resonance imaging) scan or X ray, the angle subtended by the limits of the region. For example, the lumbar region is usually convex anteriorly and concave posteriorly. Consequently, lines drawn parallel to the superior end plate of L1 and the inferior end plate of L5 intersect posteriorly. The angle of intersection is referred to as the lumbar angle or the

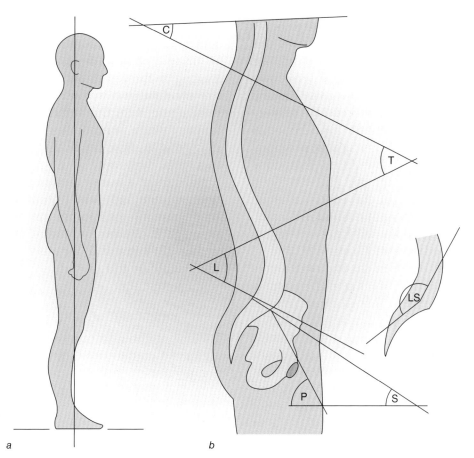

FIGURE 5.25 Anatomical descriptions of upright standing posture. *(a)* Line of action of body weight in relation to anatomical landmarks. *(b)* Angles of curvature of parts of the vertebral column: C = cervical angle; T = thoracic angle; L = lumbar angle; S = sacral angle; LS = lumbosacral angle; P = pelvic tilt angle.

angle of lumbar lordosis (*lordosis* = bent back). The angles of curvature of the cervical region and thoracic region are measured in a similar manner. The method of measuring the angles is called the Cobb method and the angles are often referred to as Cobb angles (Greenspan et al 1978). The Cobb method is also used to assess abnormal curvatures in the frontal plane (see next section).

Like the lumbar region, the cervical region is convex anteriorly and concave posteriorly. For this reason the cervical angle of curvature is sometimes referred to as the angle of cervical lordosis. In contrast to the cervical and lumbar angles, which are subtended posteriorly, the thoracic angle is subtended anteriorly because the thoracic region is usually concave anteriorly and convex posteriorly (see figure 5.25*b*). The thoracic angle is referred to as the angle of thoracic kyphosis (*kyphosis* = keel shaped). The sacral angle, lumbosacral angle, and angle of pelvic tilt also can be used in anatomical descriptions of upright posture. The sacral angle is the angle between the plane of the superior end plate of S1 and the horizontal. The lumbosacral angle is the angle of intersection between lines drawn through the geometric centers of the end plates of L5 and S1, respectively. The pelvic tilt angle is the angle between the horizontal and a line through the promontory of the sacrum and the anterior superior border of the pubic symphysis.

Whereas the angles of curvature (cervical, thoracic, and lumbar) and inclination (sacral, lumbosacral, and pelvic tilt) are relatively easy to measure from MRI scans and X rays, the relationships between the angles in terms of what constitutes correct upright posture have not yet been established (During et al 1985, Walker et al 1987, Kendall et al 2005). For example, an increase in the sacral angle will almost certainly increase lumbar lordosis. However, the actual increase in lumbar lordosis resulting from a specific increase in the sacral angle to maintain correct upright posture is not known. Similarly, the effect of an increase in the sacral angle on thoracic kyphosis and cervical lordosis is not known. Consequently, there is very little information on normal ranges for the various angles. When normal ranges

are given, they are often broad; for example, normal ranges for thoracic kyphosis and the lumbosacral angle have been reported as 20° to 40° and 128° to 160°, respectively (Drummond 1987, Kendall et al 2005). It follows that in terms of the overall shape of the vertebral column, a broad range of shapes can be considered correct upright posture with respect to the functional description given earlier.

A posture that an observer might think is anatomically abnormal may, in fact, be functionally normal. This is often the case in childhood and early adulthood. During the growth period, the vertebral column, like the rest of the skeleton, models its size, shape, and structure in response to the loads imposed on it (see chapter 11). Consequently, a person's vertebral column can develop abnormal curvature (relative to the majority of people of the same age) to continue to function normally. In such cases, and in the absence of any pathological conditions, the shape of the vertebral column will eventually stabilize; that is, the vertebral column will have a consistent shape when in an upright posture. In middle to old age, the loading on the vertebral column can change as a result of changes in muscle strength and body weight. Because the vertebral column is no longer capable of modeling in response to the change in loading, it will become functionally abnormal, depending on the change in loading.

Functional abnormality can result in joint degeneration and pain (Kendall et al 2005). After maturity, all joints are vulnerable to permanent changes in loading. However, the joints most vulnerable are anatomically abnormal weight-bearing joints (see chapter 11).

> **KEY POINT**
>
> Descriptions of correct and incorrect upright posture tend to be functional or anatomical. Functional descriptions refer to the degree of muscular effort or load on the vertebral column in relation to the maintenance of a balanced position. Anatomical descriptions refer to the orientation or curvature of the regions of the vertebral column.

Abnormal Curvature of the Vertebral Column

Abnormal curvatures of the vertebral column often occur prior to maturity in both the median and frontal planes. Whereas some forms of abnormal curvature are more common than others, the range of abnormal shapes is broad.

Abnormal Curvature in the Median Plane

The most common forms of abnormal curvature in the median plane are increased thoracic kyphosis and increased lumbar lordosis (Bullock-Saxton 1988, Kendall et al 2005). These two conditions, which can occur together, are often referred to as simply kyphosis and lordosis, but using the terms in this way can result in confusion because the terms are also used to describe normal curvature of the thoracic and lumbar regions and can be applied to

other regions of the column to describe certain abnormalities. Increased thoracic kyphosis (i.e., an abnormally large thoracic angle) results in a pronounced hump that projects posteriorly, giving rise to the descriptive terms *roundback, humpback, dowager's hump,* and, in extreme cases, *hunchback.* Figure 5.26 shows normal curvature of the vertebral column (figure 5.26*a*) together with abnormal curvatures: **roundback** (figure 5.26*b*), **hollowback** (figure 5.26*c*), combined roundback and hollowback (figure 5.26*d*), swayback (figure 5.26*e*), and flatback (figure 5.26*f*).

There are two forms of roundback: postural roundback and Scheuermann's roundback (Drummond 1987; Kendall et al 2005). Postural roundback is nonstructural and disappears when the person makes a conscious effort to hold the trunk in a normal upright posture. It is caused by slouched sitting and standing

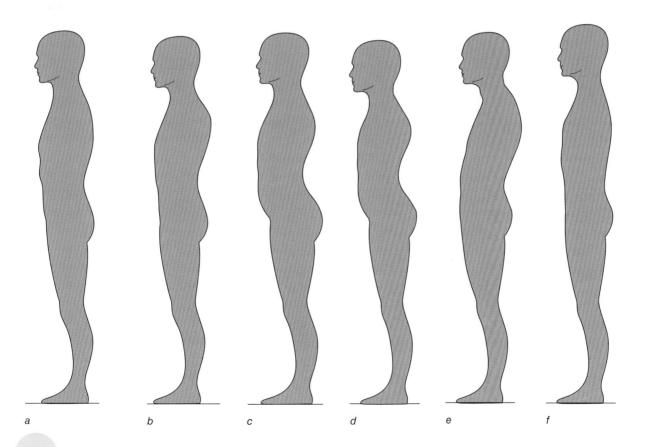

a   b   c   d   e   f

FIGURE 5.26 Normal and abnormal curvature of the vertebral column in the median plane in upright standing posture. *(a)* Normal posture. *(b)* Roundback. *(c)* Hollowback. *(d)* Combined roundback and hollowback. *(e)* Swayback. *(f)* Flatback.

postures that, over a period of time, result in abnormal lengthening of the muscles and ligaments on the posterior aspect of the trunk and abnormal shortening of the muscles and ligaments on the anterior aspect of the trunk (Bullock-Saxton 1988). This imbalance in the soft tissues maintains the postural roundback. Postural roundback is especially common in girls who start puberty early and become overly self-conscious of breast development; in this situation a girl is likely to adopt a round-shouldered posture to hide her breasts (Keim 1982, Kendall et al 2005). Normal thoracic curvature can usually be completely restored by a suitable exercise program.

In contrast to postural roundback, which is nonstructural, Scheuermann's roundback is a structural deformity involving abnormal growth of the vertebral bodies. Wedge-shaped vertebral bodies, in which the height of the bodies anteriorly is less than that posteriorly, and reduced disc space anteriorly characterize the condition. Scheuermann's roundback usually occurs between the ages of 11 and 17 years and is said to be present when at least 3 vertebrae show wedging of 5° or more (Drum-

mond 1987, Kendall et al 2005) (figure 5.27). The degree of wedging is usually greatest at the apex of the hump. In addition to the wedging of the vertebrae, the anterior longitudinal ligament is tight, resulting in a rigid hump that persists even when a conscious effort is made to flatten it.

Scheuermann's roundback is thought to be the result of abnormal ossification of the vertebral bodies, which results in abnormally weak end plates. The line of action of upper-body weight passes in front of the thoracic region so that the thoracic region is normally subjected to an anterior bending load—compression loading on the anterior aspects of the vertebral bodies and tension on the posterior aspects. In response to this normal bending load, abnormally weak end plates grow abnormally and become wedged. The wedging increases the bending load on the affected region so that the condition becomes self-aggravating. In addition, the pressure exerted by the intervertebral discs can depress the central areas of the vertebral end plates; each small depression is called a Schmorl's node (see figure 5.27b). The incidence of Scheuermann's roundback

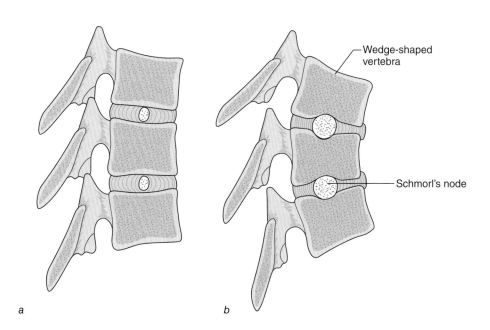

Wedge-shaped vertebra

Schmorl's node

a

b

FIGURE 5.27 Scheuermann's roundback. *(a)* Normal shape and orientation of thoracic vertebrae. *(b)* Scheuermann's roundback showing wedging of vertebrae and Schmorl's nodes.

has been reported as between 0.4% and 8.3% of the general population (Drummond 1987, Lowe and Line 2007).

In children, Scheuermann's roundback can give rise to pain, probably as a result of increased compression stress on subchondral bone and the development of Schmorl's nodes. The incidence of back pain in childhood tends to increase with increase in body weight (see case study 3). However, the incidence of pain from Scheuermann's roundback is greatest after maturity and the pain is often associated with an increase in the thoracic angle (Drummond 1987). Figure 5.28 shows a model of the development of Scheuermann's roundback and the likely sources of pain that result from it. Scheuermann's roundback usually results in compensatory changes in the curvature of the other regions of the vertebral column; for example, a marked increase in the thoracic angle is usually associated with a marked increase in the lumbar angle (see figure 5.26d).

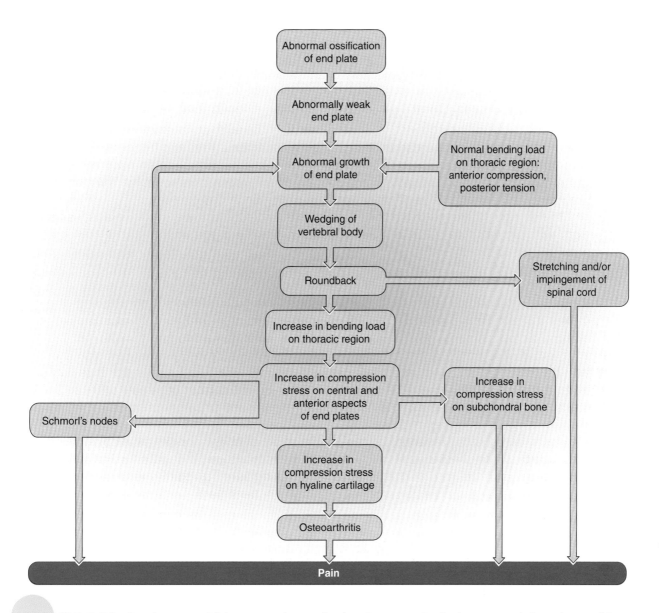

FIGURE 5.28 Development of Scheuermann's roundback and sources of pain that can result from the condition.

CASE STUDY 3   MUSCULOSKELETAL PAIN IN OBESE CHILDREN

Stovitz SD, Pardee PE, Vazquez G, Duval S, Schwimmer JB. 2008. Musculoskeletal pain in obese children and adolescents. *Acta Paediatrica* 97:489-493.

Childhood obesity is a global epidemic, and physicians are increasingly recognizing that many obese children exhibit significant musculoskeletal differences compared with nonobese children. These differences are most noticeable in the skeletal structure and alignment of the lower limbs, with a greater reported incidence, relative to nonobese children, of decreased femoral anteversion (the angle of the neck of the femur relative to shaft in the transverse plane), tibial varum (lower leg angled inward), and lower longitudinal arches of the feet (flatter feet). Whereas the evidence of a direct link between obesity and musculoskeletal abnormalities in childhood is still very limited, it would appear that obese children, like obese adults, experience musculoskeletal pain, especially joint pain, more frequently than nonobese children. The purpose of the present study was to investigate the relationship between body weight and musculoskeletal pain in the back, hips, knees, ankles, and feet in obese children and adolescents. The subjects were 135 children aged 5 to 18 years (12.3 ± 3.2 years) who were evaluated for the management of obesity in a pediatric clinic between 2004 and 2005.

Body mass index (BMI) was calculated as the weight in kilograms divided by the square of height in meters. Obesity was defined as having a BMI in the 95th percentile or greater for age- and gender-matched norms. The location (back, hips, knees, ankles, feet) and frequency (1 or more times per month, 1-3 times per week, 4-7 days per week) of musculoskeletal pain were assessed by a written questionnaire. For subjects who reported pain in more than one joint, the frequency of joint pain was taken as the level of most frequent joint pain at any one site.

Of the 135 subjects, 68 were female and 67 were male. The children were, on average, severely overweight with a mean mass of 90.5 ± 38 kg (199.5 ± 83.8 lb), mean BMI of 36.1 ± 10.4 kg/m^2, and mean BMI percentile of 98.6% ± 3%. Children with musculoskeletal pain were significantly older (12.8 ± 2.9 years) and taller (1.6 ± 0.13 m) than those subjects without pain (11.5 ± 3.5 years; 1.5 ± 0.18 m). There was no difference in prevalence of pain based on gender or ethnicity. Among the children with musculoskeletal pain (n = 83, 61%), the majority (44 of 83) complained of pain at least once a week. Back pain (n = 53, 39%) was the most prevalent form of musculoskeletal pain, followed by foot pain (n = 35, 26%), knee pain (n = 32, 24%), ankle pain (n = 20, 15%), and hip pain (n = 4, 3%). Children with pain in the back, hips, knees, or ankles weighed significantly more than children free of pain in the same joint. Similarly, children with pain in the hips, knees, or ankles had a significantly higher BMI than those children who did not have pain at those sites. Even after adjustment for age, gender, and ethnicity, the reporting of pain was significantly higher for those with a higher weight or higher BMI.

Whereas further research is needed concerning the effect of obesity on the developing skeleton and on the quality of life of the obese child, the authors recommend that parents and all professionals concerned with the care of children consider musculoskeletal pain when making recommendations for physical activity as a component of obesity treatment.

Treatment

Obese children tend to be less active and less physically fit than normal-weight children. Vigorous weight-bearing exercise like running and jumping is likely to aggravate musculoskeletal pain in an unfit, obese child. In the absence of disease, treatment should be based on gradually reducing body weight through a combination of lower food consumption and increased physical activity. To increase physical fitness to a level at which weight-bearing exercise can be tolerated, physical activity should be based on non-weight-bearing activities such as swimming and seated strength and endurance training (Bo Anderson et al 2006, Curtis and d'Hemecourt 2007).

Increased lumbar lordosis (i.e., an abnormally large lumbar angle) results in a pronounced hollow in the low back, giving rise to the descriptive term *hollowback* (see figure 5.26c). Hollowback can occur in a similar manner to Scheuermann's roundback, that is, abnormal ossification of the vertebral end plates with subsequent abnormal growth of the vertebral bodies. Wedging of the vertebrae and Schmorl's nodes can occur but these conditions do not usually occur to the same extent as in roundback (Drummond 1987, Kendall et al 2005). The most frequent causes of hollowback appear to be

- a compensatory increase in the lumbar angle secondary to the development of roundback and
- a strength imbalance between the muscles that control the position of the pelvis.

Weakness in the abdominal and buttock muscles can increase the pelvic tilt angle. This increases the sacral angle and, therefore, the lumbar angle (figure 5.29). When an increase in the lumbar angle is due to muscle imbalance, normal lumbar curvature usually can be completely restored by a suitable exercise program (Keim 1982, Kendall et al 2005). Hollowback is often associated with pain that can arise from a variety of sources (figure 5.30).

In addition to roundback and hollowback, two other distinct types of abnormal curvatures can occur in the median plane: swayback and flatback (see figure 5.26, *e* and *f*). Swayback is characterized by a backward tilt of the trunk about the hips so that the abdomen protrudes anteriorly. Swayback is usually associated with a forward head position and involves increases in cervical lordosis, thoracic kyphosis, lumbar lordosis, pelvic tilt angle, and sacral angle. Flatback is characterized by a flattening of the thoracic and lumbar curves and involves a decrease in thoracic kyphosis, lumbar lordosis, pelvic tilt angle, and sacral angle.

In the absence of disease, abnormal curvatures in the median plane are generally assumed to be compensatory postures that develop because of imbalances in the strength and length of the muscles that control the position of the pelvis (Kendall et al 2005). For example, relative to the muscle balance in normal upright posture, hollowback is usually associated with strong, shortened hip flexor muscles and weak abdominal muscles; treatment involves increasing the strength of the abdominal muscles and lengthening the hip flexor muscles. Similarly, relative to the muscle balance in normal upright posture, roundback is usually associated with strong, shortened abdominal muscles and weak trunk extensors in the thoracic region; treatment involves increasing the strength of the trunk extensors in the thoracic region and lengthening the abdominal muscles (Kendall et al 2005).

Abnormal Curvature in the Frontal Plane

In an upright posture, the vertebral column is normally straight in the frontal plane (figure 5.31a). Abnormal curvature of the vertebral column in the frontal plane is referred to as **scoliosis.** A scoliosis can be unilateral (one sided) or bilateral (both sides). There are three main forms of unilateral scoliosis: thoracic,

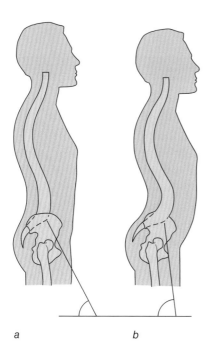

FIGURE 5.29 Relationship between degree of lumbar lordosis and pelvic tilt angle in upright standing posture. *(a)* Normal orientation of the vertebral column and pelvis. *(b)* Increased degree of lumbar lordosis associated with an increase in the pelvic tilt angle.

a   *b*

lumbar, and thoracolumbar (figure 5.31, *b-d*). In a bilateral scoliosis there are two compensatory curves, which bend in opposite directions to each other (figure 5.31*e*). Lumbar scoliosis and bilateral scoliosis are often associated with marked pelvis obliquity.

Scolioses are assessed by the Cobb method using the upper and lower vertebrae in the

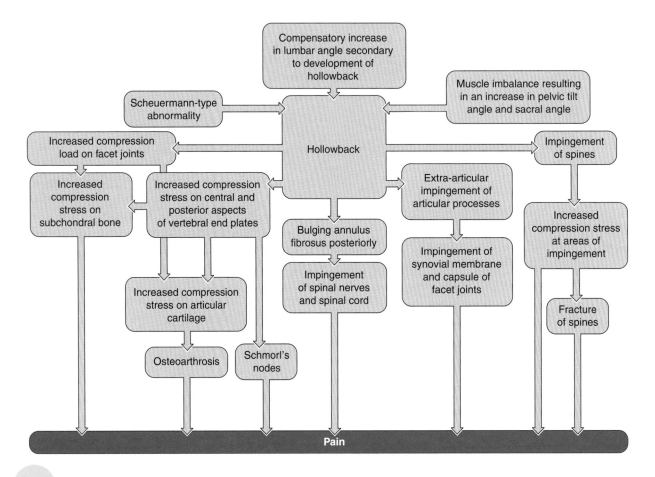

FIGURE 5.30 Development of hollowback and sources of pain that can result from the condition.

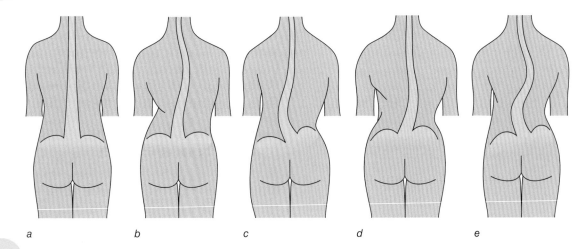

FIGURE 5.31 Forms of scoliosis. *(a)* Normal orientation of the vertebral column. *(b)* Right thoracic scoliosis. *(c)* Left lumbar scoliosis. *(d)* Right thoracolumbar scoliosis. *(e)* Bilateral scoliosis: right thoracic and left lumbar.

curve as the limits of the curve. In bilateral scoliosis both curves are assessed. A Cobb angle of less than 10° is regarded as within the bounds of normality. Approximately 2% of the adult population has some degree of scoliosis, but it exceeds 20° in less than 0.05% (Keim 1982, Kendall et al 2005). Scoliosis usually develops during the growth period; whereas the incidence of scoliosis is similar in boys and girls, the incidence of severe scoliosis—a Cobb angle of 30° to 40°—is five to eight times greater in girls than in boys (Keim 1982, Kendall et al 2005).

In some cases of scoliosis, the underlying cause can be a leg length inequality that tilts the pelvis in the frontal plane (Giles and Taylor 1981). This situation usually results in a unilateral lumbar or thoracolumbar scoliosis. In the absence of a leg length inequality, poor posture with the weight shifted to one side when standing and sitting can contribute to the development of scoliosis (Perdriolle and Vidal 1985). However, in 80% of cases of scoliosis the underlying cause is unknown—idiopathic (no known cause) scoliosis (Richter et al 1985, Kendall et al 2005). In these cases the underlying cause is thought to be one of the following:

- Abnormal ossification of the vertebral bodies, which results in abnormalities similar to those resulting from Scheuermann's roundback: wedge-shaped vertebrae and Schmorl's nodes (Keim 1982, Giles and Taylor 1982).

- An imbalance in the levels of contraction in the muscles on opposite sides of the vertebral column resulting from malfunctioning in the nervous system (Haderspeck and Schultz 1981, Schultz et al 1981, Ford et al 1988).

- A combination of skeletal abnormality and malfunctioning in the nervous system. It has been shown that the height–width ratio of the vertebral bodies in children with idiopathic scoliosis is greater than in children without the condition; children with idiopathic scoliosis have more slender vertebral columns (Skoglund and Miller 1981). The increase in slenderness makes the vertebral column more vulnerable to slight changes in loading. For example, whereas a slight imbalance in the levels of contraction in the muscles on opposite sides of the vertebral column may have no effect on a normal vertebral column, the same imbalance can result in an excessive bending load on an abnormally slender vertebral column (Schultz et al 1981).

A scoliosis tends to develop in three phases:

1. Occurrence of slight scoliosis
2. A period of rapid progression of the scoliosis
3. A period of more gradual progression of the scoliosis up to maturity, when the scoliosis stabilizes (Perdriolle and Vidal 1985)

Scolioses are classified according to the age of the person at the onset of phase 2—the period of rapid progression. There are three categories (Perdriolle and Vidal 1985):

1. Infantile scoliosis: phase 2 occurs in the period between birth and 6 years of age.
2. Juvenile–pubertal scoliosis: phase 2 occurs in the period between 6 years of age and the onset of puberty.
3. Pubertal (adolescent) scoliosis: phase 2 occurs in the period between the onset of puberty and maturity.

In cases of juvenile–pubertal and pubertal scoliosis, there can be a considerable period of time between phases 1 and 2 in which the degree of scoliosis remains at a fairly low level. The degree of scoliosis at maturity depends on the age of the person at the onset of phase 2: The earlier the onset of phase 2, the greater the amount of skeletal growth remaining and the greater the final degree of scoliosis (Bunnell 1986, Masso et al 2002). Consequently, infantile scoliosis often results in the greatest deformity and pubertal scoliosis tends to result in the least deformity. Fortunately, pubertal scoliosis is the most common form of scoliosis (Richter et al 1985, Masso et al 2002). Scoliosis often gives rise to pain, especially after maturity. Figure 5.32 summarizes the development

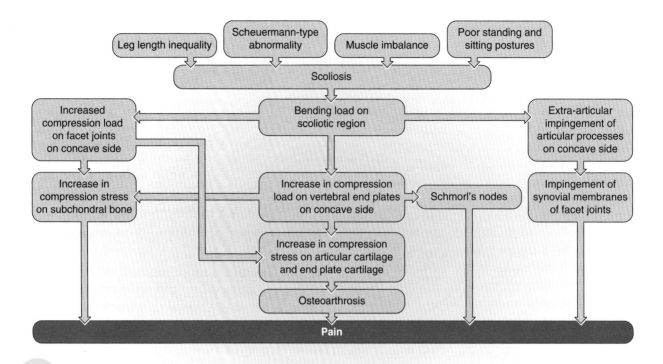

FIGURE 5.32 Development of scoliosis and sources of pain that can result from the condition.

of scoliosis and the sources of pain that arise from it.

The primary objective of scoliosis treatment is to minimize curvature progression (Weiss et al 2006). Treatment depends on the age of the person and degree of deformity at diagnosis. There are three main approaches to conservative treatment: physical therapy, intensive rehabilitation, and brace treatment (Negrini et al 2003). Physical therapy is aimed at improving trunk strength and flexibility in general. Intensive rehabilitation involves more specific strength and flexibility training for particular joints and muscles. Brace treatment, usually in combination with physical therapy or intensive rehabilitation, tends to be used when physical therapy and intensive rehabilitation on their own are ineffective, especially in young children who still have considerable potential for growth.

Key Terms

roundback An abnormally large thoracic angle resulting in a pronounced hump that projects posteriorly.

hollowback An abnormally large lumbar angle resulting in a pronounced hollow in the low back.

scoliosis Abnormal curvature of the vertebral column in the frontal plane; can be unilateral or bilateral.

Joints of the Pelvis

The axial skeleton articulates with the lower limbs via the sacroiliac articulations—the joints between the sacrum and the innominate bones (figure 2.31 and figure 5.33). The sacroiliac articulations are partly synovial and partly syndesmosis. The joints between the auricular surfaces of the sacrum and innominate bones—usually referred to as sacroiliac joints—are synovial in structure. Located immediately posterior to each sacroiliac (SI) joint is a short, strong interosseous ligament between sacrum and ilium, which constitutes a syndesmosis (figure 5.34).

The sacrum is directly responsible for transmitting loads between the vertebral column and pelvis across the lumbosacral and SI

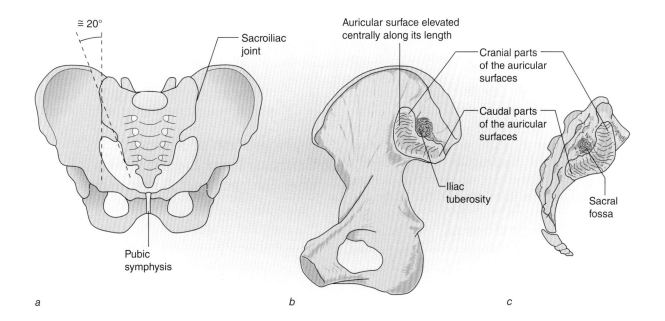

FIGURE 5.33 Sacroiliac joint. *(a)* Anterior aspect of male pelvis showing the angle of inclination of the sacroiliac joints with respect to the median plane. *(b)* Medial aspect of the right innominate bone. *(c)* Right lateral aspect of the sacrum.

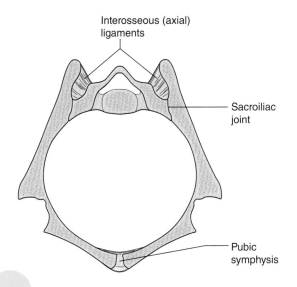

FIGURE 5.34 Oblique section through the pelvis—the pubic symphysis and angles of the sacroiliac joints—showing the interosseous ligaments.

articulations. Like adjacent motion segments in the vertebral column, the lumbosacral motion segment and SI articulations are functionally interdependent. Malfunction in one SI articulation is likely to overload not only the lumbosacral motion segment but also the other SI articulation. Similar malfunction in the lumbosacral motion segment is likely to overload the SI articulations (Grieve 1976, Wilder et al 1980, Standring 2004).

Sacroiliac Joints

The SI joints have many of the structural characteristics of a typical synovial joint—the auricular (articular) surfaces are covered with a form of articular cartilage, there is a small joint cavity containing synovial fluid, and the whole joint is lined on the inside with a synovial membrane (Bowen and Cassidy 1981). Each joint is also supported by a number of ligaments. However, there are a number of distinct differences between the SI joints and a typical synovial joint in terms of articular cartilage, articular surfaces, and range of movement.

The type of articular cartilage on the sacral surface of a SI joint is different from that on the iliac surface. The limited amount of information available suggests that the sacral articular cartilage resembles hyaline cartilage, whereas the iliac articular cartilage resembles fibrocartilage, although both can change with age (Bowen and Cassidy 1981). The sacral articular cartilage is about three times thicker than the iliac articular cartilage, 6 mm versus 2 mm in a normal adult.

At birth the articular surfaces are fairly smooth and flat, allowing sliding movements in virtually all directions. Consequently, at this stage the SI joints resemble synovial joints. During the first decade the surfaces remain fairly flat but become less smooth as the difference in texture between the sacral and iliac cartilage becomes more pronounced. After puberty, each of the articular surfaces begins developing an irregular uneven contour. Prior to the start of the third decade, the early 20s, the sacral articular surface becomes depressed along its length, whereas the iliac articular surface develops a reciprocal elevation along its length (see figures 5.33, b and c, and 5.34). The sacral depression and iliac elevation are sometimes referred to as the sacral groove and iliac ridge (Bowen and Cassidy 1981). The actual articular surfaces gradually develop a complex irregular pattern of grooves, ridges, eminences, and depressions that reciprocate with each other only in part (Grieve 1976, Wilder et al 1980, Standring 2004). The irregular and uneven articular surfaces are atypical of ordinary synovial joints. Furthermore, whereas ordinary synovial joints are usually designed to allow a large range of movement, the partial interlocking of the SI articular surfaces severely restricts movement. In old age the SI joint cavity often becomes obliterated, with a gradual increase in fibrocartilaginous adhesions, and synostosis can occur (Standring 2004).

The overall shape of the auricular surfaces varies between people, but these surfaces usually are roughly C shaped or L shaped (see figure 5.33, b and c). For descriptive purposes, each auricular surface is divided into cranial and caudal portions. With the sacrum in the anatomical position, the cranial portion points upward and, in some cases, slightly backward, whereas the caudal portion points backward and, in some cases, slightly downward (see figure 5.33, b and c). The region between the cranial and caudal portions is sometimes referred to as the angle of the auricular surface.

The sacrum is wedge shaped; it tapers from top to bottom such that on average the plane of each SI joint makes an angle of approximately 20° to the median plane when viewed from the front (figure 5.33a). The configuration of the SI joints is such that any downward force on the sacrum, such as the downward load of upper-body weight when the trunk is upright, is resisted by the ligaments supporting the SI joints, which in turn stabilizes the SI joints.

Syndesmoses and Supporting Ligaments of the Sacroiliac Articulations

The syndesmosis of each SI articulation consists of a short but broad, strong ligament called the interosseous ligament or axial ligament. The interosseous ligament spans the gap between the sacrum and ilium posterior to the SI joint (see figure 5.34). In this region the sacral surface is dominated by the large sacral fossa, and the iliac surface is dominated by the large iliac tuberosity. The area of articulation of the interosseous ligament is approximately 50% greater than that of the SI joint (22 cm^2 vs. 14 cm^2 on average in adults) and reflects the importance of the interosseous ligament to the strength of the SI articulation (Miller et al 1987). In addition to the interosseous ligament, three groups of ligaments support the SI joints. The anterior and posterior SI ligaments provide direct support, whereas the iliolumbar ligaments, above the SI joint, and the sacrospinous and sacrotuberous ligaments, below the SI joint, provide indirect support. The anterior SI ligaments run downward and medially across the front of the SI joint. They arise from the anterior aspect of the ilium adjacent to the SI joint and attach onto the upper two thirds of the anterior lateral aspect of the sacrum (figure 5.35a). The posterior SI ligaments run across the posterior aspect of the SI joint. The ligaments arise from the posterior medial aspect of the iliac crest and the adjoining notch between the posterior superior and posterior inferior iliac spines. The ligaments run medially and fan out to attach onto the tubercles of the intermediate and lateral sacral crests (figure 5.35b).

The iliolumbar ligaments run laterally downward and slightly forward from the transverse processes of L4 and L5 to the anterior medial superior aspect of the ilium above the SI joint (figure 5.35, a and b). The sacrospinous and

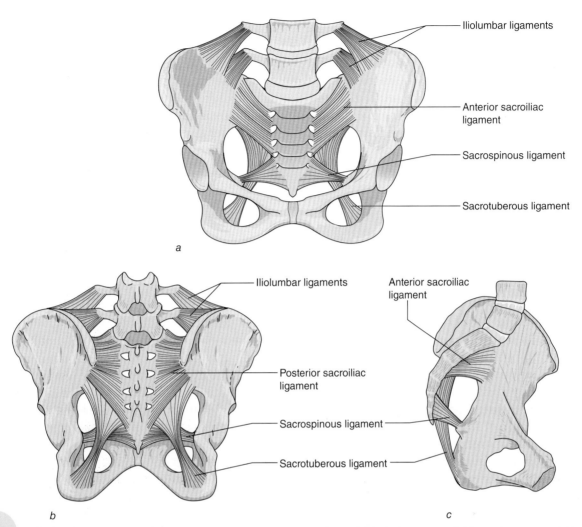

Iliolumbar ligaments

Anterior sacroiliac ligament

Sacrospinous ligament

Sacrotuberous ligament

a

Iliolumbar ligaments

Anterior sacroiliac ligament

Posterior sacroiliac ligament

Sacrospinous ligament

Sacrotuberous ligament

b

c

FIGURE 5.35   Supporting ligaments of the sacroiliac articulations. *(a)* Anterior aspect. *(b)* Posterior aspect. *(c)* Left aspect of medial section through the pelvis.

sacrotuberous ligaments join the sacrum to the ischium. The sacrospinous ligament is thin and triangular; it arises from the ischial spine and fans out medially superiorly and posteriorly to attach on to the lateral border of the coccyx and lower third of the sacrum (figure 5.35). The sacrotuberous ligament is very strong. Laterally, it is attached to the medial aspect of the tuberosity and ramus of the ischium. From this broad attachment the ligament converges to form a thick central band that passes behind the sacrospinous ligament. The central band then fans out to gain attachment to the superior medial aspect of the greater sciatic notch and lateral border of the sacrum below the SI joint. The fibers of the posterior sacroiliac, sacrospinous, and sacrotuberous ligaments blend with each other to a considerable extent.

KEY POINT

The axial skeleton articulates with the lower limbs via the sacroiliac articulations—the sacroiliac joints and the interosseous syndesmoses. The sacroiliac articulations are supported by five groups of ligaments: anterior and posterior sacroiliac ligaments, iliolumbar ligaments, sacrospinous ligaments, and sacrotuberous ligaments.

Pubic Symphysis

The pubic symphysis or interpubic joint is a symphysis joint that lies in the median plane (see figures 5.33*a*, 5.34, 5.35, 5.36). The elliptical medial ends of the pubic bones (see figure 5.33*b*) are covered with hyaline cartilage, and a disc of fibrocartilage is sandwiched between the layers of hyaline cartilage (figure 5.36). In men, the disc is approximately 4 mm thick and has a cross-sectional area in the median plane of approximately 5 cm². Consequently, the amount of movement allowed by deformation of the disc is very small. Strong ligaments on the superior, anterior, inferior, and posterior aspects of the joint further restrict movement in the joint. The superior ligament is attached to the pubic crest, that is, the region between the pubic tubercles. The posterior ligament runs across the posterior aspect of the joint and is fairly thin. The inferior or arcuate (arch) ligament, on the other hand, is thick. It attaches to the pubic arch and blends with the disc. The anterior ligament is also thick and blends with the tendons and aponeuroses of the abdominal muscles. The combination of relatively thin disc and strong ligaments produces a strong joint with a minimal range of movement.

The pubic symphysis stabilizes the pelvis and, in particular, the SI joints and provides a certain amount of shock absorption in response to loads transmitted across the pelvis. Under normal circumstances the stability of the SI joints is maintained largely by the clamping effect of the pubic symphysis anterior to the SI joints and the ligamentous support posterior to the SI joints. Instability in the pubic symphysis

or the interosseous ligaments will reduce the clamping effect on the SI joints (figure 5.34). Any reduction in the stability of the pubic symphysis results in abnormal transmission of loads across the SI joints. This, in turn, can result in pain as a consequence of impingement and stretching of spinal nerves or progressive degeneration of the pubic symphysis and SI joints.

In men and young women, the pubic symphysis and the SI joints are normally very stable. However, in adult women the stability of the pubic symphysis and SI joints is decreased during the latter part of the menstrual cycle, during menopause, and during pregnancy (Grieve 1976, Don Tigny 1985, Colliton 1996). Approximately 7 to 10 days prior to the start of menstruation, the ligaments of the pubic symphysis and SI articulations soften and lengthen as a result of hormonal changes. Consequently, the joints become less stable and more susceptible to injury. During menstruation, the ligaments regain their normal length and toughness to restore normal stability of the joints. During pregnancy the ligaments of the public symphysis and SI articulations gradually soften and lengthen. In addition, the cartilage of the joints gradually becomes thicker because of absorption of water. Consequently, the joints become much more flexible, an essential feature of normal vaginal childbirth, but the stability of the joints is considerably reduced. After delivery the stability of the joints is gradually restored over 6 to 12 weeks (Grieve 1976).

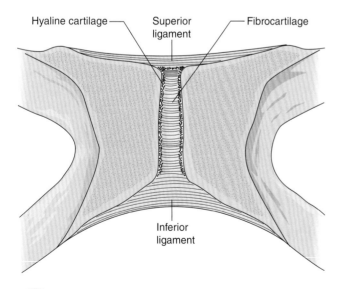

Hyaline cartilage — Superior ligament — Fibrocartilage

Inferior ligament

FIGURE 5.36 Oblique section through the pubic symphysis.

> **KEY POINT**
>
> The pubic symphysis stabilizes the pelvis, particularly the sacroiliac joints, and provides shock absorption in response to loads transmitted across the pelvis.

Movements of the Sacroiliac Joints

In men, the SI joints are normally stable. As previously described, the stability of the SI joints in women is reduced at certain times by hormonal changes. However, at other times the stability of the SI joints in women is similar to that in men. The function of the SI articulations is to transmit loads between the sacrum and ilia. This is achieved most efficiently when the SI joints are stable and have a limited range of movement. A high level of stability not only facilitates shock absorption in response to impact loading but also ensures that all of the main components of the SI articulations (articular cartilages, syndesmoses, supporting ligaments) participate fully in the transmission of load so that none of the components is overloaded. Inadequate stability of the SI joints can overload certain components of the SI articulations during load transmission.

In response to impact loads that result, for example, from heel strike during walking and running, the SI articulations act as shock absorbers in concert with the other joints of the vertebral column and the pubic symphysis to absorb the energy of the impact. Shock absorption in the SI articulations occurs in a manner similar to that in intervertebral joints—in response to impact loading a small amount of movement occurs in the SI joint, which is strongly resisted by the interosseous ligament and other supporting ligaments. Energy is absorbed by deformation of the articular cartilages and tension in the interosseous ligament and other supporting ligaments (Wilder et al 1980, Adams et al 2007). Degeneration of the SI joints and consequent loss of shock-absorbing capacity can overload the lumbar intervertebral joints; this is reflected in the increased incidence of degenerative joint disease in the lumbar region following degeneration of the SI joints (Wilder et al 1980, Adams et al 2007, Ha et al 2008).

It is generally agreed that a small amount of movement takes place in the SI joints in response to nonimpact changes in loading (Colachis et al 1963, Wilder et al 1980, Adams et al 2007). For example, in a completely relaxed recumbent posture, as shown in figure 5.37a, upward pressure on the base of the sacrum rotates the sacrum backward relative to the ilia about a mediolateral axis; the coccyx moves upward and, to a lesser extent, the sacral promontory moves downward (figure 5.37, a and b). In upright stance, the weight of the upper body (head, arms, and trunk) is exerted downward on the lumbosacral joint,

which rotates the sacrum forward about a mediolateral axis relative to the ilia; when the person moves from a recumbent position to upright standing, the sacral promontory moves forward and downward and the coccyx moves upward and backward (Kapandji 1974, Don Tigny 1985) (figure 5.37, c and d).

Authors disagree regarding the position of the axis of rotation of the sacrum. Some authors regard the interosseous ligament (axial ligament) as the axis of rotation, hence the name of the ligament, whereas others maintain that the axis of rotation passes through the angle of the auricular surfaces of the SI joint (Bowen and Cassidy 1981, Kapandji 1974). Both assertions are probably correct to a certain extent because it is likely that the rotation of the sacrum is accompanied by some linear movement (Wilder et al 1980). In this case the

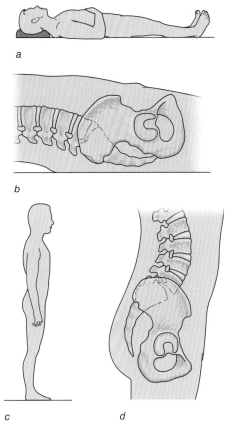

FIGURE 5.37 Movement of the sacrum relative to the ilia when moving between recumbent and standing postures. *(a)* Recumbent posture. *(b)* Position of the sacrum relative to the ilia in recumbent posture. *(c)* Standing posture. *(d)* Position of the sacrum relative to the ilia in standing posture.

axis of rotation is not fixed but tends to move in the region between the axial ligament and the angle of the auricular surfaces.

In view of the large variation in the size, shape, and contour of the SI joints among people, each person probably has a unique pattern of SI movement consisting of certain amounts of linear and angular movement. Nevertheless, it is generally agreed that forward rotation of the sacrum is the dominant feature of SI joint movement in response to the downward force of upper-body weight in the upright stance. The forward rotation of the sacrum results in movement of the innominate bones such that the posterior aspects of the ilia are drawn together and the ischial tuberosities move farther apart. The forward rotation of the sacrum and resulting movement of the innominate bones are referred to as **nutation of the pelvis** (Kapandji 1974) (*nutare* = to nod). Nutation decreases the anteroposterior diameter of the pelvic brim and increases the anteroposterior and transverse diameters of the pelvic outlet (figure 5.38). Nutation is usually severely restricted by the ligaments

that support the SI joints. **Counternutation** is the reverse of nutation—backward rotation of the sacrum resulting in separation of the posterior aspects of the ilia and drawing together of the ischial tuberosities. Counternutation occurs in the recumbent supine position with the hips extended (figure 5.37, *a* and *b*). Counternutation increases the anteroposterior diameter of the pelvic brim and decreases the anteroposterior diameter of the pelvic outlet (figure 5.38). Like nutation, counternutation is usually severely restricted by the ligaments that support the SI joints.

The degree of nutation and counternutation increases considerably during pregnancy because of the decrease in stability of the SI and pubic symphysis joints. During childbirth the increased degree of nutation and counternutation facilitates the passage of the fetus along the birth canal. During the early stages of labor, the woman is usually encouraged to adopt a recumbent prone position with the hips extended as in figure 5.37, *a* and *b*. The counternutation of the pelvis that results from this position enlarges the pelvic brim and favors the descent of the fetal head into the pelvis (Kapandji 1974). During the expulsive phase of labor the woman is often encouraged to flex the hips and knees. In this position, tension in the hamstring muscles in the back of the thighs rotates the innominate bones backward relative to the sacrum, thereby resulting in nutation of the pelvis. The effect of nutation is to enlarge the pelvic outlet (Kapandji 1974).

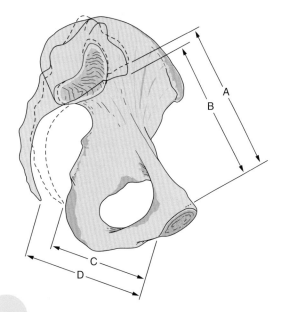

FIGURE 5.38 Medial aspect of the left innominate bone and right lateral aspect of the sacrum showing the position of the sacrum relative to the innominate bone in nutation (solid outline) and counternutation (dotted outline) of the pelvis. A and B show the anteroposterior diameter of the pelvic brim in counternutation and nutation. C and D show the anteroposterior diameter of the pelvic outlet in counternutation and nutation.

Key Terms

nutation of the pelvis Forward rotation of the sacrum about the sacroiliac joints that results in movement of the innominate bones such that the posterior aspects of the ilia are drawn closer together and the ischial tuberosities move farther apart.

counternutation of the pelvis Backward rotation of the sacrum about the sacroiliac joints that results in movement of the innominate bones such that the posterior aspects of the ilia move farther apart and the ischial tuberosities are drawn closer together.

KEY POINT

The sacroiliac joints are usually very stable. However, small amounts of movement (combination of linear and angular motion) occur in the sacroiliac joints in response to changes in loading. These movements are referred to as nutation and counternutation of the pelvis.

Abnormalities of the Sacroiliac Joints and Pubic Symphysis

The anterior branches of the fourth and fifth lumbar nerves, which form part of the sacral plexus, pass in front of the SI joints (figure 5.39). Branches from these nerves innervate the joint capsules of the SI joints. Excessive movement of an SI joint stretches the joint capsule, which in turn stretches the capsular nerves, resulting in pain. Excessive movement in an SI joint can also cause pain by stretching and impingement of the sacral spinal nerves. Pain from these sources is referred to as sacroiliitis and is often misdiagnosed as some form of low back pain (Don Tigny 1985). Sacroiliitis can occur whenever the stability of the SI joints is decreased. For reasons described earlier, adult women are much more vulnerable to sacroiliitis than men. This is reflected in a study of 222 sport-related hip and pelvic injuries in 204 patients; the incidence of sacroiliitis was found to be 15.6% and 6.4% in women and men, respectively (Lloyd-Smith et al 1985). The pelvic pain that accompanies pregnancy and especially labor and childbirth is largely due to sacroiliitis resulting from the increased range of SI joint movement (Hainline 1994, Endresen 1995).

The contours of the auricular surfaces of the SI joints are irregular and in the normal stable condition partially interlock with each other. In this situation, the movement of the SI joints that occurs in response to normal levels of impact loading and nonimpact changes of loading is usually insufficient to disrupt the normal pattern of interlocking or to cause sacroiliitis. However, in an unstable joint, excessive movement can cause the contours of the auricular surfaces to interlock in an abnormal position, which will be reinforced by tension in the supporting ligaments created by the excessive movement. This situation, a form of dislocation of the SI joint, can result in continuous pain (chronic sacroiliitis). The most common form of this condition is called anterior dysfunction of the SI joint, in which the SI joint becomes locked in an abnormal position with the innominate bone rotated anteriorly on the sacrum, that is, in a position of counternutation (Don Tigny 1985).

Anterior dysfunction occurs as a result of standing postures in which the trunk is inclined forward, for example, lifting with straight legs, making a bed, ironing, washing dishes, and shaving. In these situations the innominate bones rotate forward on the sacrum (counternutation), but the amount of rotation is usually small given the stability of the SI joints and the abdominal and gluteal muscles, which stabilize the innominate bones. However, when the SI joints are unstable, the amount of rotation can be excessive, especially if the abdominal and gluteal muscles are weak. In these circumstances one or both SI joints can become locked in an abnormal position, resulting in unilateral (one SI joint) or bilateral (both joints) anterior dysfunction. Anterior dysfunction can cause chronic sacroiliitis. In addition, because the amount of movement in an SI joint affected by anterior dysfunction is virtually nil, the ability of the joint to assist in shock absorption is

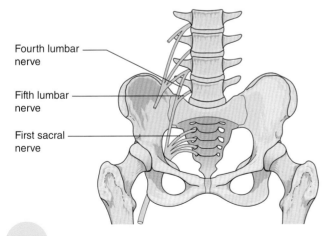

Fourth lumbar nerve

Fifth lumbar nerve

First sacral nerve

FIGURE 5.39 Anterior aspect of the pelvis showing the locations of the lumbar and sacral nerves in relation to the sacroiliac joints.

severely impaired. Consequently, the adjacent lumbar motion segments are overloaded, which increases the risk of low back pain. In some cases a change in posture can dislodge the anterior dysfunction in the affected joints. However, in other cases the anterior dysfunction may be aggravated by certain events, such as the following:

- Women are particularly susceptible to anterior dysfunction just prior to and during menstruation. If the anterior dysfunction persists after menstruation the likelihood of self-correction is considerably reduced, because the ligaments of the SI articulations and pubic symphysis shorten, tending to stabilize the anterior dysfunction.

- After anterior dysfunction has occurred, there is an increased likelihood that the sacrum will slip downward on the ilium in the affected joint. When this happens the innominate bone is prevented from rotating backward on the sacrum, which prevents self-correction of the anterior dysfunction. This condition occurs when there is a sudden increase in the vertical load exerted on the SI joint, for example, when one lifts an object, steps down heavily, or sits down heavily.

Coughing and sneezing can increase the pain of sacroiliitis when anterior dysfunction persists. Manipulation therapy is usually required to correct persistent anterior dysfunction.

Instability in the pubic symphysis reduces the stability of the SI joints and increases the likelihood of anterior dysfunction. Except for when hormonal changes in women reduce SI joint stability, the pubic symphysis is usually a very stable joint in women and men and dislocation of the pubic symphysis is rare (Kapandji 1974). Consequently, when dislocation or separation of the pubic symphysis does occur, it is, like a prolapsed intervertebral disc, a serious injury that is difficult to treat and that may not heal completely.

The pubic symphysis and ipsilateral SI joint are subjected to considerable shear loading in any activity involving a one-leg stance such as walking, running, jumping, and landing. Repetitive, high-level, shear loading on the pubic symphysis can cause microtears in the disc and supporting ligaments, giving rise to a painful condition called osteitis pubis (Hanson et al 1978, Lentz 1995, Johnson 2003). Just as sacroiliitis is often misdiagnosed as low back pain, osteitis pubis is often misdiagnosed as muscle strain. In the study of sport-related hip and pelvic injuries by Lloyd-Smith and colleagues (1985), the incidence of osteitis pubis in men and women was found to be 7.9% and 4.2%, respectively. During the second decade of life, a median cleft often develops in the rear part of the disc of the pubic symphysis, which seems to be the long-term effect of repetitive, high-level, shear loading.

KEY POINT

Excessive movement of a sacroiliac joint stretches the joint capsule and impinges sacral spinal nerves, resulting in sacroiliitis. The pelvic pain accompanying pregnancy, and especially labor and childbirth, is largely due to sacroiliitis. Excessive movement in the pubic symphysis will likely cause microtears in the disc and supporting ligaments, resulting in osteitis pubis.

Summary

The joints of the axial skeleton and pelvis are well designed—in terms of both the structure of individual joints and the shape of the vertebral column—to transmit large forces and to provide shock absorption at the expense of flexibility. The next chapter describes the structure and function of the joints of the appendicular skeleton.

Review Questions

1. Differentiate between a motion segment and an intervertebral joint.
2. Describe the structure and function of an intervertebral joint and a facet joint.
3. Describe the variation in size and shape of intervertebral discs.
4. Explain how disc thickness is affected by external loading and osmotic swelling pressure.
5. Describe the response of the annulus fibrosus and nucleus pulposus of an intervertebral disc to movement (about a horizontal axis with respect to upright posture) in a motion segment.
6. Describe the effects of flexion and extension of the vertebral column on loading in facet joints.
7. Describe the ligaments of the vertebral column.
8. Describe the factors that limit flexibility in different regions of the vertebral column.
9. Explain how back pain can arise in the absence of structural damage.
10. Describe the effects of aging and chronic loading on intervertebral discs and the sources of pain that can result.
11. Differentiate between spondylolysis and spondylolisthesis.
12. Differentiate between functional and anatomical descriptions of upright posture.
13. Describe the structure and functions of the sacroiliac articulations and the pubic symphysis.
14. Differentiate between counternutation of the pelvis and anterior dysfunction of a sacroiliac joint.

Joints of the Appendicular Skeleton

The large joints of the axial skeleton (intervertebral and pelvic joints) are mainly cartilaginous and provide shock absorption at the expense of flexibility. In contrast, virtually all of the joints of the appendicular skeleton are synovial, or mainly synovial, and are designed primarily to facilitate flexibility at the expense of shock absorption. This chapter describes the structure and function of the joints of the appendicular skeleton.

6

OBJECTIVES

After reading this chapter, you should be able to do the following:

1. Differentiate between joints and joint complexes.
2. Describe the structure and function of the shoulder complex.
3. Describe the structure and function of the elbow complex.
4. Describe the structure and function of the wrist complex.
5. Describe the structure and function of the hip joint.
6. Describe the structure and function of the knee complex.
7. Describe the structure and function of the ankle complex.

Joints and Joint Complexes

Most whole-body movements involve simultaneous movement in many joints. Whereas the range of movement patterns that people use to carry out a particular task may be broad, each movement pattern reflects a certain degree of **functional interdependence** between the joints used; movement in a particular joint is associated with simultaneous movements in other joints, especially those in the same skeletal chain such as the hip, knee, and ankle joints and the shoulder, elbow, and wrist joints. For some groups of joints, such as the intervertebral joints discussed in chapter 5, the functional interdependence is clear; that is, the movement of the vertebral column always involves simultaneous movement in a number of motion segments. For others, such as the group consisting of the shoulder, acromioclavicular, and sternoclavicular joints, the degree of functional interdependence is less clear. For example, whereas small movements of the shoulder joint, such as in writing at a desk, may occur without simultaneous movement in the acromioclavicular and sternoclavicular joints, most movements of the arm about the shoulder joint involve simultaneous move-

ment in all three joints. A group of joints that are functionally interdependent is called a **joint complex** (Peat 1986). The contribution of each joint in a joint complex to a particular movement of the body depends on the extent of body movement; the greater the movement, the greater the contribution of each joint.

With respect to the following descriptions of joints and joint complexes in the appendicular skeleton, the values given for ranges of movement refer to normal, healthy, untrained people. As described in chapter 4, the flexibility of a joint largely depends on the length range in which the muscles that control the joint normally function. Consequently, trained people may have much greater ranges of movement in certain joints than untrained people.

Key Terms

functional interdependence (of a joint complex)   The tendency of a particular group of joints to move simultaneously to achieve a particular body movement.

joint complex   A group of joints that are functionally interdependent.

Shoulder Complex

The **shoulder complex** consists of the shoulder (glenohumeral), acromioclavicular, and sternoclavicular joints, which link together the humerus, scapula, clavicle, and sternum in

a chain (figure 6.1*a*). Loads transmitted from the upper limb toward the axial skeleton are transmitted directly via the shoulder, acromioclavicular, and sternoclavicular joints and indirectly by ligaments and muscles supporting the shoulder complex and by muscles linking the shoulder complex to the axial skeleton.

The shoulder complex, in association with the muscles that link it to the axial skeleton, provides the upper limb with a range of movement exceeding that of any other joint or joint complex (Peat 1986). The large range of movement is due to four sources of relative motion in the shoulder complex: the shoulder, acromioclavicular, and sternoclavicular joints, and the **scapulothoracic gliding mechanism.** The latter refers to the ability of the scapula to glide and rotate relative to the posterior aspect of the rib cage. The scapulothoracic gliding mechanism, combined with movements in the acromioclavicular and sternoclavicular joints, enables the glenoid fossa to follow the head of the humerus and maintain maximum congruence of the shoulder joint in movements of the humerus relative to the axial skeleton; this includes all movements of the arm that involve medium- to full-range extension, flexion, abduction, and adduction of the shoulder complex. Consequently, the range of movement of the humerus relative to the axial skeleton is normally much greater than the range of movement of the shoulder joint (the range of movement of the humerus relative to the scapula). Normal movement of the humerus and, therefore, of the upper limb depends on normal functional interrelationships between the four sources of relative motion. Restriction of normal movement in any of the four sources of relative motion can impair normal function of the upper limb.

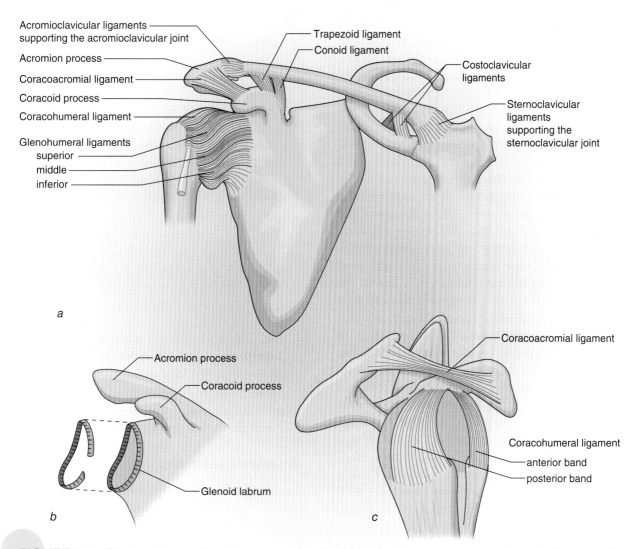

FIGURE 6.1   The shoulder complex. *(a)* Anterior aspect of the right shoulder complex showing the main ligaments. *(b)* Glenoid labrum. *(c)* Lateral aspect of the right shoulder showing coracoacromial ligament and coracohumeral ligaments.

Key Terms

shoulder complex The functionally interdependent shoulder, acromioclavicular, and sternoclavicular joints.

scapulothoracic gliding mechanism The system of muscles that enables the scapula to glide and rotate relative to the posterior aspect of the rib cage.

> **KEY POINT**
>
> The shoulder complex and the scapulothoracic gliding mechanism provide the upper limb with a range of movement exceeding that of any other joint or joint complex.

Shoulder Joint

The shoulder joint is a synovial ball-and-socket joint. The head of the humerus is almost hemispherical and articulates with the glenoid fossa, which is shallow and relatively small; the articular surface of the glenoid fossa is approximately one third the size of that of the head of the humerus (Standring 2004).

Given the relatively small and shallow contact area between the articulating surfaces of the shoulder joint and the relatively large loads transmitted across the shoulder joint in forceful actions of the upper limb, the shoulder joint is vulnerable to dislocation. Dislocations of the joint, especially anterior dislocation (dislocation of the head of the humerus forward of the glenoid fossa), are common in contact sports (Hawkins and Mohtadi 1994). The size of the area of articulation is increased by a ring of fibrocartilage called the glenoid labrum (*labrum* = lip) around the edge of the glenoid fossa (figure 6.1*b*). The glenoid labrum considerably deepens the glenoid fossa and thus increases the congruence and stability of the joint.

The joint is enclosed by a joint capsule strengthened on its anterior aspect by three capsular ligaments—the superior, middle, and inferior glenohumeral ligaments (figure 6.1*a*). The superior aspect of the joint is supported by the coracohumeral ligament, which consists of two bands linking the lateral aspect of the base of the coracoid process to the superior aspects of the lesser tuberosity (anterior band)

and greater tuberosity (posterior band) (figure 6.1*c*). Although the coracohumeral and glenohumeral ligaments, together with the rest of the capsule, help stabilize the joint in certain joint positions, the stability of the joint is largely maintained by four muscles collectively referred to as the rotator cuff (figure 6.2). The muscles originate from the subscapular fossa (subscapularis), supraspinous fossa (supraspinatus), and infraspinous fossa (infraspinatus and teres minor) and converge on the proximal end of the humerus to attach onto the lesser tuberosity (subscapularis) and greater tuberosity (supraspinatus, infraspinatus, and teres minor). In addition to stabilizing the shoulder joint, the rotator cuff assists other muscles that move the shoulder joint and shoulder girdle.

> **KEY POINT**
>
> The shoulder joint is a synovial ball-and-socket joint. The glenoid fossa is small and shallow relative to the head of the humerus, but the area of articulation is increased by the glenoid labrum. The joint capsule is strengthened on its anterior aspect by the glenohumeral ligaments and on its superior aspect by the coracohumeral ligament, but joint stability is largely maintained by the rotator cuff.

The coracoacromial arch restricts shoulder joint movements. The arch is formed by the acromion process, coracoid process, and the coracoacromial ligament linking the two bony processes (see figure 6.1, *a* and *c*). Impingement of the surgical neck of the humerus on the coracoid process and anterior aspect of the coracoacromial ligament restricts flexion. Extension is restricted by impingement of the surgical neck of the humerus on the acromion process. Abduction is restricted by impingement of the greater tuberosity of the humerus on the coracoacromial ligament.

Acromioclavicular Joint

The acromioclavicular joint is a synovial gliding joint and usually has an incomplete articular disc or one or two menisci (figure 6.3). The capsule of the acromioclavicular joint is strength-

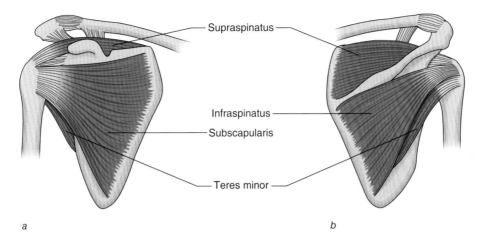

FIGURE 6.2 *(a)* Anterior aspect and *(b)* posterior aspect of the right rotator cuff.

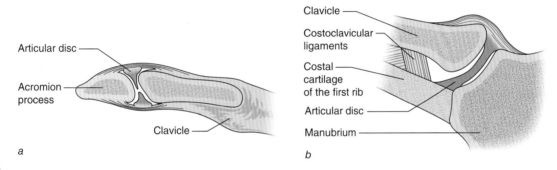

FIGURE 6.3 Coronal sections through the *(a)* right acromioclavicular joint and *(b)* right sternoclavicular joint.

ened superiorly by a capsular ligament called the superior acromioclavicular ligament. The acromioclavicular joint is indirectly stabilized by the coracoclavicular ligaments—the trapezoid and conoid ligaments, which join the clavicle to the coracoid process close to the acromioclavicular joint (see figure 6.1*a*). The fibers of the trapezoid ligament run downward and medially from the inferior aspect of the clavicle to the superior aspect of the coracoid process. The fibers of the conoid ligament run vertically from the inferior aspect of the clavicle to the superior aspect of the base of the coracoid process.

Because of the oblique orientation of the articular surfaces of the acromioclavicular joint (figure 6.3*a*), forces transmitted across the shoulder joint toward the axial skeleton often dislocate the acromioclavicular joint by driving the acromion process under the lateral end of

the clavicle. Such forces, occurring for example as a result of a punching action or falling on an outstretched hand, are normally strongly resisted by the coracoclavicular ligaments, especially the trapezoid. However, dislocation of the acromioclavicular joint is not uncommon, especially in contact sports, usually due to a fall or a hit (Taft et al 1987).

KEY POINT

The acromioclavicular joint is a synovial gliding joint. The joint is indirectly stabilized by the coracoclavicular ligaments, which resist forces transmitted across the shoulder joint that tend to dislocate the acromioclavicular joint. Dislocation of the acromioclavicular joint is not uncommon, especially in contact sports.

Sternoclavicular Joint

The sternoclavicular joint is a synovial joint that has structural and functional characteristics of both saddle and gliding joints. The sternoclavicular joint has two degrees of freedom: rotation in the coronal plane, as in raising the shoulders, and rotation in a transverse plane, as in moving the shoulders forward. The capsule of the sternoclavicular joint is strengthened on its posterior, superior, and anterior aspects by capsular ligaments. The joint is indirectly stabilized by the costoclavicular ligaments, which join the clavicle to the costal cartilage of the first rib close to the sternoclavicular joint (see figure 6.1*a* and 6.3*b*).

The fibers of the anterior band of the costoclavicular ligament run downward and medially from the inferior aspect of the clavicle to the costal cartilage and adjoining bony portion of the first rib. The fibers of the posterior band of the costoclavicular ligament run downward and laterally from the inferior aspect of the clavicle to the costal cartilage and adjoining bony portion of the first rib. The sternoclavicular joint has a complete articular disc, which is attached superiorly to the superior aspect of the medial end of the clavicle, inferiorly to the costal cartilage of the first rib, and to the joint capsule for the remainder of its circumference (Standring 2004; see figure 6.3*b*). This attachment enables the disc, assisted by the joint capsule and the posterior band of the costoclavicular ligament, to strongly resist forces transmitted along the clavicle toward the axial skeleton, which can dislocate the sternoclavicular joint by driving the medial end of the clavicle over the manubrium. The strength of this mechanism is such that dislocation of the

sternoclavicular joint is rare; fracture of the clavicle is more common (Standring 2004).

Movements During Abduction of the Upper Limb

Given the various restrictions on the movement of the joints of the shoulder complex, most upper-limb movements that involve more than 30° of abduction of the arm with respect to the trunk are brought about by a combination of movement in two or more of the four sources of movement in the shoulder complex (Peat 1986). Consequently, lack of flexibility in any of these components can impair a person's ability to abduct her arm. Abduction is an essential feature of many overhead arm actions such as reaching up to a high shelf, changing a bulb in a ceiling light, putting on your hat, washing and combing your hair, pitching a baseball, throwing a football, serving in tennis, throwing a javelin, spiking and blocking in volleyball, bowling in cricket, and all of the major swimming strokes (Miniaci and Fowler 1993, Copeland 1993).

The first 30° of abduction of the arm is brought about almost entirely by abduction of the shoulder joint (figure 6.4, *a* and *b*). The next phase of abduction of the arm—from 30° to approximately 100°—uses a combination of shoulder abduction and rotation of the scapula and clavicle as a single unit about an oblique axis passing through the sternoclavicular joint and the region of the medial end of the spine of the scapula. During this phase, shoulder abduction contributes about 30°, and rotation of the shoulder girdle (sternoclavicular joint and scapulothoracic mechanism) contributes about 40° (figure 6.4*c*).

At this point, the costoclavicular ligaments become taut and prevent further upward rotation of the clavicle about the sternoclavicular joint. The final phase of abduction of the arm—from 100° to approximately 180°—uses a combination of shoulder abduction, rotation of the scapula about an anteroposterior axis through the acromioclavicular joint, and axial rotation of the clavicle (counterclockwise with

> **KEY POINT**
>
> The sternoclavicular joint is a synovial joint with characteristics of both saddle and gliding joints. The joint has a complete articular disc, which, together with the joint capsule and costoclavicular ligaments, strongly resists forces that could dislocate the joint. Dislocation of the sternoclavicular joint is rare.

respect to the right clavicle when viewed from a lateral aspect). Rotation of the scapula about the acromioclavicular joint is limited to about

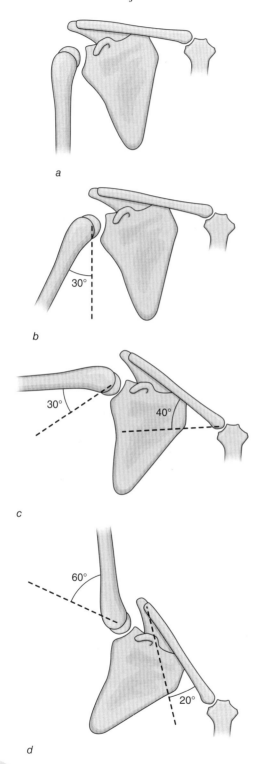

a

b

c

d

FIGURE 6.4 Movements of the right shoulder complex during abduction of the arm.

20° by the coracoclavicular ligaments, which then become taut.

The other 60° in the final phase is brought about by a combination of shoulder abduction, axial rotation of the clavicle, and scapulothoracic movement (rotation and sliding). During this final phase, shoulder abduction is accompanied by external rotation of the shoulder so that the articular surfaces of the joint remain in close contact with each other and the greater tuberosity of the humerus is prevented from impinging on the coracoacromial arch (figure 6.4d).

Impingement of any supporting structures of the shoulder joint (capsule, ligaments, and rotator cuff) on the coracoacromial arch results in a painful condition generally referred to as **shoulder impingement** (Miniaci and Fowler 1993). Shoulder impingement can occur in any form of overhead arm movement when flexibility in one or more components of the shoulder complex is limited (Brunet et al 1982, Silliman and Hawkins 1991). Whereas tight shoulder ligaments may restrict flexibility in these joints, shoulder impingement is more likely to be caused by lack of extensibility in some of the muscles of the shoulder complex, especially those concerned with the scapulothoracic mechanism. Not surprisingly, shoulder impingement is fairly common in sports such as swimming and tennis, but it can also affect everyday movements such as combing one's hair (Miniaci and Fowler 1993).

Key Term

shoulder impingement A painful condition that results from impingement of one or more of the supporting structures of the shoulder joint (capsule, ligaments, and rotator cuff) on the coracoacromial arch.

KEY POINT

Lack of flexibility in the shoulder complex and lack of extensibility in the muscles that control the scapulothoracic gliding mechanism can impair abduction of the upper limb and result in shoulder impingement.

Elbow Complex

The **elbow complex** is synovial and consists of the humeroulnar joint (between the trochlea and trochlear notch), the humeroradial joint (between the capitulum and the head of the radius), and the proximal radioulnar joint (between the head of the radius and radial notch) (see figures 2.25 and 2.26). The humeroulnar joint is a typical hinge joint designed for flexion and extension of the elbow. The proximal radioulnar joint is a pivot joint that facilitates supination and pronation of the forearm. The humeroradial joint is biaxial—it can rotate about two axes:

1. *A vertical axis,* the long axis of the radius, as in pronation and supination of the forearm. In this type of movement the superior aspect of the head of the radius spins on the capitulum as the head

of the radius pivots within the radial notch.

2. *A mediolateral axis,* as in flexion and extension of the elbow. In this type of movement the head of the radius slides on the capitulum.

The elbow complex is enclosed within a joint capsule strengthened on all aspects by capsular ligaments (figure 6.5). The fibers of the capsule blend with the annular ligament, which holds the head of the radius against the radial notch of the ulna (figure 6.5a). The anterior aspect of the capsule helps prevent hyperextension of the elbow and is strengthened by the lateral and medial oblique ligaments (figure 6.5a). The posterior aspect helps prevent hyperflexion and is strengthened by transverse, lateral oblique, and medial oblique ligaments (figure

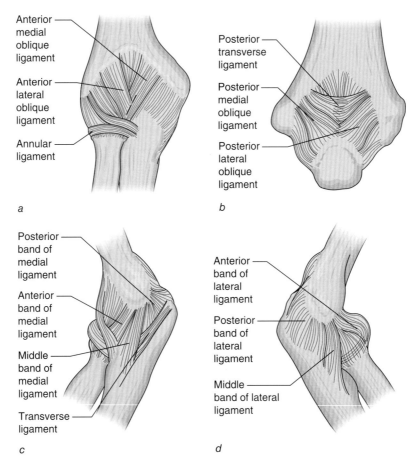

FIGURE 6.5 Capsular ligaments of the right elbow complex. *(a)* Anterior aspect. *(b)* Posterior aspect. *(c)* Medial aspect. *(d)* Lateral aspect.

6.5*b*). The normal range of flexion and extension in the elbow is approximately 140°.

Fan-shaped medial and lateral ligaments strengthen the medial and lateral aspects of the capsule, respectively. The medial ligament arises from the medial epicondyle of the humerus and is characterized by three distinct bands—the anterior, intermediate (middle), and posterior bands (figure 6.5*c*). A transverse ligament also strengthens the medial aspect of the capsule. The lateral ligament arises from the lateral epicondyle of the humerus and, like the medial ligament, is characterized by anterior, intermediate (middle), and posterior bands (figure 6.5*d*). The medial and lateral ligaments maximize congruence of the elbow complex in all positions of the flexion and extension range and thereby help prevent abduction and adduction of the elbow.

Key Term

elbow complex The functionally interdependent humeroulnar joint, humeroradial joint, and proximal radioulnar joint.

KEY POINT

The elbow complex is synovial and consists of the humeroulnar joint (uniaxial), which facilitates flexion and extension of the elbow; the humeroradial joint (biaxial), which facilitates flexion and extension of the elbow and pronation and supination of the forearm; and the proximal radioulnar joint, which facilitates pronation and supination of the forearm. Intracapsular ligaments strengthen all aspects of the elbow complex capsule.

Wrist Complex

The **wrist complex** consists of the wrist joint and the midcarpal joint. The wrist joint, a synovial ellipsoid joint, is located between the distal ends of the radius and ulna and the proximal surfaces of the scaphoid, lunate, and triquetrum (figure 6.6*a*). The radius articulates directly with the scaphoid and lunate. However, an articular disc separates the ulna from the lunate and triquetrum. The articular disc is attached to the anterior lateral aspect of the styloid process of the ulna and the inferior medial aspect of the radius just below the ulna notch. The articular surface of the radius is continuous with the lower surface of the articular disc; together they form a roughly elliptical biconcave surface that articulates with the reciprocal biconvex surface formed by the scaphoid, lunate, and triquetrum. Another articular disc, sometimes referred to as a meniscus, links the styloid process of the ulna, the triquetrum, and the pisiform. The midcarpal joint consists of the series of synovial gliding joints between the proximal and distal rows of carpals (figure 6.6*a*).

The capsule of the wrist joint follows the line of the joint and is separate from that of the midcarpal joint. The capsule of the midcarpal joint is more extensive than that of the wrist joint; it has a very irregular shape and incorporates not only the intercarpal joints of the midcarpal joint but also the intercarpal joints directly linked to the midcarpal joint.

Key Term

wrist complex The functionally interdependent wrist and midcarpal joints.

Stability of the Wrist

The wrist complex is stabilized by

- interosseous ligaments (ligaments that link adjacent carpals) between some of the carpals,
- more extensive ligaments that span various sections of the carpus, and
- muscles that move the carpus.

On the anterior aspect of the carpus, the main ligaments are the anterior (palmer) radiocarpal and ulnocarpal ligaments (figure 6.6*b*). The anterior radiocarpal ligament has two oblique bands. The inferior band passes downward and medially from the anterior inferior border of the styloid process of the radius to attach onto the capitate. The superior

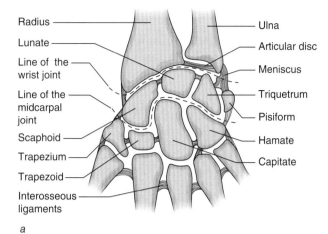

Radius
Lunate
Line of the wrist joint
Line of the midcarpal joint
Scaphoid
Trapezium
Trapezoid
Interosseous ligaments

Ulna
Articular disc
Meniscus
Triquetrum
Pisiform
Hamate
Capitate

a

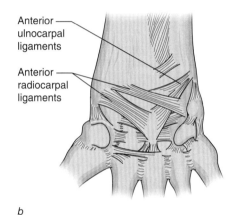

Anterior ulnocarpal ligaments
Anterior radiocarpal ligaments

b

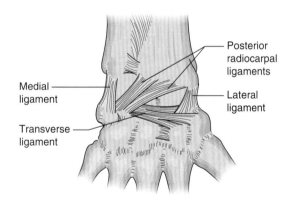

Medial ligament
Transverse ligament

Posterior radiocarpal ligaments
Lateral ligament

c

FIGURE 6.6 Ligaments of the right wrist complex. *(a)* Coronal section. *(b)* Anterior aspect. *(c)* Posterior aspect.

band passes downward and medially from the anterior lateral border of the distal end of the radius (medial to the attachment of the inferior band) to attach onto the triquetrum. The anterior (palmer) ulnocarpal ligament also has two bands, which originate from the anterior lateral

aspect of the styloid process of the ulna. The superior and inferior bands fan out to attach onto the lunate and capitate, respectively.

On the posterior aspect of the carpus, the main ligaments are the posterior (dorsal) radiocarpal, posterior (dorsal) transverse, medial, and lateral ligaments (figure 6.6c). The posterior radiocarpal ligament consists of superior and inferior bands arising from a common origin on the triquetrum. The bands fan upward and laterally to attach onto the posterior border of the distal end of the radius. The transverse ligament arises from the triquetrum and divides into three bands as it passes laterally to attach onto the scaphoid, trapezoid, and trapezium. The medial ligament originates from the inferior aspect of the styloid process of the ulna and fans out as it passes almost directly downward to attach onto the pisiform and triquetrum. The lateral ligament originates from the posterior inferior aspect of the styloid process of the radius and fans out as it passes almost directly downward to attach onto the trapezoid and trapezium.

With respect to the anatomical position, the lateral border of the wrist joint is inferior to the medial border such that lateral displacement of the carpus at the wrist joint is considerably restricted by the styloid process of the radius (figure 6.6a). In comparison, medial displacement of the carpus at the wrist joint is relatively unrestricted. Stability of the carpus in this direction appears to be one of the main functions of the obliquely oriented anterior and posterior radiocarpal ligaments, which restrict movement of the carpus medially (and downward) with respect to the wrist joint.

Movements of the Wrist

The main movements in the wrist complex are flexion, extension, abduction, adduction, and circumduction. All of these movements occur as a result of a complex combination of movements in the wrist and midcarpal joints. With respect to the anatomical position, the ranges of flexion and extension are both limited to approximately 85°. The posterior ligaments restrict flexion, and the anterior ligaments restrict extension. Abduction is restricted to approximately 15° by the oblique orientation of the wrist joint and by the medial ligament. In comparison, adduction is relatively unrestricted and has a 45° range of motion. Whereas movements of the wrist complex beyond the normal ranges can result in injury, injury occurs most often as a result of excessive abduction, usually in combination with extension. Such movements can result in fracture of the scaphoid or styloid process of the radius (Recht et al 1987).

> **KEY POINT**
>
> The wrist complex is synovial and consists of the wrist joint (basically ellipsoid) and midcarpal joint (series of gliding joints). The wrist complex is stabilized by interosseous ligaments between some of the carpals, more extensive ligaments that span various sections of the carpus, and muscles that move the carpus. The wrist complex facilitates circumduction of the hand.

Hip Joint

The hip joint is a synovial ball-and-socket joint. The head of the femur is slightly more than a hemisphere and articulates with the relatively deep acetabulum. The acetabular notch is spanned by a transverse ligament that is flush with the acetabular rim. The acetabular rim and transverse ligament form a ring-shaped articular surface with the inner border of the ring recessed with respect to the outer border like a countersunk metal washer. The recessed acetabular fossa forms the center of the ring. The acetabulum is deepened considerably by a ring of fibrocartilage around its edge and inferior border of the transverse ligament. The ring of fibrocartilage is called the acetabular labrum and is structurally and functionally similar to the glenoid labrum in the shoulder. The hip joint is enclosed within a cylindrical joint capsule that is strengthened by capsular ligaments. The anterior aspect is strengthened by three ligaments, namely, the superior iliofemoral, inferior iliofemoral, and pubofemoral ligaments (figure 6.7a). The posterior aspect of the capsule is strengthened by the ischiofemoral ligament (figure 6.7b). Unlike the shoulder joint, the hip joint has an intracapsular ligament called the ligamentum teres (figure 6.7c), which links together the head of the femur (at a point called the fovea) and the transverse ligament. As the head of the femur slides on the acetabular rim, the ligamentum teres moves within the acetabular fossa.

The shape of the articular surfaces together with the capsule, capsular ligaments, and internal ligament makes the hip joint a very stable joint. In comparison to the shoulder, the hip joint is far more stable but much less flexible. With respect to the anatomical position, all of the capsular ligaments of the hip joint run in a clockwise direction from the innominate bone to the femur. This is thought to be the result of evolution from a quadruped posture to an upright posture (Standring 2004). The orientation of the ligaments is such that they quickly become taut during hip extension and slacken during hip flexion. Consequently, the ligaments restrict hip extension but not hip flexion. However, the range of flexion and extension of the hip also depends on the angle of the knee joint. The hamstring and quadriceps muscles cross over both the hip and knee joints (see figure 1.5). With the knee extended, hip extension is limited to approximately 30° by tightness in the quadriceps as well as in the capsular ligaments, and hip flexion is limited to approximately 70° by tightness in the hamstrings. With the knee flexed, hip extension is reduced to virtually nil by tightness in the quadriceps, and hip flexion is increased to approximately 120° as a result of relaxation of both muscle groups as well as relaxation of the ligaments. The range of hip flexion is usually much greater than that of hip extension regardless of the angle at the knee. In this respect, the hip joint is similar to the shoulder joint in that the flexion range is much greater than the extension range (with respect to the anatomical position).

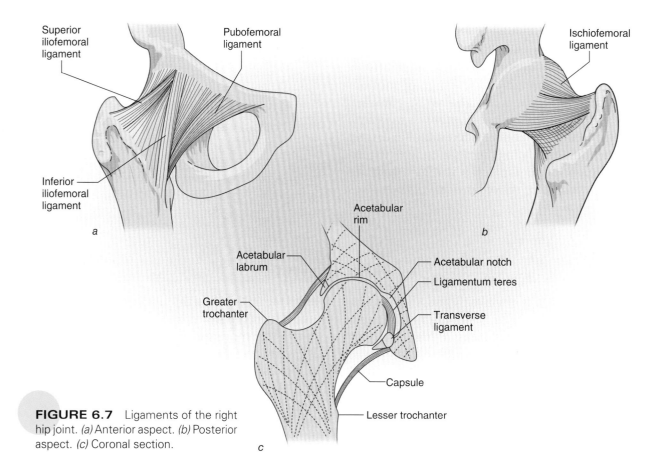

Superior iliofemoral ligament

Pubofemoral ligament

Inferior iliofemoral ligament

Ischiofemoral ligament

a

b

Acetabular rim

Acetabular labrum

Greater trochanter

Acetabular notch

Ligamentum teres

Transverse ligament

Capsule

Lesser trochanter

c

FIGURE 6.7 Ligaments of the right hip joint. *(a)* Anterior aspect. *(b)* Posterior aspect. *(c)* Coronal section.

During abduction of the hip, the pubofemoral and ischiofemoral ligaments become taut and the superior iliofemoral ligament slackens off; the normal range of hip abduction is approximately 40°. During adduction of the hip, the pubofemoral and ischiofemoral ligaments slacken and the superior iliofemoral ligament becomes taut; the normal range of hip adduction is approximately 30°. During lateral rotation of the hip, all of the anterior ligaments become taut and the ischiofemoral ligament slackens; the normal range of lateral rotation is approximately 45°. During medial rotation of the hip, all of the anterior ligaments slacken and the ischiofemoral ligament becomes taut; the normal range of medial rotation is approximately 30°.

Just as movements involving the shoulder joint are associated with movements of the shoulder girdle, movements of the hip joint are associated with movements of the vertebral column. For example, abduction of the hip is usually associated with lateral flexion of the trunk, as when a person stands on one leg with the free leg abducted (figure 6.8a). Similarly, hip flexion is usually associated with trunk flexion, as occurs during a sit-up (figure 6.8b). Consequently, reduced flexibility in the hip is partially compensated by movement in the vertebral column. To this extent, the hip joints and joints of the vertebral column can be regarded as a functional group similar to the shoulder complex.

KEY POINT

The hip joint is a synovial ball-and-socket joint, and the area of articulation is increased considerably by the acetabular labrum. The hip joint is supported internally by the ligamentum teres, externally by a cylindrical joint capsule strengthened anteriorly by iliofemoral and pubofemoral ligaments, and posteriorly by the ischiofemoral ligament. The hip joint is more stable but less flexible than the shoulder.

Injuries to the hip joint and its ligamentous supporting structures occur far less frequently than do injuries to the shoulder, knee, and ankle (Lloyd-Smith et al 1985, Weiker and Munnings 1994). This probably reflects, at least in part, the high level of stability of the hip joint relative to the shoulder, knee, and ankle. Musculoskeletal pain from the hip region is usually the result of injury to structures associated with hip joint movement rather than the hip joint itself; these injuries include sacroiliitis, osteitis pubis, stress fractures of the femoral neck, bursitis, and muscle or tendon strains (Brody 1980).

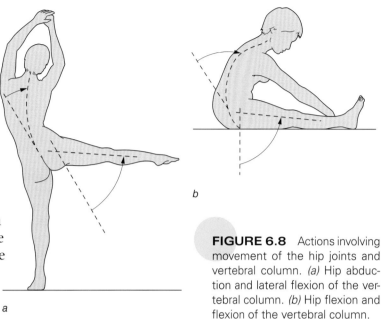

a

b

FIGURE 6.8 Actions involving movement of the hip joints and vertebral column. *(a)* Hip abduction and lateral flexion of the vertebral column. *(b)* Hip flexion and flexion of the vertebral column.

Knee Complex

The **knee complex** consists of two joints: the tibiofemoral joint and the patellofemoral joint (figure 6.9). Both joints are synovial and share the same joint capsule. The movement of the tibiofemoral joint largely determines patellofemoral joint movement. Abnormal movements in the tibiofemoral joint tend to cause abnormal movements in the patellofemoral joint. In situations involving dynamic weight bearing, the load on the two joints is usually high, and even slightly abnormal movements can overload parts of the joints (Huberti and Hayes 1984). Consequently, the incidence of injury to the knee complex is relatively high in certain groups of people, such as those involved in games and sports and in manual occupations that require lifting heavy weights (LaBrier and O'Neill 1993).

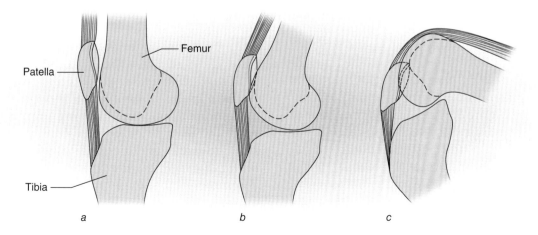

a

b

c

FIGURE 6.9 The knee complex. *(a)* Full extension. *(b)* 20° of flexion. *(c)* 140° of flexion. The patellofemoral joint is the joint between the patella and the femur. The tibiofemoral joint is the joint between the tibia and the femur.

Key Term

knee complex The functionally interdependent tibiofemoral joint and patellofemoral joint.

Tibiofemoral Joint

Whereas the tibial condyles are virtually flat, the femoral condyles have a convex pulley shape. Consequently, stability of the tibiofemoral joint is largely dependent on two menisci, strong extracapsular ligaments, and powerful muscles. The shape and orientation of the menisci and ligaments enable the tibiofemoral joint to function as a modified hinge joint, with flexion and extension as the principal plane of movement.

Menisci of the Knee

At any particular joint position, the area of contact between the articular surfaces of the femoral and tibial condyles is relatively small (see figure 6.9). However, the effective area of contact is normally increased by three to four times by the presence of two C-shaped menisci, one medial and one lateral, which form wedges between the peripheral incongruent articular surfaces of the femoral and tibial condyles (figure 6.10). The horns (or ends) of the menisci are attached to the intercondylar areas of the tibial table, that is, the superior aspect of the tibia (figure 6.11a). The anterior horn of the lateral meniscus is attached to the posterior lateral aspect of the anterior intercondylar area (figure 6.11a). The posterior horn of the lateral meniscus is attached to the anterior lateral aspect of the posterior intercondylar area. The anterior horn of the medial meniscus is attached to the anterior medial aspect of the anterior intercondylar area. The posterior horn of the medial meniscus is attached to the medial aspect of the posterior intercondylar area. The two menisci are joined anteriorly by a transverse ligament (figure 6.11b). The

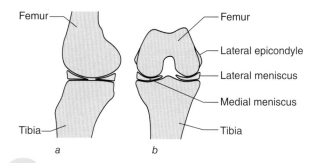

FIGURE 6.10 Menisci of the knee. *(a)* Sagittal section through the lateral tibial condyle with the tibiofemoral joint in extension. *(b)* Coronal section through the tibiofemoral joint flexed at 90°.

periphery of each meniscus is attached to the inner wall of the joint capsule. However, the articular surfaces of the menisci are free; they are in contact with the articular surfaces of the femoral and tibial condyles but free to slide on them. The ability of the menisci to slide on both articular surfaces is essential for normal joint function. The menisci have four main functions:

1. To maintain congruence between the articular surfaces in all positions of the joint. During joint movement, the menisci deform in response to the changing curvature of the femoral condyles and thus maintain congruence between the articular surfaces in all positions of the joint. The increase in congruence provided by the menisci increases joint stability and minimizes the compressive stress on the articular cartilages.

2. To provide shock absorption in the joint. The menisci deform under loading to help absorb the shock of impact loads on the joint.

3. To maintain a circulation of synovial fluid through the articular cartilages. The structure of articular cartilage is similar to that of a sponge, with a dense network of tiny interconnecting pores. During joint movement the joint reaction force compresses the articular cartilages, squeezing synovial fluid out of the cartilage. During periods of relaxation, when the joint reaction force is reduced, the viscoelasticity of the cartilage restores it to its unloaded size and shape and synovial fluid flows into the cartilage.

Even in joints that do not normally have menisci or articular discs, the articular surfaces

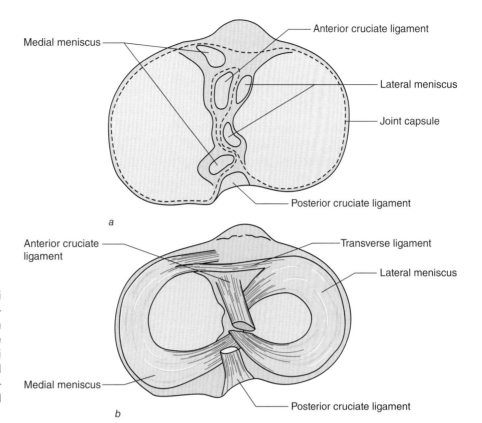

Medial meniscus

Anterior cruciate ligament

Lateral meniscus

Joint capsule

Posterior cruciate ligament

a

Anterior cruciate ligament

Transverse ligament

Lateral meniscus

Medial meniscus

Posterior cruciate ligament

b

FIGURE 6.11 Menisci and cruciate ligaments. Superior aspect of the right tibia showing the location of *(a)* the attachments of the menisci and cruciate ligaments and the outline of the joint capsule and *(b)* the menisci and cruciate ligaments.

are not perfectly congruent. Consequently, at any particular joint position some parts of the articular surfaces are under greater compressive stress than others. Furthermore, the parts of the articular surfaces under the greatest compressive stress change as the joint position changes. In this way joint movement creates a circulation of synovial fluid through the articular cartilages; this enables the cartilage cells to receive nutrients and to get rid of waste products. The menisci of the knee help to maintain the circulation of synovial fluid through the articular cartilages of the tibial and femoral condyles.

4. To help bring about normal movement between the articular surfaces. Knee flexion from full extension involves two main phases: rolling only followed by simultaneous rolling and sliding (Kapandji 1970, Norkin and Levangie 1992). The two phases are reversed when the joint is extended. For the medial condyle, the rolling-only phase corresponds approximately to the first 15° of flexion, whereas for the lateral condyle, the rolling-only phase corresponds approximately to the first 20° of

flexion. Consequently, the lateral condyle rolls farther than the medial condyle, resulting in internal rotation of the lower leg through about 20° during the rolling-only phase. The rolling-only phase at the start of knee flexion, which corresponds to the normal range of flexion and extension in ordinary walking, is sometimes referred to as unlocking the knee. The reverse movement, external rotation of the lower leg during the rolling-only phase of knee extension, is sometimes referred to as locking the knee or the screw-home mechanism.

The rolling-only phase at the start of knee flexion is followed by a phase in which rolling and sliding occur simultaneously, with sliding becoming more important as flexion progresses. The total range of knee flexion and extension is normally in the region of 160°. The ability of the menisci to deform and slide in response to the loads exerted on them during joint movement is essential to maintain the normal pattern of rolling and sliding between the articular surfaces. In turn, the normal pattern of rolling and sliding is essential for maintaining normal congruence, shock absorption, and circulation of synovial fluid through articular cartilage.

The femoral condyles are convex and the tibial condyles are fairly flat. However, their effective area of contact is normally increased three to four times by medial and lateral menisci, which have four main functions:

- To maintain congruence between the articular surfaces in all positions of the joint
- To provide shock absorption in the joint
- To maintain a circulation of synovial fluid through the articular cartilages
- To help bring about normal movement between the articular surfaces

Injuries to menisci are common in the general population and especially in activities such as dancing (modern and ballet) and sports such as football (all types), basketball, baseball, wrestling, and skiing (Baker et al 1985, Silver and Campbell 1985). The incidence of injury to the medial meniscus is approximately four times higher than for the lateral meniscus. In comparison to the movement of the lateral meniscus, the movement of the medial meniscus is rather restricted because of its firm attachment to the medial ligament via the joint capsule and the relatively large distance between its attachments to the tibial table.

This restricted range of movement is thought to be responsible for the increased incidence of injury to the medial meniscus. Damage to the menisci can result from all types of abnormal movements of the knee but especially from twisting the knee, that is, excessive axial rotation of the lower leg.

There are two types of meniscal injury:

1. Tearing the attachments of the menisci to the tibial table and joint capsule
2. Crushing the menisci between the femoral and tibial condyles, which produces circular (bucket handle) and radial tears

These injuries are illustrated in figure 6.12. A meniscus or part of a meniscus can also become trapped in the intercondylar notch of the femur, which frequently results in the knee locking—the person is unable to fully extend the knee. The only blood supply to the menisci is via the synovial membrane at the periphery. Furthermore, the menisci are usually under considerable load, especially when bearing weight, which may keep the torn parts of a damaged meniscus apart from each other. Consequently, repair of a damaged meniscus tends to be slow and may not take place at all. For this reason it used to be normal medical practice to remove a badly torn meniscus. However, improved surgical techniques and a greater awareness of the importance of the menisci to knee function have resulted in a move toward partial rather than total meniscectomy whenever possible (Yates and Jackson 1984, Silver and Campbell 1985).

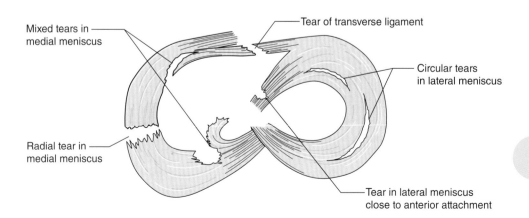

Mixed tears in medial meniscus

Tear of transverse ligament

Circular tears in lateral meniscus

Radial tear in medial meniscus

Tear in lateral meniscus close to anterior attachment

FIGURE 6.12 Types of tears in the menisci of the right knee (superior aspect).

Capsule of the Knee

The tibiofemoral and patellofemoral joints share a common capsule that has a complicated shape. The capsule is attached above to the femur, below to the tibia, and anteriorly to the patella (figure 6.13). To accommodate the extremes of full flexion and full extension of the tibiofemoral joint, the anterior part of the capsule is folded upward during extension and the posterior part of the capsule is folded downward during flexion. The upward fold, only present in extension, is referred to as the suprapatellar bursa (figure 6.13). The suprapatellar bursa intervenes between the quadriceps tendon and the anterior aspect of the femur just proximal to the patellar surface.

The downward fold, only present during flexion, is referred to as the gastrocnemius bursa. The gastrocnemius bursa intervenes between the posterior aspect of the tibial condyles and the gastrocnemius muscle. In general, a bursa is a flattened sac of synovial membrane containing synovial fluid. Bursas minimize friction between structures that slide across each other during normal movement (see chapter 7).

During flexion of the tibiofemoral joint, the suprapatellar bursa becomes progressively smaller and the gastrocnemius bursa becomes progressively larger as synovial fluid is redistributed from the suprapatellar bursa to the gastrocnemius bursa. The movement of synovial fluid is reversed during extension of the tibiofemoral joint; that is, the gastrocnemius bursa becomes progressively smaller and the suprapatellar bursa becomes progressively larger.

The suprapatellar bursa is separated from the femur by a pad of fat called the suprapatellar fat pad. Lying in the space bounded by the upper two thirds of the posterior aspect of the patellar ligament, the anterior intercondylar area of the tibia, and the anterior inferior aspect of the articular surface of the femoral condyles is a pad of fat called the infrapatellar fat pad. It is roughly triangular in sagittal cross section and is suspended superiorly from a fibro-adipose band called the infrapatellar fold. The infrapatellar fold is attached anteriorly to the inferior pole of the patella and posteriorly to the anterior border of the intercondylar notch (figure 6.13). The infrapatellar fat pad

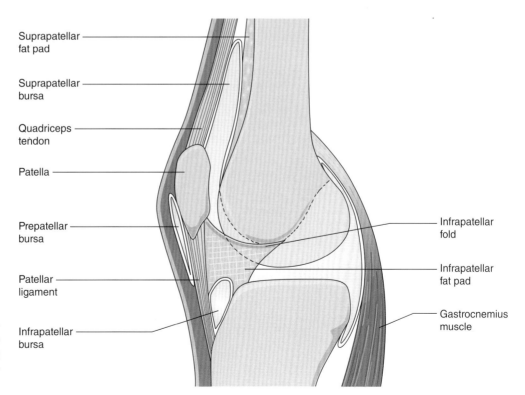

FIGURE 6.13 Sagittal section through the knee complex showing the shape of the capsule in extension.

Suprapatellar fat pad

Suprapatellar bursa

Quadriceps tendon

Patella

Prepatellar bursa

Patellar ligament

Infrapatellar bursa

Infrapatellar fold

Infrapatellar fat pad

Gastrocnemius muscle

is attached anteriorly to the posterior aspect of the patellar ligament and extends at both sides of the patellar ligament. The infrapatellar fat pad also extends narrow branches halfway up each side of the patella; the branches are referred to as alar folds (*alar* = winglike). The infrapatellar fat pad cushions the patellar ligament and lower part of the patella during movements of the knee.

Knee Ligaments

Four extracapsular ligaments support the tibiofemoral joint. The lateral ligament (also referred to as the lateral collateral ligament and the fibular collateral ligament) is attached superiorly to the lateral epicondyle of the femur and inferiorly to the head of the fibula (figure 6.14*a*). The medial ligament (also referred to as the medial collateral ligament and the tibial collateral ligament) is attached superiorly to the medial epicondyle of the femur and inferiorly to the medial aspect of the tibia below the tibial condyle (figure 6.14*b*). The posterior fibers of the medial ligament blend with the joint capsule at the level of the medial meniscus and thus the medial ligament is not completely extracapsular. The arrangement of the medial and lateral ligaments and the curvature of the femoral condyles are such that both ligaments are relatively slack when the knee is flexed but become progressively more taut as the joint extends. In a normal joint the ligaments are fully taut when the joint is fully extended. In this position of the joint both ligaments help to prevent hyperextension and lateral rotation of the tibia relative to the femur. In addition, the medial ligament and lateral ligament help to prevent abduction and adduction, respectively (see figures 4.22 and 4.23).

With the knee fully extended, the range of abduction and adduction is normally zero. Because the medial and lateral ligaments are relatively slack when the knee is flexed, a certain amount of axial rotation and abduction and adduction of the lower leg can occur when the knee is in a flexed, non-weight-bearing position. For example, when a person is sitting on a table with his lower leg hanging freely, the ranges of internal and external rotation are normally around 30° and 40°, respectively (Norkin and Levangie 1992). However, the ranges of abduction and adduction are normally very small, approximately 2° to 5°.

Injuries to the lateral and, in particular, the medial ligaments are common in sports (Reider 1996). The classic cause of medial ligament

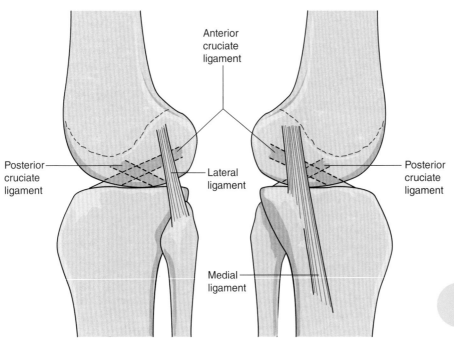

Anterior cruciate ligament

Posterior cruciate ligament

Lateral ligament

Posterior cruciate ligament

Medial ligament

a   *b*

FIGURE 6.14 Location of the main ligaments of the left knee. *(a)* Lateral aspect. *(b)* Medial aspect.

injury is a blow to the lateral aspect of the knee while the foot in contact with the ground, resulting in abduction of the knee. This situation is common in sports such as American football, soccer, and rugby. The excessive loading on the medial ligament in such situations can result in serious injury (Indelicato 1995). Medial ligament injuries of a less severe but nonetheless persistent nature are common in swimmers, especially breaststroke swimmers. The whip-kick leg action in breaststroke tends to simultaneously abduct and laterally rotate (tibia with respect to femur) the knee, resulting in strain of the medial ligament and associated medial supporting structures (Vizsolyi et al 1987).

The other two extracapsular ligaments that support the tibiofemoral joint are the anterior and posterior cruciate ligaments, which cross over each other in the intercondylar notch (*cruciate* = cross) (see figures 6.11 and 6.14). The distal end of the anterior cruciate ligament is attached to the posterior aspect of the anterior intercondylar area of the tibial table, medial to the attachment of the anterior horn of the lateral meniscus (see figure 6.11). The proximal end is attached to the posterior medial aspect of the lateral femoral condyle. The distal end of the posterior cruciate ligament is attached to the posterior aspect of the posterior intercondylar area of the tibial table, and the proximal end is attached to the anterior inferior lateral aspect of the medial femoral condyle. The anterior cruciate ligament runs laterally to the posterior cruciate ligament. Although the cruciate ligaments are located in the middle of the joint, they are extracapsular; the capsule is invaginated into the intercondylar notch where it encloses the menisci but excludes the cruciate ligaments (see figure 6.11*a*).

The cruciate ligaments bring about normal movement between the articular surfaces of the femoral and tibial condyles. For example, during the rolling–sliding phase of knee flexion, the anterior cruciate ligament pulls on the femur so that the femoral condyles slide forward as they roll backward. Similarly, during the rolling–sliding phase of knee extension, the posterior cruciate ligament pulls on the femur so that the femoral condyles slide backward

as they roll forward. The posterior cruciate ligament restricts forward movement of the femoral condyles and thus helps to prevent forward dislocation of the femur on the tibia (see figure 6.14). Similarly, the anterior cruciate ligament restricts backward movement of the femoral condyles and therefore helps to prevent posterior dislocation of the femur on the tibia. At full knee extension, both cruciate ligaments become fully taut and therefore help the medial and lateral ligaments to prevent hyperextension. In addition, the orientation of the cruciate ligaments, the anterior ligament running laterally to the posterior ligament, is such that they help to prevent medial rotation of the tibia relative to the femur in full knee extension (see figure 6.11*b*). Consequently, the cruciate ligaments work with the lateral and medial ligaments to provide rotational stability of the knee. Damage to these ligaments causes excessive laxity in the tibiofemoral joint and considerably increases the likelihood of damage to the menisci (Gollehon et al 1987). See page 115 in chapter 4 for a discussion of anterior cruciate ligament (ACL) injuries.

KEY POINT

The tibiofemoral joint is supported by four extracapsular ligaments: lateral, medial, anterior cruciate, and posterior cruciate. In a normal knee, all of the ligaments help to provide rotational stability and prevent hyperextension; in addition, the medial ligament and lateral ligament help to prevent abduction and adduction, respectively. Damage to the ligaments, especially the cruciate ligaments, causes excessive laxity and increases the likelihood of damage to the menisci.

Patellofemoral Joint

The patellofemoral joint is a synovial gliding joint in which the patella slides up and down on the patellar surface and condyles of the femur during extension and flexion of the tibiofemoral joint. In full knee extension, the superior half of the articular surface of the patella rests against the suprapatellar bursa (see figure 6.13). During knee flexion, the patella slides down the intercondylar groove formed by the patellar surface and condyles of the femur. The area of contact between the patella and femur gradually increases up to about 90° of flexion and then starts to decrease as the patella moves into the widest part of the intercondylar notch (see figure 6.9). From full knee extension to full knee flexion, the patella slides on the femur for a distance of approximately 8 cm in the adult. The movement of the patella on the femur is determined by the shape of the articular surfaces of the patella and intercondylar groove and the various fibrous attachments to the patella. The articular surface of the patella is V shaped and articulates reasonably well with the reciprocally shaped intercondylar groove of the femur for most of the range of flexion and extension of the knee. Consequently, in terms of the articular surfaces, normal movement of the patella involves sliding in a more or less sagittal plane.

There are four main fibrous attachments to the patella: quadriceps tendon, patellar ligament, and the medial and lateral retinacula (figure 6.15). The quadriceps tendon and patellar ligament are continuous with each other and enclose the nonarticular parts of the patella. The distal end of the patellar ligament is attached to the tibial tuberosity of the tibia. The four quadriceps muscles, which form the bulk of the soft tissue in the anterior aspect of the thigh, attach onto the quadriceps tendon. The main function of the quadriceps muscles is to extend the knee, and the patella increases the mechanical efficiency of the quadriceps muscles. The quadriceps can still extend the knee if, for some reason, the patella has to be removed. However, in this case knee extension is relatively weak because of a reduction in the mechanical efficiency of the quadriceps. In addition, if the patella is absent the quadriceps tendon is compressed against the intercondylar groove (figure 6.16), which not only subjects the quadriceps tendon to friction during sliding but also increases the compressive stress on the intercondylar groove because of the reduced area of contact with the quadriceps tendon relative to contact with the patella. Consequently, the patella has three main functions:

1. To increase the mechanical efficiency of the quadriceps muscle group
2. To prevent contact and friction between the quadriceps tendon and intercondylar groove
3. To minimize compression stress on the intercondylar groove

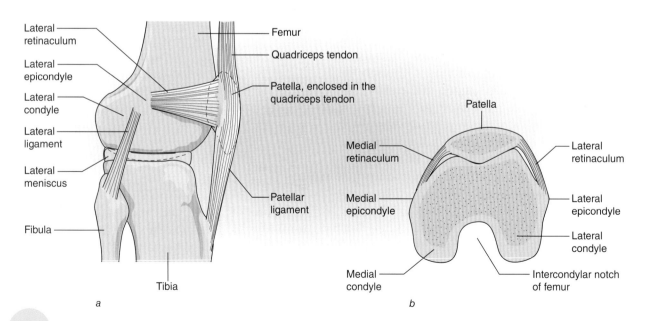

FIGURE 6.15 Fibrous attachments to the patella. *(a)* Lateral aspect of the right knee complex. *(b)* Superior aspect of a transverse section through the right patellofemoral joint.

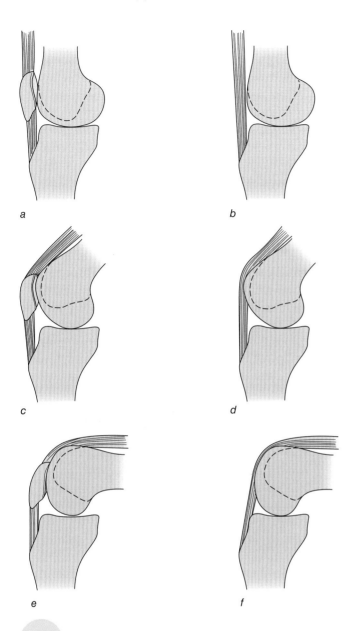

a

b

c

d

e

f

FIGURE 6.16 Effect of the patella on the line of action of the force exerted by the patellar ligament. *(a, c, e)* With the patella. *(b, d, f)* Without the patella. *(a, b)* Full knee extension. *(c, d)* Knee flexed at 45°. *(e, f)* Knee flexed at 90°.

Normal and Abnormal Tracking of the Patella

The medial retinaculum attaches the medial border of the patella to the medial epicondyle of the femur (figure 6.15*b*). The lateral retinaculum attaches the lateral border of the patella to the lateral epicondyle of the femur (figure 6.15*b*). The retinacula blend with the capsule and, together with the quadriceps tendon and patellar ligament, normally maintain maximum congruence and normal movement of the patellofemoral joint throughout the range of flexion and extension of the knee. The movement of the patella along the intercondylar groove is usually referred to as **patellofemoral tracking,** and normal tracking corresponds to the maintenance of maximum congruence (figure 6.17*a*). Any change in the normal pattern of forces exerted on the patella by the various fibrous attachments may displace the patella laterally or medially in the intercondylar groove, which is likely to reduce congruence in the patellofemoral joint and produce abnormal tracking of the patella in the intercondylar groove during knee flexion and extension (Fagan and Delahunt 2008). Lateral tracking and medial tracking refer to abnormal tracking of the patella on the lateral and medial sides of the patellofemoral joint, respectively (figure 6.17, *b* and *c*).

Lateral and medial tracking of the patella stretches the joint capsule and retinacula and increases the compressive stress on the articular surfaces that remain in contact because of the decrease in the area of contact between them. If prolonged, abnormal tracking can result in a dull pain around and beneath the patella. The pain can be exacerbated by

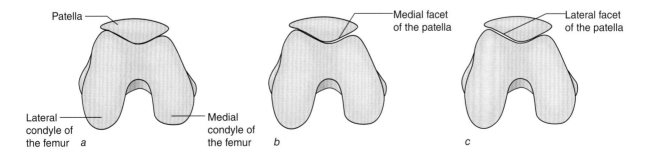

Patella

Medial facet of the patella

Lateral facet of the patella

Lateral condyle of the femur   a

Medial condyle of the femur   b

c

FIGURE 6.17 Tracking of the patella (right leg). *(a)* Normal tracking. *(b)* Lateral tracking. *(c)* Medial tracking.

exercise involving flexion and extension of the knee while weight bearing, such as running, jumping, and ascending and descending stairs (Witvrouw et al 2000, McClinton et al 2007). Pain can also occur when one sits for long periods with the knee flexed, such as in a car, an airplane, or a theater (Callaghan and Oldham 1996). In older adults the condition is often associated with damage to the articular cartilage of the patella and is referred to as chondromalacia patellae (*chondro* = cartilage, *malacia* = softening) (Casscells 1982, Standring 2004). However, in children and young adults the condition usually occurs without any obvious damage to the articular cartilage; in these cases the condition is usually referred to as **patellofemoral pain syndrome** (or a similar term such as *patellofemoral compression syndrome, patellofemoral malalignment syndrome,* or *patellagia)* (Galea and Albers 1994, Percy and Strother 1985, Fagan and Delahunt 2008). The pain is thought to be due to stretching of the fibrous supporting structures (joint capsule and retinacula) and associated synovial membrane and compressive stress on the subchondral bone. Unlike the articular cartilage, the fibrous supporting structures, synovial membrane, and subchondral bone all have a rich supply of pain receptors. The incidence of patellofemoral pain syndrome and, to a lesser extent, chondromalacia patellae is high in theatrical dancers (Reid 1987, Winslow and Yoder 1995) and participants in sports such as long-distance running and volleyball (Clement et al 1981, Ferretti et al 1990, Taunton et al 2003, Van Ghent et al 2007) and those who train by jogging (Kujala et al 1986; Thijs et al 2008). Patellofemoral pain syndrome also affects a high proportion of nonathletes (McConnell 1986, Fagan and Delahunt 2008).

Key Terms

patellofemoral tracking The movement of the patellofemoral joint. The lateral and medial facets of the patella articular surface are normally in contact with the corresponding lateral and medial parts of the patella surface of the femur during knee flexion–extension; this is referred to as normal tracking. Abnormal tracking occurs when the patella is displaced laterally during knee flexion–extension such that there is no contact between the medial patellar facet and the medial part of the patella surface of the femur; this is referred to as lateral tracking. Similarly, abnormal tracking occurs when the patella is displaced medially during knee flexion–extension such that there is no contact between the lateral patellar facet and the lateral part of the patella surface of the femur; this is referred to as medial tracking.

patellofemoral pain syndrome Persistent dull pain around the patella, frequently associated with abnormal tracking, which is thought to be caused by stretching of the fibrous supporting structures (joint capsule and retinacula) and associated synovial membrane and compressive stress on subchondral bone.

KEY POINT

The loads exerted on the patella by the quadriceps tendon, patellar ligament, and retinacula normally maintain maximum congruence and normal tracking of the patellofemoral joint. Any change in the normal pattern of these loads may result in abnormal tracking. If prolonged, abnormal tracking can result in patellofemoral pain syndrome.

Causes of Abnormal Tracking

There are four main causes of abnormal tracking that often interact with each other: skeletal abnormalities, strength imbalance in the quadriceps, strength imbalance in fibrous supporting structures, and compensatory movements of the knee in response to abnormal movement in the foot.

Skeletal Abnormalities A variety of skeletal abnormalities may contribute to abnormal tracking: the size of the Q angle, genu varum and genu valgum, patella alta and patella infera, and the size of the lateral border of the patellar surface.

• *The Q angle*. The position of the tibial tuberosity affects the line of pull of the quadriceps and therefore the tracking of the patella. The normal position of the tibial tuberosity is such that in the anatomical position the patellar ligament runs downward and slightly laterally (figure 6.18*a*). However, if the tuberosity is located more laterally than normal, there is a tendency to lateral tracking. Similarly, if it is located more medially than normal, there is a tendency to medial tracking. The orientation of the patellar ligament to the quadriceps tendon can be assessed by measuring the **Q angle**—the angle between the line joining the anterior superior iliac spine and the center of the patella and the line joining the center of the patella and the midpoint of the tibial tuberosity (figure 6.18*b*). The average Q angle in males (around 12°) has been reported to be smaller than in females (around 14°) (Aglietti et al 1983, Herrington and Nester 2004). The larger Q angle in females is thought to be largely attributable to their relatively greater hip width compared with men. Increases and decreases of the Q angle with respect to these normal values are associated with an increased incidence of patellofemoral pain syndrome (Huberti and Hayes 1984).

• *Genu varum and genu valgum*. **Genu varum (bowleg)** and **genu valgum (knock-knee)** may lead to abnormal tracking of the patella. When these conditions develop during childhood, it is usually the result of normal modeling in response to abnormal patterns of loading imposed on the knees by the combined effect of muscle forces and body weight (see chapter 11). The skeletal adaptations of genu varum and genu valgum ensure normal transmission of loads across the knee; the knees are

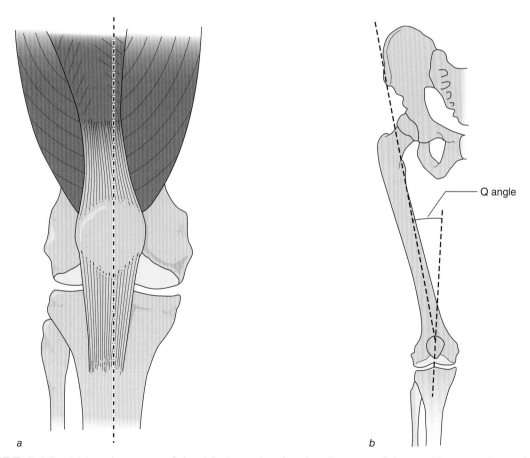

FIGURE 6.18 *(a)* Anterior aspect of the right knee showing the alignment of the quadriceps tendon and patellar ligament. *(b)* The Q angle.

functionally normal even though they appear to be abnormally aligned. The joints are only likely to become functionally abnormal when the balance between body weight and muscle forces changes. This applies to all people, but those with genu varum and genu valgum are more likely to develop functionally abnormal knees as a result of changes in the balance between muscle forces and body weight than people with normally aligned knees.

With increased age, muscle weakness is likely to increase the degree of varus in a person with genu varum and increase the degree of valgus in someone with genu valgum. When this happens the joints become incongruent, which increases compressive stress on the parts of the articular cartilage in contact—the lateral condyles of the femur and tibia in someone with genu valgum or the medial condyles of the femur and tibia in someone with genu varum. If the situation persists, the overloaded cartilage may erode. As the cartilage becomes thinner, the degree of misalignment between femur and tibia increases. Removal of either the lateral or medial meniscus may lead to the same chain of events: excessive stress on one side of the joint, gradual thinning of the cartilage, and a gradual increase in the degree of varus or valgus (Yates and Jackson 1984). Clearly, any increase in the degree of varus or valgus of the knees during adulthood can cause abnormal tracking of the patella; an increase in varus can cause medial tracking and an increase in valgus can cause lateral tracking.

- *Patella alta and patella infera.* The length of the patellar ligament largely determines the height of the patella relative to the tibiofemoral joint. Normally the ratio of the length of the patella to that of the patellar ligament is approximately 1.0; a variation of more than 20% in this ratio is regarded as abnormal (Insall and Salvati 1971). A ratio of 0.8 or less is referred to as **patella alta** (a high patella), and a ratio of 1.2 or greater is referred to as **patella infera** (a low patella). Patella alta and patella infera are associated with an increased incidence of patellofemoral pain syndrome, which may be due in part to abnormal tracking of the patella (Insall and Salvati 1971, Lancourt and Cristini 1975, Kujala et al 1987).

- *Lateral border of the patella surface.* Passive and active mechanisms can restrain the movement of the patella. The passive mechanisms are the retinacula, their capsular attachments, and the shape of the patellofemoral joint. The patellofemoral joint is V shaped in transverse section, which prevents side-to-side movement of the patella in the intercondylar groove (see figure 6.17a). The active mechanism is the appositional force exerted on the patella—the force pressing the patella into the intercondylar groove as a result of contraction of the quadriceps muscles. The greater the force exerted by the quadriceps and the greater the degree of knee flexion, the greater the appositional force (figure 6.19a). However, as the knee extends, the angle between the quadriceps tendon and patellar ligament decreases such that the appositional force on the patella decreases. At full knee extension the appositional force may be close to zero, and the patella may displace laterally (figure 6.19b). One of the main restraints that prevents lateral displacement of the patella is the lateral border of the patellar surface, which is more prominent anteriorly than the medial border (see figure 6.17b). If the lateral border of the patellar surface is less prominent than normal, the patella may dislocate laterally at full extension of the knee (Kapandji 1970). Frequent lateral dislocation of the patella is referred to as recurrent dislocation of the patella.

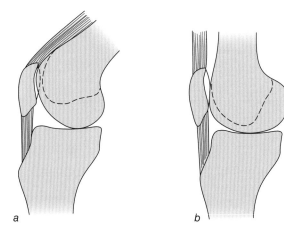

FIGURE 6.19 Effect of the knee angle on apposition of the patella. *(a)* Knee flexed at 45°. *(b)* Full knee extension.

Key Terms

Q angle   The angle between the line joining the anterior superior iliac spine and the center of the patella and the line joining the center of the patella and the midpoint of the tibial tuberosity.

genu varum (bowleg)   A deformity of the lower limbs where the horizontal distance between the knees is greater than that between the ankles when standing upright.

genu valgum (knock-knee)   A deformity of the lower limbs where the horizontal distance between the knees is less than that between the ankles when standing upright.

patella alta   Condition that occurs when the ratio of the length of the patella to that of the patellar ligament is less than or equal to 0.8.

patella infera   Condition that occurs when the ratio of the length of the patella to that of the patellar ligament is greater than or equal to 1.2.

Strength Imbalance in the Quadriceps

The four quadriceps muscles (rectus femoris, vastus lateralis, vastus intermedius, vastus medialis) pull on the patella in different directions, but normally the net effect of all four muscles is a more or less vertical pull that brings about normal tracking of the patella (figure 6.20). The tendency of the vastus lateralis and, to a lesser extent, the rectus femoris and vastus intermedius to pull the patella laterally is normally balanced by the vastus medialis. The vastus medialis is most active during the last 20° of knee extension. If the full range of knee extension is not used, a strength imbalance can develop between the vastus medialis and vastus lateralis; the vastus medialis can become relatively weaker, resulting in lateral displacement and lateral tracking of the patella. A long-distance runner who does not fully extend his knees during the propulsion phase of each stride is prone to develop this condition. Weakness of the vastus medialis is associated with an increased incidence of patellofemoral pain syndrome and recurrent lateral dislocation of the patella (Kujala et al 1986, Bose et al 1980, Cash and Hughston 1988).

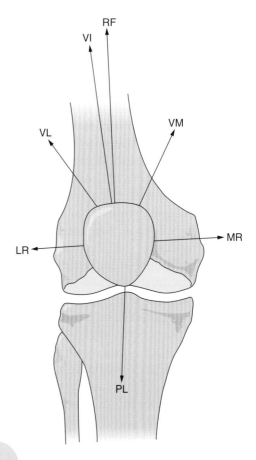

FIGURE 6.20   Active and passive forces acting on the patella. Active forces include the quadriceps: RF = rectus femoris; VL = vastus lateralis; VM = vastus medialis; VI = vastus intermedius. Passive forces include the following: PL = patellar ligament; LR = lateral retinaculum; MR = medial retinaculum.

Strength Imbalance in Fibrous Supporting Structures

The patellar ligament, the quadriceps tendon, and the medial and lateral retinacula are all joined by the joint capsule and fascia of the muscles controlling the movement of the knee. In some people the fibrous structures on one side of the joint, usually the lateral side, become thicker or shorter than those on the other side of the joint and the patella is pulled to one side, resulting in abnormal tracking (Silver and Campbell 1985, Micheli and Stanitski 1981).

Compensatory Movement in the Knee Due to Abnormal Movement in the Foot

In activities involving weight bearing, foot pronation (see next section) can result in internal rotation of the tibia with respect to the femur and foot supination can result in external rotation of the tibia with respect to the femur. Internal rotation of the tibia can result in simultaneous medial rotation and medial displacement of the patella. If foot pronation is associated with forceful activity in the quadriceps, as occurs in many sports, abnormal tracking of the patella may occur because of a combination of medial rotation and medial displacement of the patella. If such movements occur with high frequency, for example, in distance running or in jumping and landing in volleyball, patellofemoral pain syndrome can occur.

It is not surprising that patellofemoral pain syndrome is the most common injury in joggers and long-distance runners (Clement et al 1981). Most runners contact the ground with the posterior lateral aspect of the heel. Immediately after heel strike, the foot pronates and the arches of the foot depress to absorb the shock of the impact. After the impact phase the foot normally quickly recoils into a less-depressed and less-pronated position such that abnormal tracking of the patella is unlikely to occur. However, in many runners the ground contact period is associated with excessive or prolonged pronation, which considerably increases the likelihood of abnormal

tracking and the occurrence of patellofemoral pain syndrome (Cavanagh 1980, Hontas et al 1986). The excessive or prolonged pronation may be the result of extrinsic factors such as inadequate footwear or intrinsic factors such as fatigue in the muscles that control pronation.

Anterior knee pain can occur in the form of jumper's knee even when patella tracking is normal. Jumper's knee is characterized by pain at the insertion of the quadriceps tendon into the upper pole of the patella and at the insertion of the patellar ligament into the inferior pole of the patella (Ferretti et al 1990). The various sources of pain arising from jumper's knee and abnormal tracking of the patella are summarized in figure 6.21. Case study 4 discusses running and patellofemoral pain.

> **KEY POINT**
>
> There are four main causes of abnormal tracking of the patella that often interact with each other: skeletal abnormalities, strength imbalance in the quadriceps, strength imbalance in fibrous supporting structures, and compensatory movements of the knee in response to abnormal movement in the foot.

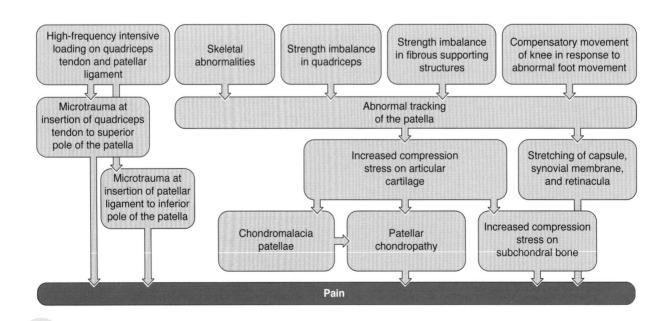

FIGURE 6.21 Sources of anterior knee pain.

CASE STUDY 4 RUNNING AND PATELLOFEMORAL PAIN

Thijs Y, De Clercq D, Roosen P, Witvrouw E. 2008. Gait-related intrinsic risk factors for patellofemoral pain in novice recreational runners. *British Journal of Sports Medicine* 42:466-471.

The growing interest in disease prevention has led to increased participation in sports in recent years. Jogging and running have emerged as some of the most popular forms of recreational exercise. As with other popular forms of exercise, increased participation in jogging and running has resulted in an increase in the number of injuries. Epidemiological studies have shown that most of the injuries sustained through running are knee injuries, and patellofemoral pain (PFP) syndrome is the most prevalent injury. PFP is differentiated from other knee injuries by pain on direct compression of the patella against the femoral condyles with the knee in full extension, tenderness of the posterior surface of the lateral or medial rim of the patella on palpation, and pain on resisted knee extension.

There is no consensus on the cause of PFP syndrome. The results of some studies suggest that abnormal kinematics of the ankle and foot are associated with PFP syndrome, but these studies have been based on runners already suffering from PFP syndrome, so it is not clear whether the alleged abnormal kinematics are a cause or a consequence of PFP syndrome. The purpose of the present study was to examine the intrinsic characteristics (foot shape during relaxed upright standing, plantar pressure patterns while running barefoot) of novice recreational runners enrolled in a 10-week "start to run" program who developed PFP syndrome as a result of participation in the program.

One hundred and twenty-nine runners (107 women, 22 men) with no history of musculoskeletal injury started the program. Twenty-seven dropped out of the program at various times because of reasons other than injury. Consequently, 102 completed the program (age 37 ± 9.5 years, mass 69 ± 15 kg, body mass index 25 ± 3 kg/m^2). Prior to the start of the running program, investigators used the foot posture index (FPI) to assess the foot shape during relaxed upright standing of all subjects. The FPI is a validated clinical tool for assessing foot posture, in particular, the degree to which a foot can be considered to be pronated, supinated, or neutral. In addition to static foot posture, plantar pressure patterns (under eight areas of the plantar surface of the foot) while running barefoot were measured for all subjects prior to the start of the program.

The 102 runners who completed the program sustained a number of injuries: PFP syndrome (17; 16 women, 1 man), shin splints (11), Achilles tendinitis (10), ankle pain (6), iliotibial band friction syndrome (4), adductor injuries (3), ankle inversion injuries (3), patellar tendinitis (1), and meniscal injury (1). There was no significant evidence of an association between foot shape and risk of developing PFP syndrome. Risk of developing PFP syndrome was negatively (inversely) related to time to peak vertical force underneath the heel and positively related to peak vertical force underneath the second metatarsal of a predisposing foot.

Application

The results of the study indicate that many (around 17%) previously nonsymptomatic novice runners participating in a low-intensity, low-distance introductory running program are likely to develop PFP syndrome in the absence of preconditioning. This finding is consistent with the results of studies that examined the effect of preconditioning on the incidence of injury arising from running training programs (Nigg 2001, Fagan and Delahunt 2008). In particular, risk of PFP syndrome in novice runners has been shown to be associated with (1) weak hip extensor muscles, (2) weak quadriceps muscles, (3) a strength imbalance between the quadriceps, or (4) weak quadriceps muscles combined with a strength imbalance between the quadriceps. Consequently, preconditioning for a recreational running training program should include strengthening of the leg, in particular the hip and knee extensors.

Rearfoot Complex

In dynamic situations the foot is required to act both as a shock absorber to cushion the impact of contact of the foot with the ground and as a propulsive mechanism to propel the body in the desired direction. The foot often performs these functions on a variety of support surfaces. The ability of the foot to function effectively in diverse environments is due to its structure, in particular its flexible arched shape and complex movement capability.

Many of the 26 bones in each foot articulate with 2 or more other bones such that there are approximately 40 joints in each foot. Consequently, most movements of the foot involve a large number of joints, and the movement of individual joints in each movement is difficult to describe. However, as in most movements of the body, there tends to be high degree of functional interdependence between the joints of the foot, especially between the ankle, subtalar, and midtarsal joints, such that movement of one joint tends to bring about fairly predictable movement in adjacent joints (Kitaoka et al 1997, Nester 1997, Singh et al 1992). The term **rearfoot complex** is frequently used to describe the functional interdependence between the ankle, subtalar, and midtarsal joints (Bowden and Bowker 1995, Downing et al 1978, Nester 1997).

Ankle Joint

The ankle joint, between the tibia, fibula, and talus, is a hinge joint that facilitates rotation about an axis of rotation that runs approximately 20° anterior–superior with respect to the horizontal in the sagittal plane and 20° anterior–medial with respect to the coronal plane in the horizontal plane with respect to the center of the joint (figure 6.22) (Singh et al 1992). Consequently the movement of the ankle joint is triplanar; that is, movement occurs simultaneously in the sagittal, coronal, and transverse planes about mediolateral, anteroposterior, and vertical axes, respectively, with movement predominantly in the sagittal plane. Sagittal plane motion of the foot about the ankle joint is usually referred to as **plantar flexion** and **dorsiflexion** (figure 6.23a). In dorsiflexion, sometimes referred to as true flexion of the ankle, the dorsal (superior) surface of the foot is drawn closer to the shin. In plantar flexion, sometimes referred to as extension of the ankle, the dorsal surface of the foot is moved away from the shin (pointing the toes). Transverse plane motion of the foot about the ankle is usually referred to as abduction and adduction (figure 6.23b), and coronal plane motion of the foot about the ankle is usually referred to as eversion and inversion (figure 6.23c).

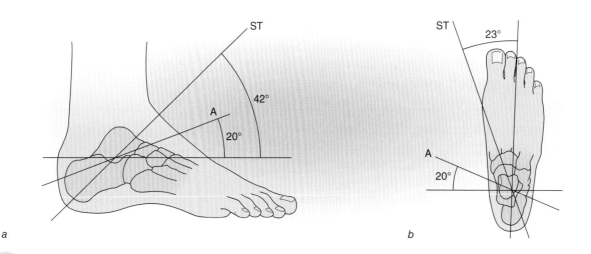

a

b

FIGURE 6.22 Orientation of the axes of rotation of the ankle (A) and subtalar (ST) joints.

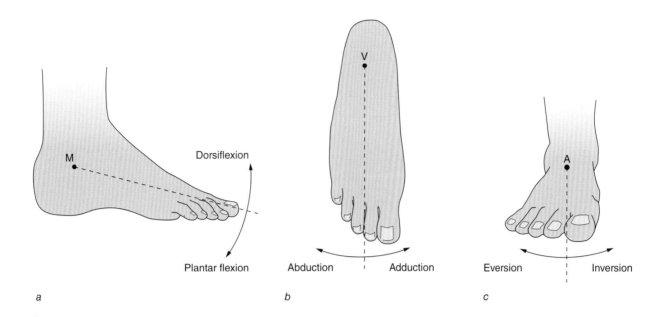

Dorsiflexion

Plantar flexion

Abduction | Adduction

Eversion | Inversion

a

b

c

FIGURE 6.23 Components of triplanar movement of the foot about the axis of the rearfoot complex. M = medio-lateral component of the axis of the rearfoot complex; V = vertical component of the axis of the rearfoot complex; A = anteroposterior component of the axis of the rearfoot complex.

Ankle Joint Ligaments

The articular surfaces of the distal tibia (trochlear surface and medial malleolus) and distal fibula (lateral malleolus) form a mortise (a slot-shaped space) that articulates with superior (trochlear surface), lateral (lateral facet), and medial (medial facet) articular surfaces of the talus (figure 6.24a). The mortise restricts side-to-side movement of the talus in the ankle joint. The mortise is maintained by the integrity of distal tibiofibular joint, which is a syndesmosis formed by the downward extension and thickening of the interosseous membrane between the shafts of the tibia and fibula (figure 6.24b). The syndesmosis is supported by strong anterior and posterior tibiofibular ligaments. The anterior tibiofibular ligament runs downward and laterally from the anterior lateral aspect of the tibia just above the ankle joint to the anterior aspect of the lateral malleolus. The posterior tibiofibular ligament runs downward and laterally from the posterior lateral aspect of the tibia just above the ankle joint to the posterior aspect of the lateral malleolus (figure 6.24b). The inferior aspect of the anterior tibiofibular ligament overlaps the anterior lateral aspect of the ankle joint, and the inferior aspect of the posterior tibiofibular

ligament overlaps the posterior lateral aspect of the ankle joint.

When viewed from above, the trochlear surface of the talus is wedge shaped with the broad end in front (figure 6.25). Consequently, the intermalleolar space slightly increases during ankle dorsiflexion, which tightens the mortise (because of strain on the syndesmosis and the anterior and posterior tibiofibular ligaments) and increases the side-to-side stability of the ankle joint. However, during ankle plantar flexion, the intermalleolar space slightly decreases, which loosens the mortise and tends to reduce the side-to-side stability of the ankle joint.

The medial aspect of the ankle joint is supported by the deltoid ligament (also referred to as the medial collateral ligament; figure 6.26). The deltoid ligament is a strong ligament that fans out from the anterior, medial, and posterior aspects of the medial malleolus to attach onto a more or less continuous arc formed by the tuberosity of the navicular, the plantar calcaneonavicular ligament (which spans the gap between the tuberosity of the navicular and the sustentaculum tali of the calcaneus and supports the head of the talus), the sustentaculum tali, and the medial tubercle of the

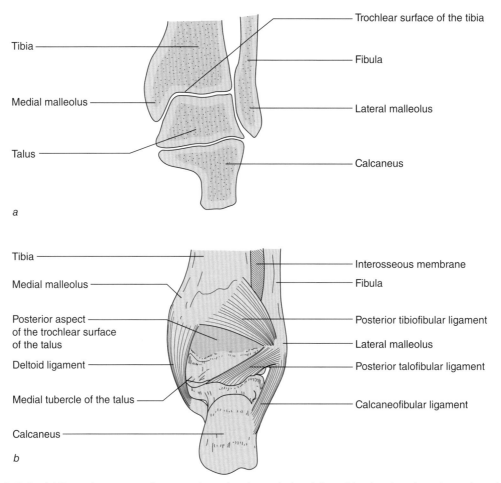

FIGURE 6.24 *(a)* Posterior aspect of a coronal section through the right ankle showing the orientation of the tibia, fibula, talus, and calcaneus. *(b)* Posterior aspect of the right ankle showing the supporting ligaments.

talus. The deltoid ligament as a whole strongly resists eversion of the ankle and subtalar joints. The fibers of the anterior band become taut at the limit of plantar flexion, and the fibers of the posterior band become taut at the limit of dorsiflexion.

The lateral aspect of the ankle joint is supported by the anterior talofibular ligament, the lateral talocalcaneal ligament, the calcaneofibular ligament, and the posterior talofibular ligament (figure 6.27). The anterior talofibular ligament runs medially and slightly forward and downward from the anterior inferior aspect of the lateral malleolus to the anterior lateral aspect of the talus immediately in front of the lateral articular surface of the talus. The lateral talocalcaneal ligament runs downward and slightly backward from the lateral process of the talus (figure 6.27) to the adjacent lateral aspect of the calcaneus. The calcaneofibular ligament runs medially downward and back-

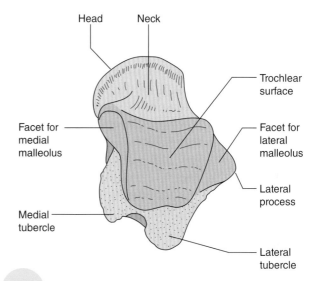

FIGURE 6.25 Superior aspect of the right talus.

ward from the inferior aspect of the lateral malleolus to the lateral aspect of the calcaneus. The posterior talofibular ligament runs medi-

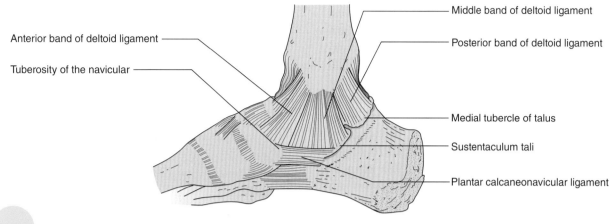

Anterior band of deltoid ligament

Tuberosity of the navicular

Middle band of deltoid ligament

Posterior band of deltoid ligament

Medial tubercle of talus

Sustentaculum tali

Plantar calcaneonavicular ligament

FIGURE 6.26 Medial ligaments of the right ankle.

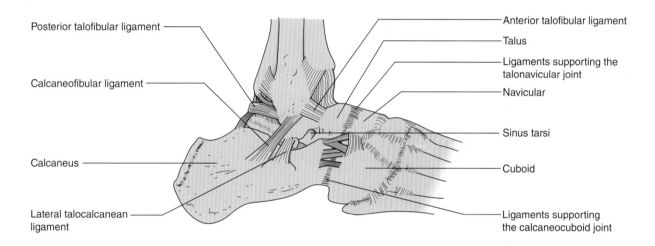

Posterior talofibular ligament

Calcaneofibular ligament

Calcaneus

Lateral talocalcanean ligament

Anterior talofibular ligament

Talus

Ligaments supporting the talonavicular joint

Navicular

Sinus tarsi

Cuboid

Ligaments supporting the calcaneocuboid joint

FIGURE 6.27 Lateral ligaments of the right ankle.

ally and slightly backward from the posterior inferior aspect of the lateral malleolus to the lateral tubercle of the talus (figure 6.24). The posterior talofibular ligament becomes taut at the limit of dorsiflexion, and the anterior talofibular ligament becomes taut at the limit of plantar flexion.

Ankle Injuries

Ligament injuries of the ankle are common, especially in sports (Boruta et al 1990). The most common ankle injury is an inversion sprain, which results in damage to one or more of the lateral ligaments and in some cases fracture of the medial malleolus or the lateral malleolus (Lassiter et al 1989) (figure 6.28). Chronic (i.e., persistent) lateral ankle instability occurs in 10% to 20% of people fol-

lowing a severe inversion sprain (Perlman et al 1987, Peters et al 1991). Most inversion sprains involve plantar flexion as well as inversion. As described previously, the ankle joint is least stable in plantar flexion because of loosening of the tibiofibular mortise. Furthermore, the ligament likely to be under the most strain during plantar flexion and inversion is the anterior talofibular ligament, which is reported to be the weakest of the lateral ligaments (Siegler et al 1988, Giza et al 2003, Fong et al 2007). Consequently, it is not surprising that the ankle is frequently injured during movements of the foot that involve rapid and forceful plantar flexion or inversion; these movements may occur, for example, when walking, running, or landing on uneven surfaces (McKay et al 2001, McHugh et al 2006).

In addition to causing ligament injuries, movements involving extreme plantar flexion can result in impingement injuries, in particular impingement of the posterior lateral border of the trochlear surface of the tibia on the lateral tubercle of the talus or impingement of the posterior border of the medial malleolus on the medial tubercle of the talus (figure 6.29, *a* and *b*). This is a common cause of injury in theatrical dancers (Hardaker et al 1985, Hopper and Robinson 2008). Extreme dorsiflexion can result in impingement of the anterior border of the trochlear surface of the tibia on the neck of the talus (figure 6.29, *a* and *c*).

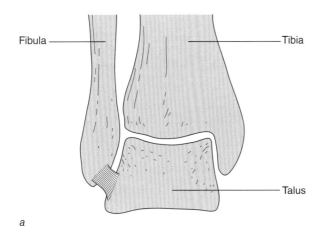
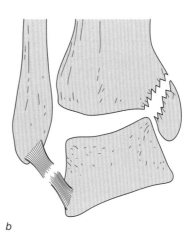

FIGURE 6.28 Inversion sprain injury. *(a)* Normal orientation of the tibia, fibula, and talus. *(b)* Torn lateral ligaments and fractured medial malleolus resulting from excessive and forceful inversion.

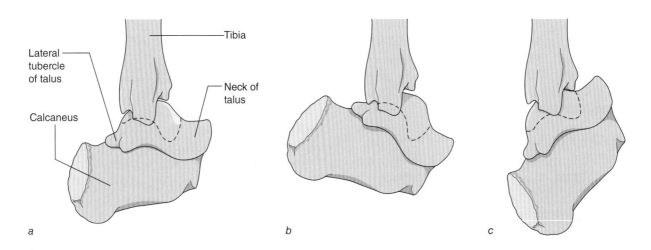

FIGURE 6.29 Medial aspect of the left foot. Impingement of the tibia and talus. *(a)* Neutral orientation of the tibia and talus. *(b)* Impingement of the posterior border of the trochlear surface of the tibia on the lateral tubercle of the talus resulting from extreme plantar flexion. *(c)* Impingement of the anterior border of the trochlear surface of the tibia on the neck of the talus resulting from extreme dorsiflexion.

Subtalar Joint

The subtalar joint, between the talus and calcaneus, is part synovial and part syndesmosis. The anterior synovial part of the joint is separated from the posterior synovial part of the joint by a funnel-shaped channel called the sinus tarsi (figure 6.27). The syndesmosis part of the subtalar joint is a broad interosseous talocalcaneal ligament consisting of medial and lateral bands that run downward from the sulcus tali (superior part of the sinus tarsi) to the sulcus calcanei (inferior part of the sinus tarsi).

Like the ankle joint, the subtalar joint has triplanar movement. Inman (1976) showed that the orientation of the axis of the joint varies considerably between people, with a mean orientation of approximately 42° anterior–superior with respect to the horizontal in the sagittal plane and 23° anterior–medial with respect to the median plane in the horizontal plane with respect to the center of the joint (figure 6.22).

Pronation and Supination of the Rearfoot Complex

Little empirical information is available concerning the movement of the midtarsal joint, which consists of a biplanar–biaxial saddle joint (calcaneocuboid) and a triplanar–triaxial ball-and-socket joint (talonavicular). However, it is clear that the rearfoot complex facilitates triplanar movements of the foot. These movements are usually referred to as **pronation** and **supination** (figure 6.30) (Kitaoka et al 1997, Nester 1997).

Pronation involves simultaneous abduction, dorsiflexion, and eversion (figure 6.30, a and b). Similarly, supination involves simultaneous adduction, plantar flexion, and inversion (figure 6.30, b and c). The orientation of the rearfoot complex axis varies considerably, with a mean orientation of approximately 51° anterior–superior with respect to the horizontal in the sagittal plane and 18° anterior–medial with respect to the median plane in the horizontal plane with respect to the center of the rearfoot complex (Downing et al 1978).

The movements of supination and pronation as described here refer to movements of the rearfoot complex when the foot is not bearing weight. When the foot is bearing weight, these movements are constrained depending on the magnitude and distribution of the ground reaction force acting on the plantar (inferior) surface of the foot. Under weight-bearing conditions, the most noticeable movements of the foot occur about an anteroposterior axis through the foot (similar to inversion and eversion). For this reason, in describing the movement of the foot under weight-bearing conditions, the terms *supination* and *inversion* are sometimes used synonymously, as are the terms *pronation* and *eversion*. However, the actual movements of the foot under weight-bearing conditions are modifications of supination and pronation and, as such, involve simultaneous triplanar movement in all the joints of the rearfoot complex (Lundberg and Svensson 1988, Stacoff et al 1991).

Key Terms

rearfoot complex The functionally interdependent ankle, subtalar, and midtarsal joints; facilitates pronation and supination.

plantar flexion Rotation of the foot about a mediolateral axis through the ankle joint in which the plantar surface of the foot is pushed away from the shin (pointing the toes).

dorsiflexion Rotation of the foot about a transverse axis through the ankle joint in which the dorsal surface of the foot is drawn closer to the shin.

pronation Simultaneous abduction, dorsiflexion, and eversion of the foot.

supination Simultaneous adduction, plantar flexion, and inversion of the foot.

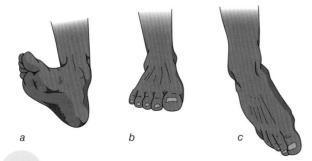

a   b   c

FIGURE 6.30 Pronation and supination. *(a)* Full pronation. *(b)* Neutral position. *(c)* Full supination.

Summary

In contrast to the joints of the axial skeleton—which primarily provide shock absorption—the joints of the appendicular skeleton primarily facilitate relatively large ranges of movement; this design is reflected in the wide variety of synovial joints present in the appendicular skeleton. All of the joints in the appendicular skeleton have a certain degree of functional interdependence with other joints, especially those in the same joint complex or skeletal chain, and therefore abnormal or restricted movement in a joint can result in abnormal movement in other joints in the joint complex or skeletal chain. Chapters 2 to 6 have been largely concerned with the structure and function of the skeletal system. Chapter 7 covers the neuromuscular system, which actively controls the movement of the skeletal system.

Review Questions

1. Describe the coracoacromial arch and its effect on movements of the shoulder joint.
2. Describe the role of the coracoclavicular and costoclavicular ligaments in the stability of the acromioclavicular and sternoclavicular joints.
3. Describe the movements of the shoulder complex during abduction of the upper limb to the overhead position.
4. Describe the ligaments of the elbow complex and their functions.
5. Explain why the normal range of wrist adduction is approximately three times that of wrist abduction.
6. Compare and contrast the structure and functions of the shoulder and hip joints.
7. Describe the structure and functions of the menisci of the tibiofemoral joint.
8. Describe the changes in the shape of the capsule of the knee complex during flexion and extension of the knee.
9. Describe the structure and functions of the collateral ligaments and cruciate ligaments of the knee.
10. Explain why knee flexion increases the range of hip flexion and decreases the range of hip extension.
11. Describe the functions of the patella.
12. Describe the main causes of abnormal tracking of the patella.
13. Describe the possible sources of pain characterized by patellofemoral pain syndrome.
14. Describe the effect of dorsiflexion and plantar flexion on the stability of the ankle joint.
15. Describe the types of ankle injury likely to result from extreme dorsiflexion and plantar flexion.

The Neuromuscular System

The muscular system is the interface between the nervous and skeletal systems. The muscles produce the forces that determine the movements of the joints, but the nervous system determines the level and timing of the muscle forces. The nervous system constantly monitors and interprets information from the various senses concerning body position and body movement, including information from muscles and joint supporting structures, and on the basis of this information sends instructions to the muscles to coordinate body movement. Those parts of the nervous and muscular systems responsible for bringing about coordinated movement are referred to as the neuromuscular system. This chapter describes the structure and function of the neuromuscular system.

7

OBJECTIVES

After reading this chapter, you should be able to do the following:

1. Differentiate the cerebrospinal nervous system, autonomic nervous system, central nervous system, and peripheral nervous system.

2. Describe the events resulting in transmission of an impulse along a nerve fiber and from one nerve fiber to another.

3. Describe the general organization of nerve tissue in the brain, spinal cord, and spinal nerves.

4. Describe the structure of a skeletal muscle fiber and the organization of fibers in pennate and nonpennate muscles.

5. Differentiate kinesthetic sense and proprioception.

6. Describe the length–tension relationship in a sarcomere, muscle fiber, and muscle–tendon unit.

7. Describe the force–velocity relationship and stretch-shorten cycle in skeletal muscle.

Body movement is brought about by the musculoskeletal system under the control of the nervous system. The nervous system continuously monitors and interprets information from the various senses and uses this information to program muscular activity to bring about voluntary and involuntary (reflex) movements.

With regard to voluntary movements, improvements in effectiveness—the extent to which the objective of a particular movement is achieved—and efficiency—the metabolic energy cost of performing the movement (expenditure of chemical energy by the cells)—depend on changes in the operation of the neuromuscular system. The ability of a physical education teacher, coach, or therapist to improve effectiveness and efficiency depends to a considerable extent on her understanding of the structure and function of the neuromuscular system.

Nervous System

The nervous system consists of approximately 13,000 million nerve cells called **neurons** and an equally large number of specialized connective tissue cells called **glial cells.** Neurons are specialized to conduct electrical impulses rapidly throughout the body to coordinate all the essential biological functions.

With regard to structure, the cells of the nervous system are organized into the **central nervous system** and the **peripheral nervous system** (figure 7.1). The central nervous system consists of the brain and spinal cord. The peripheral nervous system consists of 43 pairs of nerves (bundles of nerve fibers) that arise from the base of the brain and the spinal cord. Twelve of the pairs of nerves arise from the base of the brain and are called **cranial nerves** (not visible in figure 7.1). The other 31 pairs arise from the spinal cord and are called **spinal nerves** (figure 7.1). The cranial and spinal nerves convey information between the central nervous system and the rest of the body.

With regard to function, the cells of the nervous system are organized into the **cerebrospinal nervous system** and the **autonomic nervous system.** The cerebrospinal nervous system, also known as the somatic, craniospinal, or voluntary nervous system, is under voluntary control except for reflex movements. A reflex movement provides protection by rap-

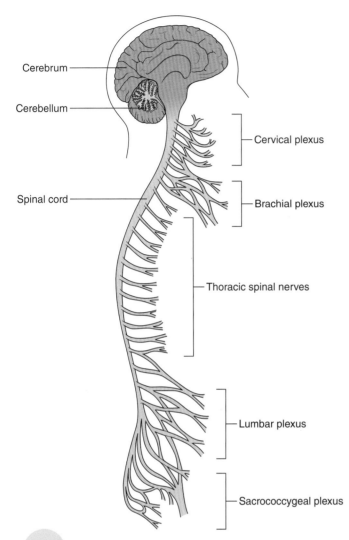

Cerebrum

Cerebellum

Spinal cord

Cervical plexus

Brachial plexus

Thoracic spinal nerves

Lumbar plexus

Sacrococcygeal plexus

FIGURE 7.1 The central nervous system consists of the brain (cerebrum and cerebellum) and spinal cord. The peripheral nervous system consists of the cranial and spinal nerves. Many of the spinal nerves link together to form plexuses. All of the spinal nerves and plexuses branch profusely throughout the regions of the body that they innervate.

idly removing part of the body from a source of danger without conscious effort. The cerebrospinal nervous system includes those parts of the nervous system concerned with consciousness and mental activities and control of skeletal muscle, as in voluntary movement of the arms and legs. The autonomic nervous system, also known as the visceral or involuntary nervous system, is not under voluntary control; it includes those parts of the nervous system that control the visceral muscles (smooth muscle, as in the arteries and digestive tract), the heart, and the exocrine and endocrine glands. The

functions of the cerebrospinal and autonomic nervous systems take place in different parts of the central and peripheral nervous systems.

Neurons

The nervous system consists of neurons that conduct electrical impulses rapidly throughout the body to coordinate essential biological functions. Neurons differ in size and shape, but they all have three common features: a cell body, processes of varying length that extend from the cell body, and specialized sites for communicating with other neurons, with specialized receptors such as pain receptors, or with specialized effectors such as motor end plates in muscle. Neurons can be classified by the direction in which they conduct impulses in relation to the brain. **Sensory neurons,** or afferent neurons, conduct impulses toward the brain; **motor neurons,** or efferent neurons, conduct impulses away from the brain.

Nerve Fibers, Dendrites, and Axons

The processes that extend from the cell bodies of neurons are called **nerve fibers.** Nerve fibers vary in length from a few millimeters to more than one meter. There are two types of nerve fibers: dendrites and axons. **Dendrites,** or afferent fibers, conduct impulses toward the cell body. **Axons,** or efferent fibers, conduct impulses away from the cell body. In addition to the sensory and motor classification, neurons are classified on the basis of the number of processes arising from the cell body into pseudounipolar, bipolar, and multipolar.

In a pseudounipolar neuron, there appears to be one process arising from the cell body that quickly divides into an afferent fiber and an efferent fiber (figure 7.2a). Sensory neurons in the peripheral nervous system are pseudounipolar neurons. A bipolar neuron has two distinct processes, one afferent and one efferent (figure 7.2b). Bipolar neurons are found in the sensory areas of the eye, ear, and nose. Multipolar neurons have numerous relatively short dendrites with a single axon that may branch at various points (figure 7.2c). Most of the neurons in the brain and spinal cord are multipolar neurons.

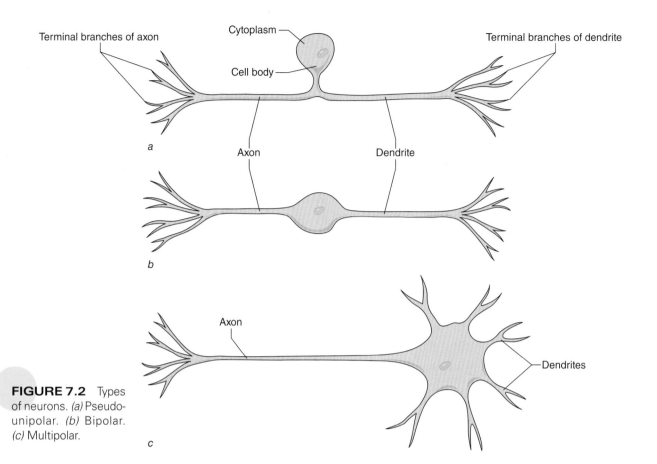

FIGURE 7.2 Types of neurons. *(a)* Pseudounipolar. *(b)* Bipolar. *(c)* Multipolar.

Neurons differ in size and shape, but they all have three common features: a cell body, processes that extend from the cell body, and specialized sites for communicating with other neurons.

Key Terms

neuron   A nerve cell.

glial cell   A specialized connective tissue cell found only in the nervous system.

central nervous system   One of two structural divisions of the nervous system; consists of the brain and spinal cord.

peripheral nervous system   One of two structural divisions of the nervous system; consists of 43 pairs of nerves (bundles of nerve fibers) that arise from the base of the brain and the spinal cord and branch extensively throughout the body.

cranial nerves   Twelve pairs of peripheral nerves that arise from the base of the brain.

spinal nerves   Thirty-one pairs of peripheral nerves that arise from the spinal cord.

cerebrospinal nervous system   One of two functional divisions of the nervous system; concerned with consciousness and mental activities and control of skeletal muscle.

autonomic nervous system   One of two functional divisions of the nervous system; concerned with the control of the visceral muscles, the heart, and the exocrine and endocrine glands.

sensory neuron   A neuron that conducts impulses toward the brain; also called an afferent neuron.

motor neuron   A neuron that conducts impulses away from the brain; also called an efferent neuron.

nerve fibers   The processes that extend from the cell body of a neuron.

dendrite   A nerve fiber that conducts impulses toward the cell body.

axon   A nerve fiber that conducts impulses away from the cell body.

Myelinated and Nonmyelinated Nerve Fibers

Glial cells provide mechanical and metabolic support to neurons. In the central nervous system there are a variety of glial cells including astrocytes, which provide support for blood vessels, and oligodendrocytes, which provide support for nerve fibers. In the peripheral nervous system there is only one type of glial cell, the **Schwann cell** (Gamble 1988). All nerve fibers of the peripheral nervous system are enveloped by Schwann cells, which provide the same type of mechanical support as the oligodendrocytes provide for nerve fibers in the central nervous system. The Schwann cells around some fibers produce a fatty substance called myelin, which is deposited around the fibers as a multilayered myelin sheath. The sheath is in the form of a spiral with up to 100 regularly spaced layers of myelin separated by folds of Schwann cell membrane. The outer fold of the Schwann cell membrane is referred to as the neurilemma (figure 7.3*a*).

The greater the number of layers of myelin in the sheath, the faster the speed of nerve transmission along the nerve fiber. Nerve fibers that have a myelin sheath are referred to as **myelinated nerve fibers** (or medullated nerve fibers), and those nerve fibers that do not have a myelin sheath are referred to as **nonmyelinated nerve fibers** (or nonmedullated nerve fibers). Each myelinated nerve fiber is enclosed by a chain of Schwann cells. Each Schwann cell envelops approximately 1 mm of nerve fiber, and successive Schwann cells are separated by a gap of approximately 1 μm (1 μm = 1 micrometer = one millionth of a meter) so that the nerve fiber is exposed; these regions are referred to as nodes of Ranvier (figure 7.3*a*). The nodes of Ranvier facilitate intercellular exchange between the nerve fibers and the surrounding extracellular fluid, which is important for the nutrition of the nerve fiber and for the transmission of impulses along the nerve fiber. Myelinated nerve fibers in the central nervous system are different from those in the peripheral nervous system in that they

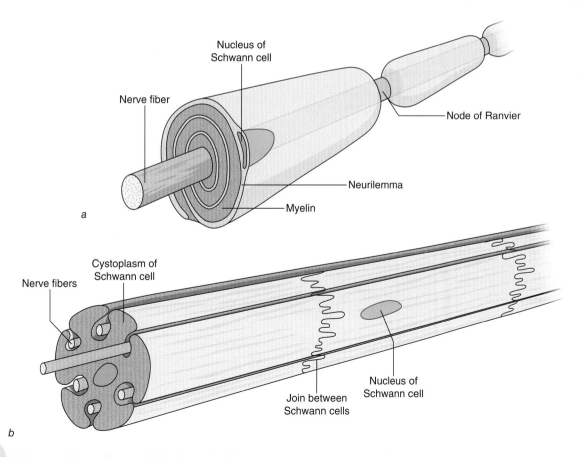

FIGURE 7.3   Nerve fibers. *(a)* Myelinated. *(b)* Nonmyelinated.

are not surrounded by Schwann cells (there are no Schwann cells in the central nervous system) and have no neurilemma. It is thought that oligodendrocytes are responsible for the formation of the myelin sheath around these fibers at the embryonic stage (Standring 2004).

Nonmyelinated nerve fibers of the peripheral nervous system are also enclosed within continuous chains of Schwann cells. However, in comparison to myelinated fibers, there are no nodes of Ranvier in nonmyelinated fibers and there may be as many as nine nerve fibers enveloped within the folds of the Schwann cells in the chain (figure 7.3b).

KEY POINT

In myelinated nerve fibers, the greater the number of layers of myelin in the sheath, the faster the speed of nerve transmission along the nerve fiber.

Nerve Fiber Endings

The dendrites and axons of all types of neurons have a large number of terminal branches or nerve endings devoid of Schwann cells and myelin sheath. There are three types of nerve endings:

1. **Sensory nerve ending,** where the nerve ending is in contact with a specialized receptor organ such as a pain receptor

2. **Motor nerve ending,** where the nerve ending is in contact with a specialized effector organ such as a motor end plate in muscle

3. **Synapse,** where the nerve ending is in contact with another neuron

Sensory and motor nerve endings are referred to as **end organs.** Neurons that only have synapses at their nerve endings are called **association neurons** or interneurons.

Key Terms

Schwann cell The only type of glial cell in the peripheral nervous system.

myelinated nerve fiber A nerve fiber that has a myelin sheath.

nonmyelinated nerve fiber A nerve fiber that does not have a myelin sheath.

sensory nerve ending A nerve ending in contact with a specialized receptor organ such as a pain receptor.

motor nerve ending A nerve ending in contact with a specialized effector organ such as a motor end plate in muscle.

synapse Where a nerve ending is in contact with another neuron.

end organs Sensory and motor nerve endings.

association neuron A neuron that only has synapses at its nerve endings; also called an interneuron.

Electrical Impulse Transmission

The cytoplasm of a neuron and the extracellular fluid surrounding the neuron contain many ions (electrically charged atoms). These include positively charged inorganic ions such as sodium (Na^+) and potassium (K^+), negatively charged inorganic ions such as chloride (Cl^-), and various organic anions (negatively charged amino acids and proteins [A^-]).

Like most membranes in the body, a nerve fiber membrane is semipermeable; it has a large number of tiny holes through which ions and small molecules (aggregations of ions) pass from one side of the membrane to the other. The movement of ions through the nerve fiber membrane depends on the permeability of the membrane (the number and size of the holes in the membrane) and the force tending to drive the ions through the membrane. The driving force has electrical, chemical, and, in the case of Na^+ and K^+, mechanical components. The electrical component depends on the polarity of the ions; like charges repel each other and unlike charges attract each other. The chemical component depends on the concentration of ions in different regions; ions move from an area of high concentration to areas of lower concentration. The mechanical component results from specialized regions of the membrane collectively referred to as the Na^+-K^+ pump. The Na^+-K^+ pump transports Na^+ out of the cytoplasm and into the extracellular fluid and transports K^+ in the opposite direction.

Under resting conditions, when the nerve fiber is not transmitting an impulse, the net effect of the membrane permeability and the driving forces on the various ions is that the electrical charge on the outside of the fiber membrane is approximately 70 mV (millivolts) higher than on the inside; the potential difference across the membrane is approximately 70 mV, with the inside negative with respect to the outside (figure 7.4a). This resting potential difference is referred to as resting membrane potential (RMP) (Enoka 1994).

The arrival of a stimulus at a nerve fiber in a state of RMP alters the permeability of the fiber membrane to Na^+ and K^+ so that Na^+ flows into the cell and K^+ flows out of the cell. The flow of Na^+ into the cell is initially greater than the flow of K^+ out of the cell so that the potential difference across the membrane decreases. If the decrease in potential difference, which

depends on the strength of the stimulus, reaches a critical level of approximately 60 mV (with the inside of the membrane negative with respect to the outside), the membrane will be depolarized; the potential difference across the fiber membrane rapidly changes from 70 mV, with the inside of the membrane negative with respect to the outside, to approximately 30 mV, with the inside of the membrane positive with respect to the outside; the change in potential difference is approximately 100 mV (figure 7.4a).

This change in potential difference constitutes an **action potential,** which results in the flow of electrical current—called local current—between the depolarized region of the cell membrane and the adjacent unpolarized regions (both sides) (figure 7.4, b and c).

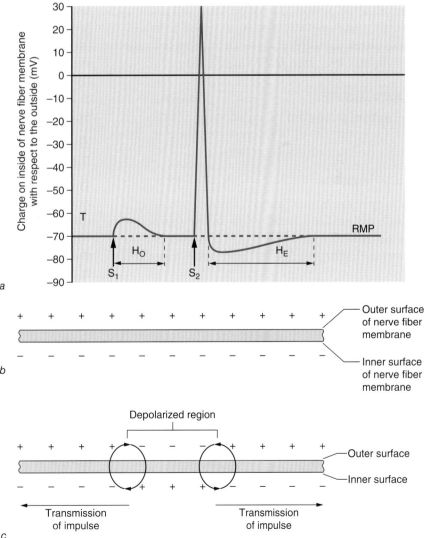

FIGURE 7.4 Resting membrane potential (RMP) and change in membrane potential. (a) Change in membrane potential in response to stimulation: T = level of hypopolarization necessary to trigger depolarization (T ≈ −60 mV); S_1 = a stimulus that results in hypopolarization but not depolarization; H_O = period of hypopolarization following S_1; S_2 = a stimulus that results in depolarization; H_E = period of hyperpolarization following repolarization. (b) Electrical charge on nerve fiber membrane during RMP. The charge on the inside of the membrane is approximately −70 mV with respect to the outside. (c) Electrical charge on depolarized region of nerve fiber membrane; the charge on the inside of the membrane is approximately +30 mV with respect to the outside.

The establishment of local current results in progressive (rapid wave) depolarization of the rest of the cell membrane so that the impulse is transmitted along the whole length of the fiber. After depolarization, the membrane is rapidly repolarized such that the action potential appears as a spike in a graph of the change in membrane potential with time (figure 7.4*a*).

The duration of the action potential spike (depolarization and repolarization) is less than 1 ms (Gamble 1988). Repolarization is due to a rapid decrease in the flow of Na^+ into the cell (caused by reduced permeability of the membrane to Na^+ ions) and continued flow of K^+ out of the cell. Following repolarization there is usually a period (15-100 ms) of hyperpolarization in which the potential difference across the membrane is slightly greater than RMP as RMP is gradually restored (figure 7.4*a*). A stimulus not strong enough to cause depolarization results in a period of hypopolarization prior to restoration of RMP, that is, a period in which the potential difference across the membrane is slightly less than RMP (figure 7.4*a*).

Key Term

action potential Depolarization of the cell membrane of a neuron resulting in the transmission of an impulse.

Synapses

Every branch of an axon terminates in an end bulb (or end foot) that rests on the surface of a neighboring neuron to form a synapse—a specialized region that facilitates one-way communication between the two neurons (figure 7.5). Each neuron synapses with hundreds or thousands of other neurons (Standring 2004). The most common type of synapse is axodendritic (between an axon and a dendrite), but synapses may also be axosomatic (between an axon and a cell body) and axoaxonic (between two axons). A synapse consists of a presynaptic membrane (the base of the end bulb), a postsynaptic membrane (the corresponding region of the adjacent neuron), and an intervening synaptic cleft.

The end bulb contains a number of synaptic vesicles full of neurotransmitters. Some of the vesicles contain excitatory transmitters and some contain inhibitory transmitters. The arrival of an action potential at a synapse results in the release of neurotransmitters into the synaptic cleft. If excitatory neurotransmitters are released, the postsynaptic membrane will be depolarized, resulting in an action potential and transmission of the impulse. If inhibitory transmitters are released, the postsynaptic membrane will be hyperpolarized, thereby preventing the development of an action potential so that the impulse will not be transmitted.

Speed of Impulse Transmission

The speed of transmission or conduction of impulses is directly proportional to fiber diameter and the thickness of the myelin sheath. Nonmyelinated fibers do not have a myelin sheath and thus have much lower conduction speeds than myelinated fibers. Nerve fibers of the peripheral nervous system are classified on the basis of conduction speed and fiber diameter (table 7.1). There are five main categories of afferent fibers, Ia, Ib, II, III, and IV, and five main categories of efferent fibers, Aα, Aβ, Aγ, B, and C. Skeletal muscle is innervated by the fastest afferent fibers (Ia and Ib) and the fastest efferent fibers (Aα, Aβ, and Aγ).

Nerve Tissue Organization in the Brain

The brain is the largest and most complex aggregation of neurons in the nervous system. It consists of the cerebrum and the cerebellum

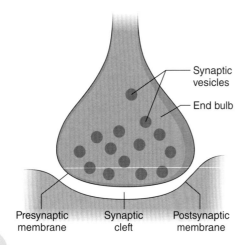

Synaptic vesicles

End bulb

Presynaptic membrane

Synaptic cleft

Postsynaptic membrane

FIGURE 7.5 A synapse.

Table 7.1 Classification of Peripheral Nerve Fibers

Class	Speed (m/s)	Innervation
Efferent (motor) fibers		
Aα	65–120	Fast twitch extrafusal fibers
Aβ	40–80	Slow twitch extrafusal fibers, muscle spindle intrafusal fibers
Aγ	10–50	Muscle spindle intrafusal fibers
B	4–25	Presynaptic autonomics
C	0.2–2.0	Postsynaptic autonomics
Afferent (sensory) fibers		
Ia	65–130	Muscle spindle intrafusal fibers
Ib	65–130	Golgi tendon organs
II	20–90	Muscle spindle intrafusal fibers, pressure receptors
III	12–45	Temperature and pain receptors
IV	0.2–2.0	Viscera, pain receptors

Adapted from J. Gamble, 1988, *The musculoskeletal system* (New York: Raven Press), 121. By permission of J. Gamble.

(see figure 7.1). The cerebrum occupies most of the cranium and is bigger than the cerebellum. The region of the cerebrum close to and including its surface is called the cerebral cortex and consists of **gray matter**—the cell bodies of neurons and their processes, which are largely nonmyelinated, together with their synapses and supporting glial cells. The cerebrum is heavily convoluted with fissures of varying depth. The largest fissure is the longitudinal central fissure that divides the cerebrum in the median plane into right and left cerebral hemispheres.

The cerebellum, which occupies the posterior inferior aspect of the cranium, is separated from the cerebrum by the transverse fissure. Like the cerebrum, the region of the cerebellum close to and including its surface consists of gray matter and is called the cerebellar cortex. The cerebellum is not convoluted but is traversed by numerous small furrows. The convolutions of the cerebrum and furrows of the cerebellum significantly increase their surface areas and thus the volume of gray matter. The inner parts of the cerebrum and cerebellum consist of **white matter,** largely myelinated nerve fibers organized into groups

in which the fibers are parallel to each other. There are five groups of fibers, resembling an intricate system of electrical wiring:

1. Fibers linking the parts of the cerebral cortex
2. Fibers linking the parts of the cerebellar cortex
3. Fibers linking the parts of the cerebral cortex with the parts of the cerebellar cortex
4. Fibers passing between the cerebral cortex and spinal cord
5. Fibers passing between the cerebellar cortex and spinal cord

Key Terms

gray matter The cell bodies of neurons of the central nervous system and their fibers, which are largely nonmyelinated, together with their synapses and supporting glial cells.

white matter Myelinated nerve fibers in the central nervous system.

KEY POINT

The brain is the largest and most complex aggregation of neurons in the nervous system. It consists of the cerebrum and the cerebellum. The cerebral cortex and cerebellar cortex consist of gray matter. The inner parts of the cerebrum and cerebellum consist of white matter.

Nerve Tissue Organization in the Spinal Cord and Spinal Nerves

In transverse section, the spinal cord is roughly oval shaped with an anterior median fissure and a posterior medial septum (figure 7.6). The central area is dominated by a roughly H-shaped mass of gray matter, with the rest of the cord consisting of white matter in which the groups of fibers run parallel with the spinal cord. The posterior projections (or horns) of the gray matter are continuous with afferent fibers that enter the spinal cord via the left and right dorsal (posterior) roots of the corresponding left and right spinal nerves. The cell bodies

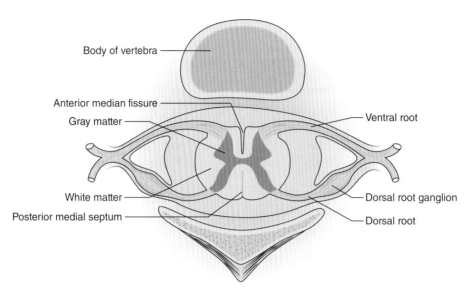

Body of vertebra

Anterior median fissure

Gray matter

Ventral root

White matter

Posterior medial septum

Dorsal root ganglion

Dorsal root

FIGURE 7.6 Transverse section through the spinal cord at the level of a pair of spinal nerves.

of the afferent fibers are located in the dorsal root ganglia; a ganglion is an aggregation of cell bodies outside the spinal cord.

The anterior projections of the gray matter are continuous with efferent fibers that leave the spinal cord via the left and right ventral (anterior) roots of the corresponding spinal nerve. A spinal nerve is formed by the aggregation of the afferent and efferent fibers as they pass through the corresponding intervertebral foramen. Each spinal nerve divides into an anterior and posterior branch just outside the intervertebral foramen; the anterior branch is usually much larger than the posterior branch. In each branch the individual nerve fibers are enveloped in a thin layer of connective tissue called the endoneurium, which supports a blood capillary network. The fibers are grouped together in bundles called fasciculi or funiculi, and each fasciculus is sheathed within a layer of connective tissue called the perineurium. The fasciculi are grouped together and sheathed within another layer of connective tissue called the epineurium to form the complete spinal nerve (figure 7.7). Each spinal nerve consists of a mixture of myelinated and nonmyelinated nerve fibers.

In all regions of the spinal cord apart from most of the thoracic region, the spinal nerves link up with each other on each side of the spinal cord to form networks called **plexuses** (figure 7.1). The upper four pairs of cervical

nerves form the left and right cervical plexuses that innervate the head and upper part of the neck. The other cervical spinal nerves combine with the first pair of thoracic spinal nerves to form the left and right brachial plexuses that innervate the arms. The twelfth thoracic spinal nerves and the spinal nerves of the lumbar and sacral regions combine to form the left and right lumbosacral plexuses that innervate the legs. The second to the eleventh thoracic spinal nerves, which innervate the trunk, do not form plexuses.

KEY POINT

In the spinal cord, the posterior horns of the gray matter are continuous with afferent fibers of the corresponding spinal nerves, and the anterior horns are continuous with efferent fibers of the corresponding spinal nerves. Each spinal nerve consists of myelinated and nonmyelinated nerve fibers.

Key Term

plexus Network of spinal nerves that innervate a specific region of the body.

Voluntary and Reflex Movements

The brain interprets sensory information from the various receptors and using this information brings about appropriate responses via

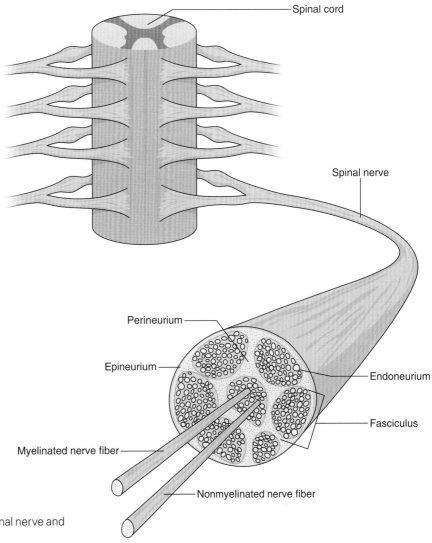

FIGURE 7.7 Structure of a spinal nerve and its relationship to the spinal cord.

effectors to ensure normal bodily functioning. Whereas the autonomic nervous system operates largely at a subconscious level, the cerebrospinal nervous system normally operates at the level of consciousness; that is, the individual consciously processes information, such as visual, auditory, and kinesthetic (sense of movement) information, and consciously brings about appropriate responses. For example, consider a child just about to touch a hot coffeepot resting on a stove.

1. As the child moves his hand close to the pot, the heat radiating from the pot may excite the heat receptors in the skin of his hand.

2. This information is transmitted by sensory neurons in the peripheral nervous system to the spinal cord, where the nerve endings of the axons of the sensory neurons synapse with many other neurons.

3. Some of these neurons relay the information to the brain center responsible for heat sensation in the hand, and the child experiences the sensation of heat.

4. Having sensed the heat from the coffeepot, the child decides to move his hand away from the pot to avoid the danger.

5. Impulses are sent from the brain to motor neurons in the spinal cord that innervate the muscles of the arm.

6. The impulses are relayed to the muscles of the arm resulting in movement of the hand away from the coffeepot.

The preceding sequence, illustrated in figure 7.8, is typical of all voluntary movements. If the child had not sensed the danger and had actually touched the hot coffeepot, he would have jerked his hand away from the pot with lightening speed before he was even conscious of the heat. This extremely rapid involuntary reaction is called a **reflex action** and is one of a host of similar reflexes that result in instant reactions to protect the body from potentially harmful stimuli. In this particular case, the reflex action results in the child's hand being in contact with the coffeepot for only a fraction of the time it would have taken for the child to voluntarily (and, therefore, consciously) take his hand away from the pot. Consequently, the child may sustain a relatively slight burn rather than a serious burn, which would have resulted from more prolonged contact with the coffeepot.

The increased speed of a reflex action compared with a voluntary action results from a decrease in the distance over which the impulses travel from receptor to effector. In our example, the heat from the coffeepot is sensed by the heat receptors in the hand, which send impulses to the spinal cord. This is basically the same as in a voluntary action—stages 1 and 2 in figure 7.8. However, in a reflex action the impulses are transmitted directly across the spinal cord to the motor neurons that innervate the muscles of the arm—stage 3A in figure

7.8. The impulses are relayed to the muscles of the arm, and the reflex action is completed. Consequently, a reflex action is faster than a voluntary action because no time is spent in transmitting impulses up the spinal cord to the brain, making a decision, and transmitting impulses back down the spinal cord—stages 3, 4, and 5 in figure 7.8 are omitted. A reflex action is triggered when the intensity of the impulse output from the receptors is above a certain threshold indicating extreme danger to the body. In our example, the hotter the coffeepot, the greater the intensity of output from the heat receptors and the greater the likelihood of a reflex action being triggered.

As the impulses are transmitted across the spinal cord (stage 3A), the impulses are simultaneously transmitted to the brain as in a voluntary action (stage 3). However, stage 3 (transmission of impulses to the brain and sensation of heat) takes longer than stages 3A and 6 such that a few tenths of a second pass after the boy jerks his hand away from the coffeepot before he perceives any pain in his hand.

Key Term

reflex action An involuntary movement that provides protection by rapidly removing part of the body from a potential source of danger without conscious effort.

KEY POINT

The autonomic nervous system operates at a subconscious level. The cerebrospinal nervous system normally operates at the level of consciousness but may produce involuntary reflex actions.

Nerve Fiber Injuries

Injuries to peripheral nerves, which can result from any kind of abnormal movement of the body or abnormal contact with the body, usually occur as a result of compression or traction (stretching) or a combination of the two (Kleinrensink et al 1994). Peripheral nerves originate at the spinal cord and branch progressively with increased distance from the spinal

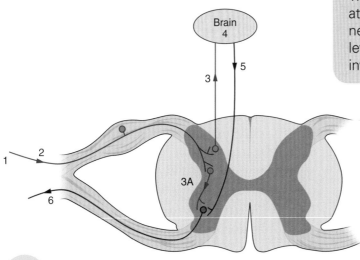

FIGURE 7.8 Pathway of impulses in voluntary and reflex movements (see text for description of stages).

cord. Consequently, an injury to a spinal nerve that is close to its origin is likely to have more serious consequences than an injury farther along the spinal nerve because more functions, sensory and motor, are likely to be affected. Compression and traction may damage any or all of the main structures: connective tissue sheaths, Schwann cells, the myelin sheath (when present), and nerve fibers. Compression and traction damage the nerve structures directly by crushing and tearing, respectively. Prolonged compression also may damage the nerve structures indirectly as a result of ischemia—a disruption of the local blood supply (due to compression of the local capillary network), which results in a deficiency of oxygen and nutrients to the affected tissues.

Injuries to peripheral nerves are classified on the basis of degree of structural and functional damage into neuropraxia, axonotmesis, and neurotmesis (Seddon 1972). Neuropraxia is the lowest level of damage. It involves damage to Schwann cells and myelin sheaths but little or no damage to nerve fibers or endoneurium. Neuropraxia is characterized by a disruption to impulse transmission, which is often associated with pain and tingling in the areas innervated by the affected nerves. Recovery of a nerve fiber from neuropraxia usually occurs within 10 to 14 days. In axonotmesis, the nerve fiber and Schwann cell covering or myelin sheath are severed and the severed part of the fiber degenerates; the endoneurium remains intact. Recovery begins when fiber sprouts emerge from the severed end of the part of the fiber still attached to the cell body. One of the sprouts eventually dominates and gradually grows along the tube formed by the endoneurium. Growth occurs at a rate of approximately 2.5 cm per month, and function gradually returns to normal as the fiber and myelin sheath (when present) return to normal. In neurotmesis, the nerve fiber, Schwann cell covering or myelin sheath, and endoneurium are all severed. The severed part of the fiber degenerates, and without surgical repair it is unlikely that much recovery will occur. With surgical repair some recovery occurs that is similar to the way in which a fiber recovers from axonotmesis, but functional outcome is often less than satisfactory.

KEY POINT

Compression and traction can damage nerves directly by crushing and tearing, respectively. Prolonged compression may damage the nerve structures indirectly as a result of ischemia.

Skeletal Muscle Structure

The composition and basic function of the muscular system are described in chapter 1. This section describes the macrostructure of skeletal muscle.

Origins and Insertions

Most of the muscles are attached to the skeletal system by tendons or aponeuroses (as described in chapters 1 and 3). However, one or both attachments of some muscles attach directly to bone without an intervening tendon or aponeurosis. These muscles are often located adjacent to bones. For example, the brachialis, a muscle that flexes the elbow, arises directly from a large area on the lower half of the anterior aspect of the humerus and is inserted via a thick and broad tendon onto the coronoid process of the ulna (figure 7.9, a and c). Most of the muscles of the upper and lower limbs are arranged in line with the direction of the long bones. For descriptive purposes, the proximal and distal attachments of each of these muscles are referred to as the origin and insertion of the muscles, respectively. For example, the origin of the brachialis is on the humerus and the insertion is on the ulna. Some muscles have more than one site of origin and more than one site of insertion. For example, the biceps brachii has two sites (or heads) of origin and one site of insertion (figure 7.9, a and b). The origins and insertions of the muscles of the trunk and the muscles that link the trunk

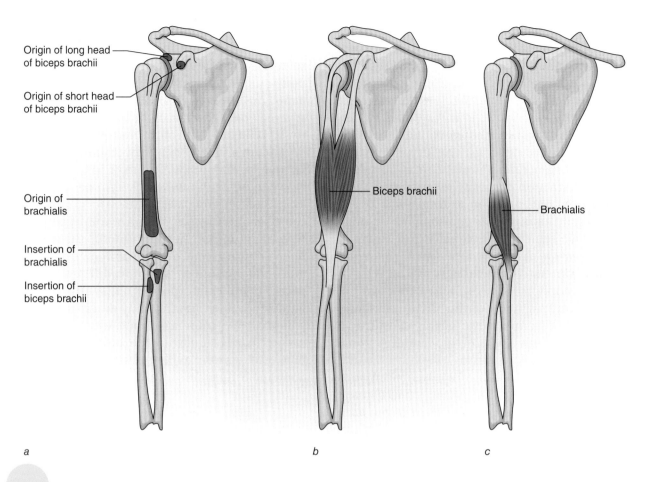

Origin of long head of biceps brachii

Origin of short head of biceps brachii

Origin of brachialis

Insertion of brachialis

Insertion of biceps brachii

Biceps brachii

Brachialis

a

b

c

FIGURE 7.9 Attachments of the biceps brachii and brachialis muscles.

to the limbs tend to be medial and lateral, respectively, but there are many exceptions. The superficial muscles of the body are shown in the appendix together with descriptions of their origins and insertions.

KEY POINT

The origins and insertions of the muscles of the limbs tend to be proximal and distal, respectively. The origins and insertions of the muscles of the trunk and the muscles that link the trunk to the limbs tend to be medial and lateral, respectively.

Pennate and Nonpennate Muscles

A skeletal muscle is made up of skeletal muscle cells bound by various layers of connective tissue that support extensive networks of nerves and blood vessels. Muscle cells are long and thin and, as such, they are usually

referred to as **muscle fibers.** Each muscle fiber is approximately 50 µm (1 µm = one millionth of a meter) wide. All the fibers in an individual muscle are about the same length. In some muscles the fibers are relatively short; for example, in the muscles that move the eyes, the fibers are 2 to 4 mm. In contrast, the fibers in some other muscles are very long; for example, the fibers of the sartorius are approximately 30 cm. The muscle fiber length in most muscles is between these two extreme values.

The fibers in all muscles are organized into bundles (as described later), and the fibers in each bundle run parallel with each other. However, the arrangement of the bundles of fibers with respect to the origin and insertion of the muscle is either pennate or nonpennate. In a **pennate muscle,** the fibers run obliquely with respect to the origin and insertion so that the line of pull of the fibers is oblique to the line of pull of the muscle (figure 7.10). Pennate muscles have a featherlike appearance and are classified according to the number of groups of fibers into unipennate, bipennate, and multipennate. A unipennate muscle contains one group of fibers that insert onto the sides of two tendons (or one bony attachment and one tendon) (figure 7.10*a*). The flexors and extensors of the fingers are unipennate muscles. A bipennate muscle, such as the gastrocnemius, contains two groups of fibers that insert onto the opposite sides of a central tendon (figure 7.10*b*). A multipennate muscle such as the deltoid is, in effect, two or more bipennate muscles combined into a single muscle (figure 7.10*c*).

In a **nonpennate muscle** the fibers run in line with the line of pull of the muscle. There are five main types: quadrilateral, strap, spiral, fusiform, and fan shaped (figure 7.11). In quadrilateral and strap muscles, the length of the fibers and the presence of tendinous intersections reflects the muscle's function as stabilizer (short fibers and tendinous intersections, figure 7.11, *a* and *c*) or mover (long fibers, figure 7.11*b*). In a spiral muscle, the muscle curves around other muscles or bones. In a fusiform (or spindle-shaped) muscle, the fibers are gathered at each end to attach onto long, relatively narrow, tendons. The biceps brachii is a fusiform muscle (see figure 7.9*b*). In a fan-shaped muscle, the fibers converge from a broad origin to a relatively small insertion. The effect of pennate and nonpennate arrangements on muscle function is described later in the chapter.

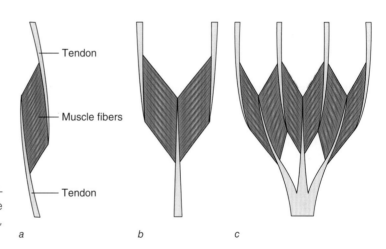

FIGURE 7.10 Pennate muscles. *(a)* Unipennate (e.g., finger flexors). *(b)* Bipennate (e.g., gastrocnemius). *(c)* Multipennate (e.g., deltoid).

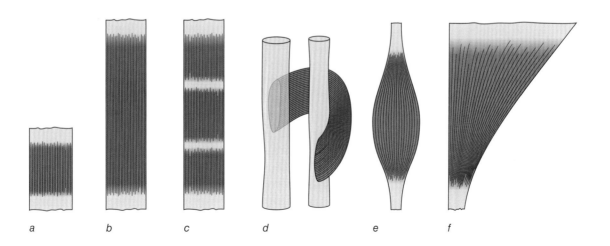

FIGURE 7.11 Nonpennate muscles. *(a)* Quadrilateral (e.g., pronator quadratus). *(b)* Strap (e.g., sartorius). *(c)* Strap with tendinous intersections (e.g., rectus abdominis). *(d)* Spiral (e.g., supinator). *(e)* Fusiform (e.g., biceps brachii). *(f)* Fan shaped (e.g., pectoralis major).

211

Key Terms

muscle fiber A muscle cell.

pennate muscle A muscle in which the fibers run obliquely with respect to the origin and insertion so that the line of pull of the fibers is oblique to the line of pull of the muscle.

nonpennate muscle A muscle in which the fibers run in line with the line of pull of the muscle.

Fusiform Musculotendinous Units

Figure 7.12 shows the structure of a fusiform muscle–tendon unit. Each muscle fiber is enveloped in a layer of areolar tissue called the endomysium, which helps to bind the muscle fibers together and provides a supporting framework for blood capillaries and the terminal branches of nerve fibers. The muscle fibers are grouped together by irregular connective tissue (a mixture of collagenous and elastic) into bundles of up to 200 fibers; each bundle is called a fasciculus (or funiculus), and the connective tissue sheath is called the perimysium. The fasciculi are bound together to form the belly of the muscle by a layer of irregular collagenous connective tissue called the epimysium. The muscle fibers gradually taper at each end. The tapering of the fibers accompanies a gradual thickening in the epimysium and perimysium layers and a change in the composition of the epimysium and perimysium layers from irregular collagenous and irregular elastic connective tissue to regular collagenous connective tissue as the epimysium and perimysium layers merge to form a tendon or aponeurosis (see figure 3.1).

Each muscle receives one or more nerves (collections of sensory and motor nerve fibers), which usually enter the muscle together with the main blood vessels (arteries enter, veins leave) at a region of the muscle that does not move a great deal during normal movement; this region is referred to as the neurovascular hilus (Standring 2004). The blood vessels and nerves branch through the epimysium and perimysium layers down to the endomysium of the individual muscle fibers.

> **KEY POINT**
>
> A skeletal muscle is made up of skeletal muscle fibers bound together by connective tissue that supports extensive networks of nerves and blood vessels. The muscle fibers are organized into bundles that have a pennate or nonpennate arrangement.

Fascial and Osseofascial Compartments

The bundles of muscle fibers that comprise a whole muscle are bound together by epimysium (figure 7.12). Similarly, any group of muscles that function as a single unit, such as the quadriceps muscles, is usually separated from other muscles by a thickening in those parts of the epimysium that surrounds the group of muscles. As the thickened epimysium forms a continuous sheath around the group of muscles, it is usually referred to as fascia (or deep fascia; see chapter 3) and the space enclosed by the fascia, occupied by the muscles, is usually referred to as a fascial compartment. If one or more of the muscles in the group is adjacent to bone, the **compartment** enclosing the muscles is referred to as an osseofascial compartment. The quadriceps are confined within an osseofascial compartment formed by the medial, anterior, and lateral aspects of the femur and the merging of the epimysial layers of the vastus lateralis, rectus femoris, and vastus medialis (figure 7.13). Conse-

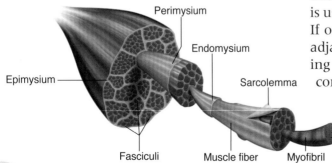

FIGURE 7.12 Structure of a fusiform muscle.

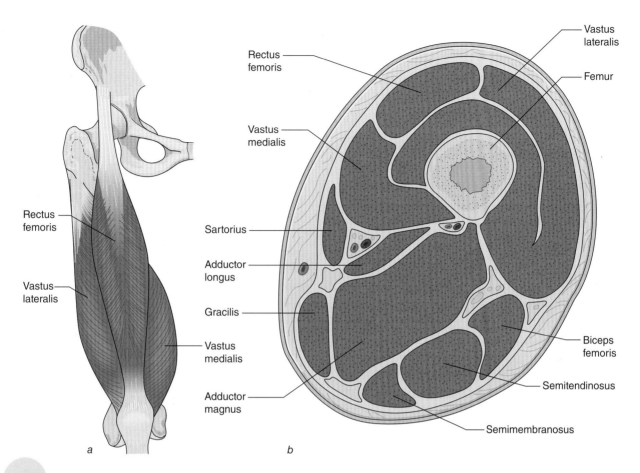

FIGURE 7.13 Muscle compartments of the right upper leg. *(a)* Anterior aspect of the upper leg showing three of the four quadriceps muscles. *(b)* Superior aspect of a transverse section through the middle of the upper leg.

quently, the quadriceps muscles are normally closely packed together, with sufficient space for the passage of blood vessels and nerves. However, any decrease in the size of the osseofascial compartment or any increase in the volume of the contents of the osseofascial compartment can increase the pressure on the blood vessels and nerves within the compartment. **Compartment pressure syndrome** occurs when the pressure within a fascial or osseofascial compartment surrounding a particular muscle or group of muscles becomes elevated to the extent that blood flow within all or part of the compartment is reduced to an abnormally low level, resulting in ischemia of muscle and nerve cells (Segan et al 1988; Englund 2005). Ischemia results in a variety of symptoms including persistent pain, increased pain with stretching, and decreased sensitivity to light touch and pinprick. If severe compart-

ment pressure syndrome is not treated, localized tissue necrosis (degeneration and death of cells) may occur.

Increased pressure within a compartment can occur because of increased stiffness of the fascial binding following injury. However, it is more likely to occur because of an increase in the volume of the contents of the compartment. There are three main causes of an increase in volume: muscle hypertrophy, muscle injury, and prolonged high-intensity activity. Strength training can increase the cross-sectional area of the muscles, which will increase the pressure inside a compartment if the fascial bindings do not adapt to the increased volume of muscle. Muscle injury results in leakage of blood from torn blood vessels, which may cause an increase in pressure within a compartment (Martinez et al 1993). Prolonged high-intensity activity in sports such as kayaking and swimming, in which certain muscle groups are more or less constantly active, can result in compartment pressure syndrome during the activity if the associated fascial bindings are too tight (Ryan et al 1987).

Key Terms

compartment A fascial or osseofascial enclosure surrounding a group of muscles that normally function as a single unit.

compartment pressure syndrome A painful condition resulting from increased pressure within a compartment.

Bursas and Synovial Sheaths

Most of the structures of the musculoskeletal system are closely packed together; the compartmentalization of muscle groups is an example of this. Because of the close packing there are many situations where adjacent structures move relative to each other, that is, slide over each other, during the course of normal movements. Sliding results in **friction**—a force parallel to the surfaces in contact that opposes the sliding movement (Watkins 2007). Friction generates heat, and too much heat can damage or wear the surfaces. In machines, the surfaces of components that slide over each other are usually highly polished and lubricated to minimize friction. Similar mechanisms exist within the musculoskeletal system—bursas and syno-

vial sheaths—and the key component of both mechanisms is synovial fluid.

A **bursa** is a flattened sac of synovial membrane containing a capillary film of synovial fluid. Bursas develop between structures that move relative to each other, providing complete separation (apart from the capillary film of synovial fluid) and freedom of movement over a limited range. Depending on their location, bursas are described as subcutaneous, subtendinous, submuscular, and subfascial. Subcutaneous bursas are interspersed between the deeper layers of the skin and underlying bone, ligament, or tendon; the prepatellar bursa is a subcutaneous bursa (see figure 6.13). A blister that develops on the skin in response to unaccustomed friction is a form of subcutaneous bursa. A blister is a short-term safety mechanism that separates the superficial and deep layers of the skin. Subtendinous bursas are interposed between tendons and bone, ligaments, or other tendons. The suprapatellar and infrapatellar bursas are subtendinous bursas (figure 6.13). Submuscular bursas are interposed between muscles and bones, tendons, ligaments, or other muscles. The subacromial bursa between the deltoid and supraspinatus muscles is a submuscular bursa (figure 7.14). Subfascial bursas are interposed between aponeurotic areas and bone. The majority of bursas are close to joints, and some bursas, such as the suprapatellar bursa and gastrocnemius bursa, are continuous with the

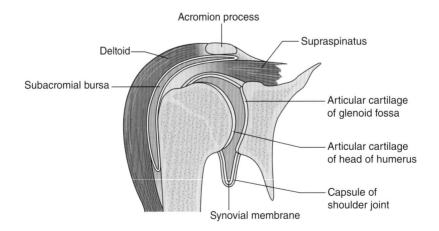

FIGURE 7.14 Coronal section through the right shoulder joint showing the subacromial bursa. The joint is shown distracted to highlight the joint structures.

joint capsule. Excessive repetitive or prolonged pressure on a bursa may inflame it, resulting in a painful condition referred to as **bursitis.** For example, subacromial bursitis, inflammation of the subacromial bursa, is a common injury in competitive swimmers.

Synovial sheaths cover tendons that pass under retinacula (bands of regular collagenous connective tissue; see figure 3.14) or within osseofibrous tunnels (passages formed by bone and fibrous tissue). For example, the tendon of the long head of the biceps brachii runs within an osseofibrous tunnel formed by the bicipital groove and the transverse humeral ligament (figure 7.9, *a* and *b*, and figure 7.15). A **synovial sheath** consists of a flattened sac of synovial membrane containing a capillary film of synovial fluid wrapped around the tendon.

The inner (or visceral) layer of the sheath is attached to the tendon, and the outer (or parietal) layer is attached to the osseofibrous tunnel (figure 7.15).

The median nerve, which innervates part of the hand and fingers, and the tendons of the wrist and finger flexor muscles run in an osseofibrous tunnel formed by the carpals and the flexor retinaculum of the wrist (figure 7.16). This tunnel is usually referred to as the carpal tunnel. All of the tendons are enclosed within synovial sheaths. Unaccustomed or prolonged repetitive flexion and extension of the wrist can inflame the synovial sheaths resulting in a painful condition, called **tenosynovitis,** in which the synovial sheaths become swollen. Tenosynovitis of the wrist may occur, for example, as a result a long, strenuous game

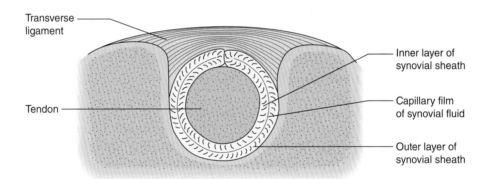

FIGURE 7.15 Horizontal section through the bicipital groove showing the synovial sheath around the tendon of the long head of the biceps brachii muscle.

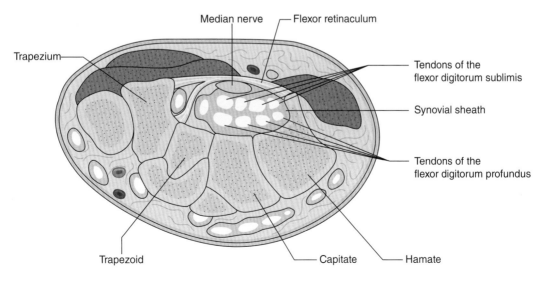

FIGURE 7.16 Transverse section through the left wrist showing the osseofibrous carpal tunnel.

of squash or badminton in a person unaccustomed to such activity. It may also occur as a result of prolonged, repetitive tasks in the food-processing industry, such as in preparing fish and poultry with a knife. The swelling increases the pressure within the carpal tunnel. The greater the pressure, the greater the possibility of neuropraxia of the median nerve, which is usually referred to as carpal tunnel syndrome (Steyers and Schelkun 1995). Carpal tunnel syndrome is characterized by pain and numbness or tingling in the sensory distribution of the median nerve—the palmar surfaces of the thumb, index finger, middle finger, and radial half of the third finger.

Key Terms

friction The force exerted parallel to the surfaces of two objects in contact with each other, which opposes sliding, or the tendency to slide, of the surfaces on each other.

bursa A closed sac comprised of synovial membrane containing synovial fluid that is interposed between tissues that slide on each other to prevent or minimize friction between them.

bursitis A painful condition resulting from inflammation of a bursa.

synovial sheath A closed flattened sac comprised of synovial membrane containing a capillary film of synovial fluid; forms a protective sleeve around a tendon or ligament to prevent or minimize friction between the tendon or ligament and adjacent bony or fibrous structures.

tenosynovitis A painful condition resulting from inflammation of a synovial sheath.

Muscle Fiber Structure and Function

A muscle fiber consists of hundreds or thousands of **myofibrils** embedded in sarcoplasm (muscle cytoplasm) and enclosed by a cell membrane called the sarcolemma (figure 7.17, a and b). Each muscle fiber has a large number of nuclei that lie just beneath the sarcolemma. Each myofibril is approximately 1 μm wide; the myofibrils are arranged parallel to each other and run the whole length of the muscle fiber. Each myofibril exhibits a characteristic pattern of alternate light and dark transverse bands due to the way that the components of the myofibril reflect light (under an electron microscope). The light and dark bands are referred to as I (isotropic) and A (anisotropic) bands, respectively. Because the light and dark bands coincide in adjacent myofibrils, the muscle fiber has a striped or striated appearance; thus, skeletal muscle is often referred to as striated muscle.

Each myofibril has two types of filaments arranged in a highly ordered way that gives rise to the light and dark bands. One type of filament is thicker than the other. The thicker filaments occupy the A bands and are composed of the protein **myosin;** these filaments are usually referred to as myosin filaments or A filaments (figure 7.17c). The thinner filaments—**actin** filaments or I filaments—occupy the I bands and are composed largely of the protein actin (Edman 1992). Each I band is divided by a transverse Z disc. The section of a myofibril between two successive Z discs is called a **sarcomere**—the basic structural unit of a muscle fiber. A myofibril consists of a chain of sarcomeres. The actin filaments project from each side of the Z discs and reach into the A bands of the corresponding sarcomeres, where they interdigitate with the myosin filaments (figure 7.17c). The region between the ends of the two groups of actin filaments in the middle of a sarcomere is referred to as the H zone.

Each actin filament consists of two strands of actin molecules wound together longitudinally in a helical manner (figure 7.17d). Each myosin filament is composed of myosin molecules. Each myosin molecule is a clublike structure consisting of two adjacent globular heads attached by a relatively short curved neck to a long shaft (figure 7.17e). In a myosin filament the myosin molecules are packed such that the shafts form the main body of the filament with the heads projecting from the main body at regular intervals (figure 7.17f). The two halves

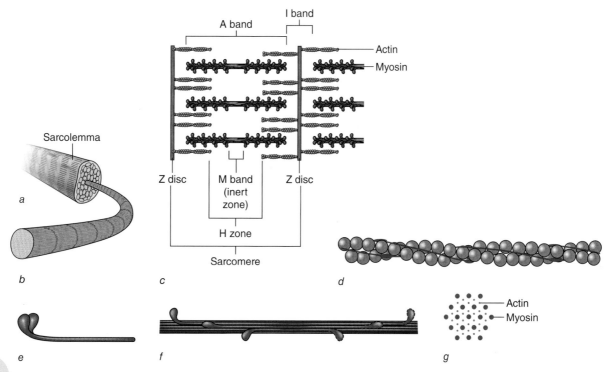

FIGURE 7.17 Structure of a muscle fiber. *(a)* Muscle fiber. *(b)* Myofibril. *(c)* Arrangement of actin and myosin filaments in a sarcomere. *(d)* Part of an actin filament. *(e)* Myosin molecule. *(f)* Part of a myosin filament. *(g)* Hexagonal arrangement of actin and myosin filaments in a sarcomere.

of each myosin filament are mirror images of each other; that is, the myosin molecules in one half of the filament are oriented in the opposite direction to the myosin molecules in the other half. This arrangement, significant in the functional interaction between the actin and myosin filaments, is such that no myosin heads project from the central region of each filament; this central region is sometimes referred to as the inert zone, M band, or M line. Each myosin filament is surrounded by six actin filaments in a regular hexagonal arrangement (figure 7.17*g*). In each half of a myosin filament the myosin molecules are arranged in groups of six; a line joining the heads of the molecules in each group forms a spiral around the myosin filament. The corresponding head in each group faces the same actin filament.

Key Terms

sarcomere   The basic structural unit of a muscle fiber that contains the contractile apparatus.

myofibril   A chain of sarcomeres.

myosin   A protein that largely comprises the thick filaments within a sarcomere.

actin   A protein that largely comprises the thin filaments within a sarcomere.

Sliding Filament Theory of Muscular Contraction

During all types of muscular contraction, the actin and myosin filaments stay the same length, but in isotonic contractions the degree of interdigitation between the two sets of filaments changes as the length of the muscle fibers changes; the width of the A bands stays the same, but the width of the I bands and H zones varies. As the muscle fibers shorten, the region of interdigitation increases, and the width of the I bands and H zones decreases. As the muscle fibers lengthen, the region of interdigitation decreases and the width of the I bands and H zones increases. These observations led to the formulation of the **sliding filament theory** of muscular contraction (Huxley and Hanson 1954). The essential features of the theory are as follows:

1. When a muscle contracts, force is generated by the formation of cross bridges between the heads of the myosin molecules in the myosin filaments and the actin filaments.

2. Flexion of the myosin molecule necks while the heads are in contact with the actin filaments exerts a pulling action on the actin filaments, which causes the actin filaments to slide relative to the myosin filaments, so that the muscle shortens.

3. After exerting their pulling action, the heads of the myosin molecules detach (or decouple) from the actin filaments and swing back to reattach (or recouple) onto the actin filaments further along the actin filaments.

4. The coupling and decoupling of cross bridges occurs at different times so that tension can be maintained while the muscle fiber shortens.

Whereas details of the processes involved have still to be discovered, the sliding filament theory has now gained general acceptance (Huxley 2000).

Types of Muscular Contraction

When a muscle fiber contracts, the heads of the myosin molecules attach to special sites on the actin filaments in the regions where the actin and myosin filaments overlap. The attachments are referred to as **cross bridges**, and each cross bridge generates a certain amount of tension. When a muscle fiber relaxes, the myosin heads detach from the actin and no tension is exerted between them.

When a muscle contracts, the tension exerted by the muscle is directly proportional to the number of cross bridges (the larger the number of cross bridges, the greater the tension). When a muscle contracts it tends to shorten, that is, it tends to pull its origin and insertion closer together. However, the muscle

may shorten, lengthen, or stay the same length depending on the external load on the muscle (the load tending to lengthen the muscle). If the muscle force is greater than the external load, the muscle will shorten; this type of contraction is called a **concentric contraction.** The lifting phase of a biceps curl exercise (elbow flexion) is an example of concentric contraction of the elbow flexor muscles. If the muscle force is less than the external load, the muscle will lengthen; this type of contraction is called an **eccentric contraction.** The lowering phase of a biceps curl exercise (elbow extension) is an example of eccentric contraction of the elbow flexor muscles. Concentric and eccentric contractions are often referred to as **isotonic contractions**—contractions that involve a change in the length of the muscle. If the muscle force is equal to the external load, as in holding the weight stationary in the middle of the flexion–extension range in a biceps curl exercise, the length of the muscle will not change; this type of contraction is called an **isometric contraction.**

Key Terms

sliding filament theory The generally agreed-upon mechanism of muscle contraction involving coupling and decoupling and sliding between actin and myosin filaments.

cross bridge The formation of a bond, which exerts a tension load, between the head of a myosin molecule in a myosin filament and an actin filament during muscle contraction.

concentric contraction A contraction during which the muscle shortens.

eccentric contraction A contraction during which the muscle lengthens.

isotonic contraction A contraction involving a change in the length of the muscle, that is, a concentric contraction or an eccentric contraction.

isometric contraction A contraction during which the length of the muscle does not change.

Motor Units

The functional unit of skeletal muscle is the **motor unit.** A motor unit consists of a motor neuron with an Aα axon (sometimes referred to as an alpha motoneuron), together with all of the terminal branches of the axon and the muscle fibers that they innervate (Gamble 1988) (figure 7.18). The number of muscle fibers innervated by a single alpha motoneuron is referred to as the innervation ratio. The innervation ratio of motor units varies considerably, from approximately 1:4 in the muscles that move the eyes to approximately 1:2,000 in the large back extensor and leg extensor muscles. When an alpha motoneuron transmits an action potential, all of the fibers in the motor unit contract. Consequently, muscles associated with very fine motor control, such as the muscles that move the eyes, have motor units with low innervation ratios so that the amount of force produced can be finely controlled.

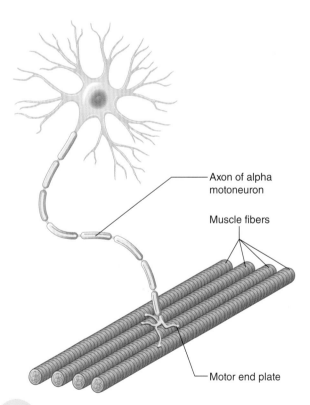

FIGURE 7.18 Composition of a motor unit: an alpha motoneuron with all the terminal branches of the axon and the muscle fibers that they innervate.

- Axon of alpha motoneuron
- Muscle fibers
- Motor end plate

Muscles associated with forceful movements are made up of motor units with relatively high innervation ratios. The fibers of a single motor unit are usually mixed with the fibers of other motor units but are grouped within a relatively small area of the muscle.

Key Term

motor unit   The functional unit of skeletal muscle consisting of a motor neuron with an Aα axon, together with all the terminal branches of the axon and the muscle fibers that they innervate.

KEY POINT

The innervation ratio is the number of muscle fibers in a motor unit. Muscles associated with fine motor control have motor units with low innervation ratios, whereas muscles associated with forceful movements are made up of motor units with relatively high innervation ratios.

A muscle fiber contracts when an action potential is generated in the sarcolemma at the junction with a motor end plate. The action potential is transmitted along and across the muscle fiber (by a system of tubules continuous with the sarcolemma), resulting in contraction. A single action potential produces a muscle twitch—a force that rapidly peaks and then equally rapidly dies away. If a series of action potentials is generated at a high enough frequency, the twitches fuse to produce tetanus, a sustained level of force.

Slow and Fast Twitch Muscle Fibers

Whereas the basic structure and function of all muscle fibers are the same, muscle fibers and motor units vary in relation to the following:

- Activation threshold—the level of stimulus required to generate an action potential
- Contraction time—the time from force onset to peak force
- Resistance to fatigue

Muscle fibers are classified into slow twitch fibers and fast twitch fibers on the basis of their contraction times. Slow twitch fibers, also referred to as Type I and red fibers, have contraction times of 100 to 120 ms. Fast twitch fibers, also referred to as Type II fibers and white fibers, have contraction times of 40 to 45 ms (Gregor 1993; Gamble 1988). The metabolism of slow twitch fibers is basically aerobic, and therefore they are resistant to fatigue. Fast twitch fibers are subdivided on the basis of their metabolic characteristics into Type IIa (aerobic and fatigue resistant) and Type IIb (anaerobic and fatigue sensitive). The muscle fibers in a particular motor unit have the same functional characteristics, and thus motor units can be classified into three categories:

- Slow contracting, fatigue resistant (S)
- Fast contracting, fatigue resistant (FR)
- Fast contracting, fatigable (FF)

Individuals differ in the proportions of the types of muscle fibers in their muscles. The average person has approximately 50% Type I, 25% Type IIa, and 25% Type IIb fibers in her calf muscles. In comparison, elite distance runners have a much higher proportion of Type I fibers, and elite sprinters have a much higher proportion of Type IIb fibers (Gamble 1988).

Whereas the classification of muscle fibers into Types I, IIa, and IIb is widely used, the categories represent ranges of metabolic and functional characteristics rather than discrete categories (Sargeant 1994). The metabolic and functional characteristics of muscle fibers appear to be influenced considerably by the type of innervation the fibers receive. It has been shown in experimental animals that altering the type of innervation changes the metabolic and functional characteristics of muscle fibers over time (Noth 1992; MacIntosh et al 2006). Table 7.2 lists the characteristics of slow and fast twitch motor units.

KEY POINT

Muscle fibers are classified into slow twitch and fast twitch fibers on the basis of their contraction times. The muscle fibers in a particular motor unit have similar functional characteristics, and so motor units are classified into three categories: slow contracting, fatigue resistant; fast contracting, fatigue resistant; and fast contracting, fatigable.

Table 7.2 Characteristics of Slow and Fast Twitch Motor Units

	Slow twitch, fatigue resistant (S)	Fast twitch, fatigue resistant (FR)	Fast twitch, fatigable (FF)
Activation threshold of muscle fibers	Low	Moderate	High
Contraction time of fibers (ms)	100-120	40-45	40-45
Innervation ratio of motor units	Low	Moderate	High
Type of muscle fibers	I	IIa	IIb
Type of axon	Aβ	Aα	Aα
Diameter of axon (μm)	7-14	12-20	12-20
Speed (m/s)	40-80	65-120	65-120
Duration and size of force	Prolonged low force	Prolonged relatively high force	Intermittent high force
Type of activities	Long-distance running and swimming	Kayaking and rowing	Sprinting, throwing, jumping, weightlifting

Based on Gamble 1988, Noth 1992, and Gregor 1993.

Kinesthetic Sense and Proprioception

The central nervous system constantly receives sensory information from a wide variety of sources concerning the various aspects of physiological functioning. Awareness of body position and body movement is provided by a range of sensory organs, in particular, those concerned with the sensations of effort and heaviness, timing of the movement of individual body parts, the position of the body in space, joint positions, and joint movements.

The input from these sources contributes to what is referred to as **kinesthetic sense** (or kinesthesia) (figure 7.19). Some aspects of kinesthetic sensitivity, such as a sense of effort and heaviness and a sense of timing of actions, are generated by sensory centers that monitor the motor commands sent to muscles. Other aspects of kinesthetic sensitivity are generated largely by input from peripheral receptors that monitor the execution of motor commands—the actual movements. For example, input from the eyes and ears is responsible for generating a sense of the position of the body in space. The sense of the position of a joint and movement of a joint is generated by a group of receptors located in the skin and musculoskeletal tissues (ligaments, capsules, tendons, muscles). These receptors are called proprioceptors, and the sensation they provide is called **proprioception. Proprioceptors** are sensory receptors that are sensitive to tension, compression, and shear load and thus are referred to as mechanoreceptors (Rossi-Durand 2006). Proprioceptors, like all sensory receptors, synapse with sensory neurons and in response to tension strain, compression strain,

or shear strain convey information to the central nervous system about muscle lengths, muscle forces, joint positions, and joint movements (speed and direction).

Key Terms

kinesthetic sense Awareness of body position and body movement.

proprioception The sense of the position of a joint and movement of a joint; proprioception is part of kinesthetic sense.

proprioceptor A sensory receptor that is sensitive to tension, compression, or shear load. In response to tension strain, compression strain, or shear strain, proprioceptors convey information to the central nervous system about muscle lengths, muscle forces, joint positions, and joint movements.

Types of Proprioceptors

The existence of proprioceptors in the skin, muscles, tendons, joint capsules, and ligaments is well established, but the precise roles of the proprioceptors and the interrelationships between them are less clear (Grigg 1994). However, it appears that proprioceptors in joint capsules and ligaments are largely responsible for generating a sense of joint position and joint

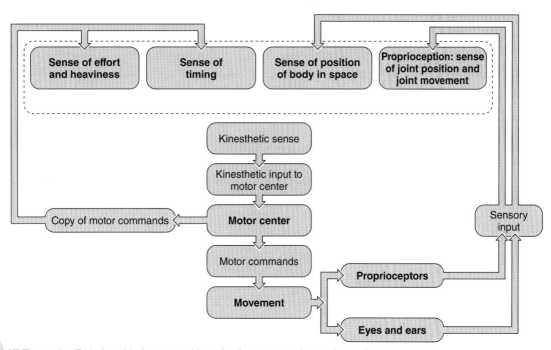

FIGURE 7.19 Relationship between kinesthetic sense and proprioception.

movement in the end ranges of joint movements—where joint capsules and ligaments may become taut. Between the end ranges it seems unlikely that the proprioceptors in joint capsules and ligaments provide much sensory information because they are unlikely to be under sufficient tension to excite them.

Figure 7.20 shows a model of proprioception. Sensory information from proprioceptors is continuously being sent to the sensory center for proprioception in the brain. The flow of information from the proprioceptors to the sensory center for proprioception increases considerably as a result of movement (Winter et al 2005; Windhorst 2007). The sensory center for proprioception integrates the proprioceptive information and sends the integrated information to the motor center in the brain. The motor center integrates the information from the sensory center for proprioception with the information from the other sensory centers (figure 7.19) to program subsequent movement.

Joint and Ligament Proprioceptors

There are two main types of proprioceptors in joint capsules: Ruffini end organs (or Ruffini corpuscles) and Pacinian corpuscles (figure 7.21). A Ruffini end organ contains a small number of main terminal branches with a profuse system of small branches on the end of each main branch. Each main branch and its smaller branches are enclosed within a connective tissue covering and located between the collagenous fibers of the capsules. Ruffini end organs appear to be mainly responsive to tension. A Pacinian corpuscle is a more complex structure than a Ruffini end organ. A Pacinian corpuscle consists of a single terminal branch surrounded by several concentric layers of Schwann cells, all enclosed in a connective tissue cover. Pacinian corpuscles, which appear to be responsive to compression, are widely distributed between the collagenous fibers of the joint capsule and surrounding fascia.

The proprioceptors in ligaments and skin are similar to those found in joint capsules. However, there are few proprioceptors in ligaments and skin compared with the number in joint capsules. In addition, because tension in ligaments and skin can be caused by movement in a number of directions, it is unlikely that the proprioceptors can provide information on movement in specific directions. For these reasons, it is thought that the contribution of the proprioceptors in ligaments and skin to proprioception is relatively small (Grigg 1994). However, the proprioceptors may make a significant contribution to joint stabilization in reflexive muscular activity.

Muscle and Tendon Proprioceptors

Whereas proprioceptors in joint capsules and, to a lesser extent, in ligaments appear to generate information on joint position and joint movement in the end ranges of joint movements, proprioceptors in muscles and tendons seem to be responsible for generating this type of information for movement between end ranges. Tendons contain proprioceptors called

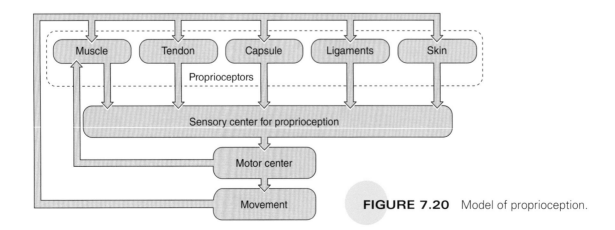

FIGURE 7.20 Model of proprioception.

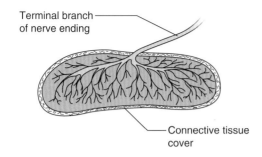

Terminal branch of nerve ending

Connective tissue cover

a

Terminal branch of nerve ending

Schwann cells

Connective tissue cover

b

FIGURE 7.21 *(a)* Ruffini corpuscle. *(b)* Pacinian corpuscle.

Golgi tendon organs that are responsive to tension. Golgi tendon organs are similar in structure to Ruffini end organs and are located between the collagenous fibers of the tendon. Muscles contain specialized proprioceptors called **muscle spindles,** which, like Golgi tendon organs, are responsive to tension.

KEY POINT

Awareness of body position and body movement is provided by a range of sense organs, in particular, those concerned with the sensations of effort and heaviness, timing of the movement of individual body parts, the position of the body in space, joint positions, and joint movements. The input from all of these sources provides what is referred to as kinesthetic sense.

Each muscle spindle consists of a number of tiny muscle fibers enclosed within a spindle-shaped connective tissue capsule (figure 7.22). The muscle spindle fibers are referred to as intrafusal (inside the spindle) to distinguish

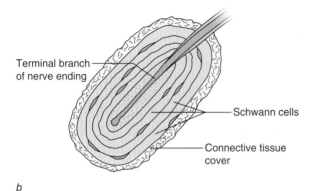

Extrafusal fibers

Muscle spindle made up of intrafusal fibers

a

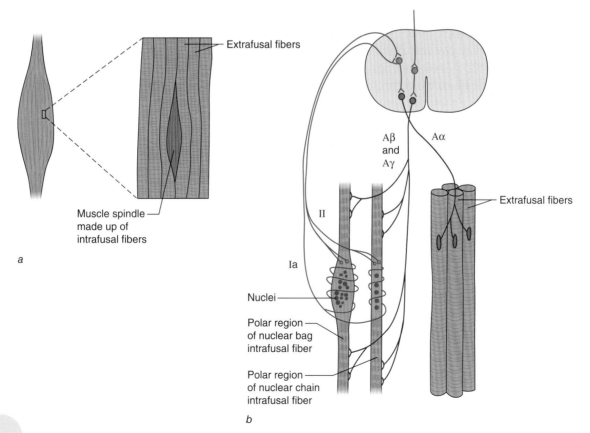

Aβ and Aγ

Aα

II

Extrafusal fibers

Ia

Nuclei

Polar region of nuclear bag intrafusal fiber

Polar region of nuclear chain intrafusal fiber

b

FIGURE 7.22 Structure of a muscle spindle. *(a)* Location of a muscle spindle. *(b)* Innervation of a muscle spindle.

them from the larger and more numerous extrafusal fibers that make up the vast majority of the muscle. Muscle spindles are embedded between extrafusal fibers. Intrafusal fibers are classified on the arrangement of their nuclei into nuclear bag fibers and nuclear chain fibers. In both types of fiber the nuclei are located in the central region of the fiber, which is devoid of myofibrils. In nuclear bag fibers the nuclei cluster in a group, whereas in nuclear chain fibers the nuclei are arranged in a line parallel to the long axis of the fiber. The regions of the fibers on each side of the central region contain a large number of myofibrils and are referred to as the polar regions (Enoka 1994).

Nuclear bag fibers and nuclear chain fibers are usually about 8 mm and 4 mm long, respectively, and a typical muscle spindle contains two bag fibers and four chain fibers (Roberts 1995). The central regions of both types of fiber are supplied with Type Ia and Type II sensory nerve endings. The endings of Type Ia nerve fibers spiral around the intrafusal fibers, and the endings of Type II nerve fibers consist of a number of branches with end bulbs; the spirals and end bulbs adhere closely to the sarcolemma of each fiber.

The polar regions of the intrafusal fibers are supplied with motor nerve endings from Aβ and Aγ motor nerve fibers, sometimes referred to as the fusimotor nerves. Stimulation of the intrafusal fibers via these nerves results in contraction of the polar regions, which stretches the central regions and excites the spiral and end bulb nerve endings. This results in sensory discharge via the Type Ia and II sensory nerve fibers. The Type Ia and II fibers synapse in the spinal cord directly with the Aα motor neurons that supply the extrafusal muscle fibers of the same muscle, resulting in contraction of the extrafusal fibers. The level of contraction of the extrafusal fibers depends on the level of activation, which in turn depends on the degree of tension in the intrafusal fibers. There is always a certain amount of tension in the intrafusal muscle fibers (activated by the sensory centers of the brain via the Aβ and Aγ fibers), which results in a certain amount of tension in the extrafusal muscle fibers. The resting level of tension in the intrafusal and extrafusal muscle fibers is called **muscle tone;** the level of muscle tone in the extrafusal fibers is determined by the level of muscle tone in the intrafusal fibers. Muscle tone can be regarded as the sensitivity of a muscle to a change in its length (Kandel et al 2000).

Sensory output from muscle spindles occurs when the central regions of the intrafusal fibers are stretched. For the purpose of setting muscle tone, the central regions are stretched by contraction of the polar regions via stimulation by the Aβ and Aγ motor nerve fibers. The central regions of the intrafusal fibers can also be stretched by stretching the muscle as a whole, because this will stretch the muscle spindles and excite the spiral and end bulb nerve endings. Low-velocity stretching of active musculotendinous units is an essential feature of normal movements; as a muscle group shortens, its antagonist partner lengthens. It is thought that the sensory information provided by muscle spindles and Golgi tendon organs as a result of stretching provides a sense of joint position and joint movement, especially during midrange movements (Gandevia et al 1992).

Whereas low-level stretch of a musculotendinous unit seems to be important in generating proprioceptive information concerning joint movement and joint position, rapid stretch results in reflex contraction of the musculotendinous unit (via the spindle afferent to muscle efferent loop) to prevent subluxation of the associated joints (Mynark and Koceja 2001). These **stretch reflexes** are important in protecting joints from injury. For example, recurrent inversion sprains of the ankle are associated with deficient stretch reflex of the everters and dorsiflexors (Garn and Newton 1988). Excitation of muscle spindles appears to be mainly responsible for initiating reflex muscle contractions. However, there is evidence that proprioceptors in joint capsules and ligaments also contribute to the initiation of such reflex contractions (Matthews 1988; Hall et al 1994; Melnyk et al 2007).

Key Terms

muscle spindle A tension receptor in muscle.

muscle tone The resting level of tension in muscle; the level of muscle tone in the extrafusal fibers is determined by the level of muscle tone in the intrafusal fibers.

stretch reflex The involuntary contraction of a muscle in response to a sudden unexpected increase in its length; the reflex is brought about by the spindle afferent to muscle efferent loop.

KEY POINT

Proprioceptors in joint capsules and ligaments (Ruffini end organs and Pacinian corpuscles) appear to generate information on joint position and joint movement in the end ranges of joint movements. Proprioceptors in muscles (muscle spindles) and tendons (Golgi tendon organs) appear to be responsible for generating this type of information for movement in between end ranges of joint movements.

Role of Proprioceptors

The precise roles of the various types of proprioceptors are not yet clear. However, there is general agreement that proprioceptive information aids coordination and balance and, in particular, maintains joint congruence (Grigg 1994, Wilkerson and Nitz 1994, Winter et al 2005, Windhorst 2007). It appears that injury to muscles, ligaments, and capsules can damage proprioceptors, resulting in long-term proprioceptive deficits, which contribute to the development of degenerative joint diseases such as osteoarthritis (Freeman and Wyke 1967, Garn and Newton 1988, Hall et al 1995, Melnyk et al 2007). There is evidence that proprioceptive information from muscle–tendon units can be enhanced by specific exercises that emphasize activation of muscle spindles and thereby improve muscle tone. Such enhancement may compensate for proprioceptive deficits in other structures such as joint capsules and ligaments (Beard et al 1994, Skinner et al 1986, Steiner et al 1986).

Mechanical Characteristics of Musculotendinous Units

The amount of force generated by a musculotendinous unit depends on the length of the unit at the time of stimulation and the speed with which the unit changes length in the ensuing contraction. In this regard, an isometric contraction is simply one point on the continuum between maximum velocity of shortening (concentric contraction) and maximum velocity of lengthening (eccentric contraction).

Length–Tension Relationship in a Sarcomere

The amount of tension generated in a sarcomere depends on the number of cross bridges attached between the actin and myosin filaments: the greater the number of cross bridges, the greater the force. The actual number of cross bridges depends on the following:

- The degree of interdigitation between the actin and myosin filaments. Too

much and too little interdigitation decreases the number of myosin heads that are in a position to attach to form cross bridges.

- The level of stimulation (activation) applied to the sarcomere. The higher the stimulation, the greater the number of cross bridges that attach and, therefore, the greater the force.

Figure 7.23 shows the isometric **length–tension relationship** for a sarcomere; the sarcomere was maximally stimulated, but not allowed to shorten, at various sarcomere lengths and the force was recorded at each length (Gordon et al 1966). When the sarcomere is extended to the point where there is no interdigitation between the actin and myosin filaments, no tension is generated because there are no myosin heads in a position to attach to form cross bridges. This situation is represented by the point L5 in figure 7.23. As

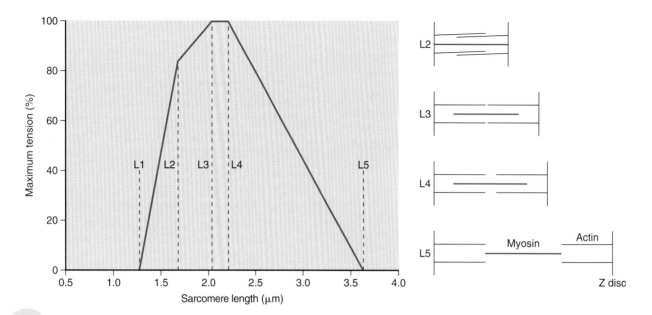

FIGURE 7.23 Isometric length–tension relationship in a sarcomere.

the sarcomere shortens and the degree of interdigitation gradually increases, there is a linear increase in the amount of force generated up to the point L4 where the length–tension relationship levels off. At L4 the maximum number of cross bridges have been formed, and tension is maximum.

The region of the length–tension relationship between L5 and L4 is referred to as the descending limb. At L4 the H zone corresponds to the inert zone of the myosin filaments (see figure 7.17c). As the sarcomere shortens between L4 and L3, the tension stays the same because no more cross bridges can be formed (the inert zones of the myosin filaments do not have myosin heads to form cross bridges). The region of the length–tension relationship between L4 and L3 is referred to as the plateau region. At L3 the ends of the actin filaments in each half of the sarcomere come together—the H zone is zero. As the sarcomere shortens between L3 and L2, the actin filaments progressively overlap each other, resulting in a progressive drop in tension because the overlapping actin filaments interfere with cross bridge formation in the region of overlap. The region of the length–tension relationship between L3 and L2 is referred to as the shallow ascending limb. At L2 the ends

of the myosin filaments abut the Z discs. As the sarcomere shortens between L2 and L1 there is a rapid and progressive drop in tension due to progressive overlap of the actin filaments and progressive longitudinal compression of the myosin filaments, both of which reduce the number of myosin heads available to form cross bridges. At L1 no cross bridges can be formed, and consequently no tension is generated. The region of the length–tension relationship between L2 and L1 is referred to as the steep ascending limb.

Length–Tension Relationship in a Musculotendinous Unit

The length–tension relationship in a musculotendinous unit is different from that in a sarcomere because of the musculotendinous unit's connective tissue, which exerts passive tension when stretched. Consequently, the tension produced by a musculotendinous unit will be the sum of the tension produced by the contractile (muscle) component and the passive tension exerted by the connective tissue components. Figure 7.24 shows the isometric length–tension relationship (or curve) for a musculotendinous unit. The contributions of the contractile and connective tissue components to total tension at any particular length are shown in the separate curves. Some of the connective tissue components are parallel with the muscle fibers and some are arranged in series; this has given rise to the terms *parallel elastic component* and *series elastic component* (Huijing 1992). The parallel elastic component consists of sarcolemma, endomysia, perimysia, and epimysium. The series elastic component consists of tendons and aponeuroses and strands of the protein titin, which connect the ends of the myosin filaments to the Z discs in each sarcomere. In addition to contributing to passive tension when stretched, the titin strands stabilize the hexagonal arrangement of the actin and myosin filaments (Lieber and Bodine-Fowler 1993).

In the absence of stimulation and any external load, the musculotendinous unit assumes a rest length at which the tension in the unit is zero, with no tension in the contractile component and no tension or stretch in the connective tissue component. Figure 7.24 shows that rest length corresponds to optimal overlap of actin and myosin filaments, that is, the length at which isometric tension is maximum.

As in a sarcomere, the shorter the musculotendinous unit, the lower the tension generated. Tension tends to reduce to zero when the unit shortens to approximately 60% of its rest length (see figure 7.24). However, this is unlikely to occur in practice given the arrangement of the musculotendinous units on the skeleton. When a musculotendinous unit contracts in a very shortened state and, as such, generates low force, it is said to be in a state of **active insufficiency** (Elftman 1966). This is more likely to occur with muscle–tendon units that cross more than one joint. For example, the hamstrings extend the hip and flex the knee. However, the muscle fibers in the hamstrings are not long enough to fully extend the hip and flex the knee simultaneously. If a person stands on one leg and attempts to fully extend the hip and fully flex the knee of the other leg at the same time, she will be unable to flex the knee much more than 90°; that is, the hamstrings will be in a state of active insufficiency. In contrast, if the person flexes her hip and then tries to fully flex her knee, she will find that the knee flexes to approximately 140°. This is possible because the hamstrings operate closer to rest length and are therefore able to exert more force (Elftman 1966).

Figure 7.24 shows that as a musculotendinous unit lengthens beyond rest length, the isometric tension generated is fairly constant between 100% and 150% of rest length and then increases to a maximum at approximately 175% of rest length. The change in isometric tension between rest length and maximum tension is associated with a gradual decrease in the amount of tension produced by the contractile component and a gradual increase in the amount of passive tension exerted by progressive stretch of the connective tissue components. Tension in the contractile component

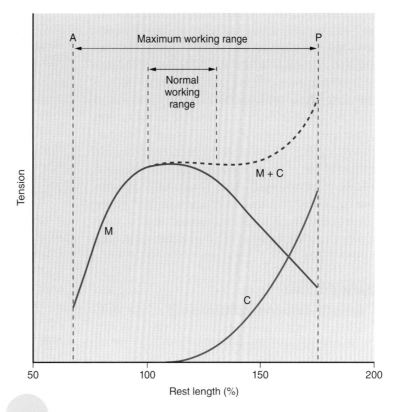

FIGURE 7.24 Isometric length–tension relationship in a musculotendinous unit. M = tension exerted by the contractile component; C = tension exerted by the connective tissue component; M+C = total tension; P = passive insufficiency; A = active insufficiency.

would be reduced to zero if the musculotendinous unit were lengthened to approximately 210% of rest length. However, the parallel elastic connective tissue components ensure that this situation does not arise by limiting the maximum length of the muscle–tendon unit to approximately 175% of rest length. In this situation the muscle–tendon unit is said to be in a state of **passive insufficiency** (Elftman 1966).

When a musculotendinous unit lengthens, all of the sarcomeres do not lengthen to the same extent. Theoretically a situation could occur where some of the sarcomeres in a muscle fiber were fully extended, that is, no interdigitation between the actin and myosin filaments, whereas other sarcomeres were not fully extended. If the muscle were stimulated to contract in this state, the fully stretched sarcomeres would not be able to contract, whereas the other sarcomeres would be able to contract. This would result in further stretching and damage to the already fully stretched sarcomeres. This theoretical condition has been referred to as muscle instability and can be prevented, at least in part, by passive insufficiency (Alexander 1989). Consequently, the maximum working length range of a musculotendinous unit is determined by the lengths at which active insufficiency and passive insufficiency occur, that is, between approximately 60% and 175% of rest length. However, it is likely that the normal working range is between approximately 100% and 130% of rest length. This range incorporates the region of the length–tension curve where contractile tension is maximum and allows maximum flexibility in tension generation (see figure 7.24).

Key Terms

length–tension relationship The relationship between length and tension in a musculotendinous unit.

active insufficiency The lower limit, minimum length, of the working range of a musculotendinous unit.

passive insufficiency The upper limit, maximum length, of the working range of a musculotendinous unit.

Force–Velocity Relationship in Muscle

The everyday physical tasks that people perform are usually well within the strength capability of the muscles used. In such movements, the muscles generate just enough tension to overcome the external load acting on them so they can move the external load. The external load may simply be the weight of a limb segment, such as the forearm in a movement involving elbow flexion. At other times the external load consists of the weight of the limb segments together with any additional load that is being moved, such as something held in the hand.

When the amount of force produced by a muscle just matches the external load, the muscle contracts isometrically. The maximum load the muscle can sustain isometrically is called the isometric strength of the muscle. When the external load is less than isometric strength, the muscle is able to contract concentrically. The speed of shortening in a concentric contraction depends on how much force the muscle needs to produce to move the external load. The greater the external load, the greater the force needs to be, and the greater the force (as a proportion of isometric strength), the slower the speed of shortening. A muscle can shorten at maximum speed when the external load on the unit is zero. When the external load on a muscle is greater than the isometric strength of the muscle, it is forced to lengthen, that is, contract eccentrically.

In an eccentric contraction a muscle resists the stretching load. The attached cross bridges are themselves stretched, adding to the overall tension such that the force produced by the muscle is greater than the isometric strength of the muscle. The force produced by a muscle during eccentric contraction depends on the speed of lengthening, which depends on the size of the external load. The greater the external load (in relation to the isometric strength of the muscle), the greater the speed of lengthening. The greater the speed of lengthening, the greater the effect of the stretch reflex, and therefore the greater the force produced by the muscle. When the external force exceeds the maximum strength of the muscle, the muscle, its tendons, or both will be injured. The relationship between speed of shortening or speed of lengthening and muscle force is referred to as the **force–velocity relationship** (figure 7.25). Figure 7.26 shows the effect of the force–velocity relationship on the length–tension relationship of a musculotendinous unit. The figure shows that at any particular length, the greater the speed of shortening (as in a concentric contraction), the lower the tension, and the greater the speed of lengthening (as in an eccentric contraction), the higher the tension.

Key Term

force–velocity relationship The relationship between speed of shortening or speed of lengthening and tension in a muscle.

> **KEY POINT**
>
> The amount of force generated by a muscle depends on the length of the muscle at the time of stimulation (length–tension relationship) and the speed with which it changes length in the ensuing contraction (force–velocity relationship).

Action and Contraction in Musculotendinous Units

The contractile (muscle) and noncontractile (connective tissue) components of a musculotendinous unit function as a single unit. During contraction of the muscle component, the length of the musculotendinous unit may increase, decrease, or stay the same. An increase in length could be due to an increase in both the muscle (eccentric contraction of the muscle) and connective tissue components (elastic stretching), but it also could be caused by an increase in the connective tissue component while the muscle component stays the same length. In this situation, the muscle *contracts* isometrically, but the musculotendinous unit *acts* eccentrically. This is called **eccentric action**. A musculotendinous unit acts eccentrically when it lengthens while its muscle component is contracting. A decrease in the length of a musculotendinous unit in which the muscle component is contracting could be caused by a decrease in both the muscle (concentric contraction of the muscle) and connective tissue components (recoil following

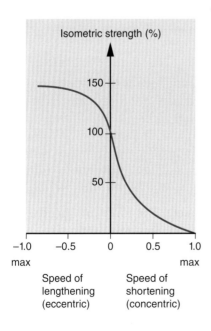

FIGURE 7.25 The force–velocity relationship in skeletal muscle. Positive speed indicates shortening (concentric), and negative speed indicates lengthening (eccentric).

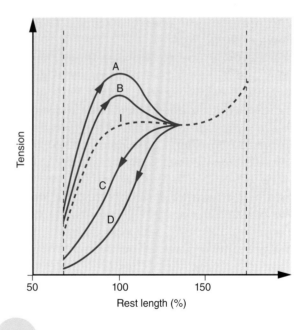

FIGURE 7.26 The effect of speed of shortening and speed of lengthening on the length–tension relationship in skeletal muscle. A and B show eccentric contractions: the speed of lengthening in A > B; C and D show concentric contractions: the speed of shortening in C < D; I shows the isometric length–tension curve.

elastic stretch), but it also could be caused by a decrease in the connective tissue component while the muscle component stays the same length. In this situation, the muscle *contracts* isometrically, but the musculotendinous unit *acts* concentrically. This is called **concentric action**. A musculotendinous unit acts concentrically when it shortens while its muscle component is contracting. The functional relationship between the contractile and noncontractile components of a musculotendinous unit is clearly very complex (Ishikawa et al 2005, Kawakami and Fukunaga 2006, Spanjaard et al 2006). It is thought that one of the main functions of the muscle component is to continuously adjust the stiffness of the tendons to maximize energy conservation, that is, minimize energy expenditure (see later section on the stretch-shorten cycle).

Key Terms

eccentric action   Lengthening of a musculotendinous unit when the muscle component is contracting.

concentric action   Shortening of a musculotendinous unit when the muscle component is contracting.

Muscle Architecture and Function

All muscles are made up of muscle fibers. However, the length and the orientation of the fibers (pennate or nonpennate) have a considerable effect on the function of the muscles. The fundamental relationships between muscle architecture and muscle function are that excursion (the distance that the muscle can shorten) and velocity of shortening are proportional to fiber length, and force is proportional to total physiological cross-sectional area of the muscle fibers (Lieber and Bodine-Fowler 1993).

All muscle fibers consist of similar sarcomeres, and the number of sarcomeres determines the length of a muscle fiber. Each sarcomere in a muscle fiber is capable of shortening to the same extent as all the other sarcomeres in the muscle fiber. Consequently, the excursion of a muscle fiber is the sum of the excursions of all the individual sarcomeres; the greater the number of sarcomeres, the longer the muscle fiber and the greater the possible excursion. Excursion and velocity of shortening are directly related because velocity of shortening is the rate of change of excursion—the rate of change in length of the muscle. The longer the muscle fiber (in terms of number of sarcomeres), the greater its excursion and velocity of shortening.

Theoretically, the ideal muscle (in terms of force and excursion capabilities) has a large cross-sectional area and very long fibers. However, such a muscle would be bulky and create considerable packing problems given its girth and areas of attachment to the skeletal system. Because there are no muscles with both of these characteristics, it is reasonable to assume that the architecture of the muscular system has evolved to provide the best compromise between structure and function. The muscles of the body represent a broad range of combinations of force and excursion capability (Lieber 1992), and it is not surprising that most movements of the body involve simultaneous activity in a number of muscles with each muscle performing a particular role.

KEY POINT

The fundamental relationships between muscle architecture and muscle function are that excursion (the distance that the muscle can shorten) and velocity of shortening are proportional to fiber length, and force is proportional to total cross-sectional area of the muscle fibers.

Roles of Muscles

With regard to control of joint movements, muscles perform various roles including stabilizer, agonist, prime mover, assistant mover, antagonist, synergist, and neutralizer. Each muscle that contributes to a particular movement can have more than one role, and the relative importance of each role may change during the movement.

Joint stabilization—the maintenance of joint congruence—is a major function of the muscular system. The extent to which a muscle contributes to joint stabilization depends largely on the line of pull of the muscle in relation to the center of the joint the muscle crosses. Generally, the closer the line of pull of the muscle to the center of the joint, the greater the stabilizing effect of the muscle (see chapter 9).

An agonist is a muscle that moves a body segment in the intended direction. For example, the deltoid and supraspinatus are both agonists in abduction of the shoulder joint, with the deltoid as prime mover and the supraspinatus as the assistant mover (figure 7.27, a-c). An antagonist is a muscle that acts in the direction opposite to that of an agonist. For example, in abduction of the shoulder joint, the latissimus dorsi and pectoralis major are antagonists (figure 7.27, d and e).

A synergist assists the action of the prime mover. For example, in abduction of the shoulder, the trapezius and serratus anterior rotate and abduct the scapula (scapulothoracic gliding mechanism, chapter 6) so that complete abduction of the arm (180° with respect to the anatomical position) can be achieved (figure 7.27, f and g).

A neutralizer prevents unwanted action of a muscle. For example, the tendency would be for the flexors of the fingers to simultaneously flex the wrist as they flex the fingers (figure 7.28). However, if the finger flexors performed both movements at the same time, the finger flexors would experience active insufficiency. Consequently, the wrist extensors neutralize the action by preventing wrist flexion. The length of the muscle fibers of the finger flexors is approximately 100 mm. To flex the wrist and hold a tight position, the finger flexors need to shorten by about 27 mm. To flex the fingers into a tight fist, the finger flexors need to shorten by about 37 mm. The finger flexors can perform both movements separately because 27 mm to 37 mm is well within the normal working range of the muscles. However, the muscles cannot perform both movements simultaneously because to do so they would need to shorten by about 64 mm (27 + 37 mm), which would produce active insufficiency (Alexander 1992). The finger flexors can only produce a tight fist when the wrist is held straight so that the tendency of the finger flexors to flex the wrist is neutralized. This is achieved by the wrist extensors (figure 7.28).

Muscle Fiber Arrangement and Force and Excursion

Figure 7.29, a and c, shows two musculotendinous units with the same rest length and the same muscle mass (same volume of myofibrils). One muscle is a nonpennate parallel-fibered muscle and the other is a unipennate muscle. The physiological cross-sectional area of each muscle is the cross-sectional area of all of the muscle fibers perpendicular to their line of pull. Assuming the length of the muscle fibers in the pennate muscle is only half that of the fibers in the nonpennate muscle, it follows that the physiological cross-sectional area of the pennate muscle is double that of the nonpennate muscle (because the muscle mass is the same in both muscles). Consequently, the pennate muscle can exert double the force of the nonpennate muscle.

However, because the fibers in the pennate muscle are oblique to the line of pull of the musculotendinous unit, not all of the force is in the line of pull of the unit. In a pennate muscle the angle of the fibers with respect to the line of pull of the musculotendinous unit is usually 30° or less (Alexander 1968), such that 90% or more of the force exerted by the muscle is directed in the line of pull of the unit. Thus, pennate muscles are normally capable of exerting far more force in the line of pull of the musculotendinous unit than nonpennate muscles of the same muscle mass.

KEY POINT

The muscles of the body represent a broad range of combinations of force and excursion capability, and most body movements involve simultaneous activity in a number of muscles with each muscle performing a particular role including stabilizer, agonist, prime mover, assistant mover, antagonist, synergist, and neutralizer.

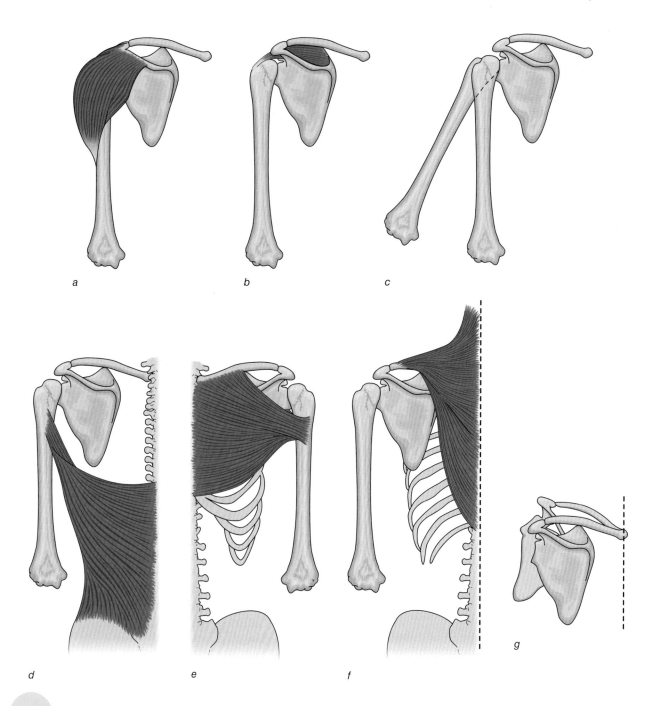

FIGURE 7.27 Roles of muscles in abduction of the arm. *(a)* Posterior aspect of the left shoulder showing the deltoid muscle (agonist and prime mover: shoulder abduction). *(b)* Posterior aspect of the left shoulder showing the supraspinatus muscle (agonist and assistant mover: shoulder abduction). *(c)* Action of the deltoid and supraspinatus (shoulder abduction). *(d)* Posterior aspect of the left shoulder and trunk showing the latissimus dorsi muscle (antagonist). *(e)* Anterior aspect of the left shoulder showing the pectoralis major muscle (antagonist). *(f)* Posterior aspect of the left shoulder showing the trapezius muscle (synergist: rotation of the clavicle and scapula). *(g)* Posterior aspect of the left shoulder showing the action of the trapezius muscle (rotation of the clavicle and scapula).

In contrast to their capacity to generate force, the excursion range of nonpennate musculotendinous units is usually much greater than that of pennate musculotendinous units of similar muscle mass. The increased excursion range of nonpennate muscles is the result of (1) increased fiber length and (2) line of pull directly in line with the line of pull of the mus-

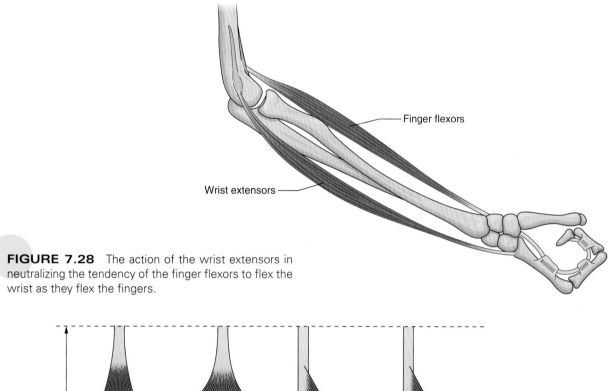

FIGURE 7.28 The action of the wrist extensors in neutralizing the tendency of the finger flexors to flex the wrist as they flex the fingers.

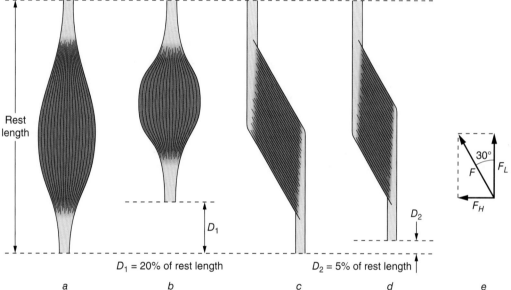

FIGURE 7.29 Effects of muscle structure on force and excursion. *(a)* Nonpennate muscle at rest length. *(b)* Nonpennate muscle in *(a)* with fibers shortened to 70% of rest length. *(c)* Pennate muscle at rest length. *(d)* Pennate muscle in *(c)* with fibers shortened to 70% of rest length. *(e)* Components of the muscle force F in the line of pull of the musculotendinous unit, F_L, and perpendicular to the line of pull F_H.

culotendinous unit. In figure 7.29, *b* and *d*, the muscles are shown with their fibers shortened to 70% of their rest length. The corresponding shortening of the musculotendinous units is 20% and 5%, respectively, in the nonpennate and pennate muscles.

Biarticular Muscles

Many of the musculotendinous units of the body, especially those in the upper and lower limbs, pass over more than one joint (see figure 7.28). These muscles are usually referred to as **biarticular muscles,** because muscles that pass over more than two joints function in the same way as muscles that pass over just two joints (Lieber 1992). Biarticular muscles are too short to fully flex or fully extend simultaneously all of the joints over which they cross. For example, the hamstrings are two-joint muscles that can extend the hip and flex the knee, but these actions (as explained earlier in the chapter) cannot be performed maximally at the same time. Indeed, hip extension is usually associated with knee extension, and hip flexion is usually associated with knee flexion, as in walking and running. In this way, the length of the muscle stays within the normal working range of approximately 100% to 130% of rest length. The main advantages of biarticular muscles are that tension is produced in one muscle rather than two (or more), which conserves energy, and working within the 100% to 130% range allows maximum flexibility in tension generation (Lieber 1992, Van Ingen Schenau et al 1994).

Case study 5 discusses resistance training in older adults, especially the development of muscular power so as to perform activities of daily living.

Key Term

biarticular muscle   A muscle that passes over more than one joint.

CASE STUDY 5   RESISTANCE TRAINING IN OLDER ADULTS

Henwood TR, Riek S, Taaffe DR. 2008. Strength versus muscle power-specific resistance training in community-dwelling older adults. *Journal of Gerontology* 63A(1):83-91.

Muscle strength decreases with age, which contributes to a decrease in the ability to perform activities of daily living and consequently to loss of independence and quality of life. Resistance training can increase muscle strength throughout life and can considerably reduce the loss of muscle strength that normally occurs with aging. The most effective way to increase muscle strength is via high-resistance exercise, that is, loads 75% or greater of 1RM (1RM = 1 repetition maximum, which is the maximum load for a given exercise, such as a bench press or biceps curl, that a person can move through a full range of motion with correct technique for one repetition). However, exercises (repetitions) involving relatively high loads inevitably involve relatively low-velocity movements. Consequently, high-resistance exercise tends to significantly increase strength but not speed of movement.

Many activities of daily living, such as rising from a chair or climbing stairs, are characterized by fairly rapid, explosive muscle contractions. Not surprisingly, many interventions that include high-resistance training in older people have resulted in increases in strength but little or no change in the ability to perform activities of daily living. Whereas muscle strength reflects the ability of a muscle to produce force, muscle power, the product of force and velocity, reflects the ability to produce force quickly. The purpose of the present study was to compare the effect of high-resistance, low-velocity training and moderate-resistance, high-velocity training on muscle function and physical performance in older adults.

Sixty-seven healthy, independent, community-dwelling (living in their own homes) older adults aged 64 to 84 years were randomized to one of three groups: a high-velocity group (HV; 11 men, 12 women), a strength training group (ST; 10 men, 12 women), and a control (no training) group (CO; 10 men, 12 women). Following 2 weeks of conditioning, participants trained twice weekly for an additional 22 weeks using 6 exercises (chest press, supported row, biceps curl, leg press, prone leg curl, leg extension). In each training session the HV group performed 3 sets of 8 repetitions, the first set at 45% 1RM, the second set at 60% 1RM, and the third set at 75% 1RM. The participants were instructed to perform the concentric phase of each repetition as rapidly as possible and then return through the eccentric phase at a slow and controlled pace. In each training session the ST group performed 3 sets of 8 repetitions at 75% 1RM at 3 s per repetition. Loads in the first week were based on 1RM measures determined during the conditioning period. To maintain high-level loading, 1RM loads were reassessed when participants could complete 10 or 11 repetitions in the final set.

Muscle function (dynamic strength, isometric strength, movement velocity, muscle power, muscle endurance) and functional performance (floor rise to standing; stair climb; usual, fast, and backward 6 m walk; repeated chair rise to standing five times; 400 m walk; functional reach) were assessed at baseline (after conditioning) and after training. Muscle strength increased significantly and similarly in the HV and ST groups compared with the control group (HV 51% ± 9.0%; ST 48.3% ± 6.8%; CO 1.2% ± 5.1%). Peak muscle power also increased significantly and similarly with training (HV 50.5% ± 4.1%; ST 33.8% ± 3.8%; CO –2.5% ± 3.9%). Training also significantly and similarly improved three of the functional performance measures (repeated chair rise to standing, stair climb, fast 6 m walk) with no significant changes in the other measures and no changes in any of the measures in the control group.

The authors concluded that because the HV training program resulted in similar increases in muscle strength, muscle power, and functional performance as the ST training program, but with less total work performed per training session (because of lower loads during the first two sets), HV training may be less strenuous and more acceptable than ST training to older people as a method of improving muscle function.

Application

The results of the study indicate that power training (high-velocity movements using moderate to high resistance, 45-75% 1RM) was as effective as strength training (relatively low velocity movements using high resistance, 75% 1RM) in increasing peak power, strength, and some components of activities of daily living (ADLs). Because ADLs depend more on power than strength and given that power declines more rapidly with age than does strength in 65- to 90-year-olds (Skelton et al 1995), it is recommended that older adults engage in resistance training 2 to 3 times per week using one set of 8 to 10 exercises (to cover all of the main muscle groups), 8 to 12 repetitions per exercise, with resistance of approximately 60% 1RM to emphasize speed of movement during the concentric phases of the repetitions (Pollock et al 1998, Cress et al 2004).

Stretch-Shorten Cycle

When a concentric contraction occurs without prestretch, the initial phase of contraction takes up the slack in the series elastic connective tissue (SEC) components; it is only when the slack in the SEC components has been taken up, that is, when the SEC components have become taut, that the force produced by the muscle is transmitted to the skeleton. The need for the muscle to expend energy in taking up the slack in the SEC components before force can be transmitted to the skeleton is not efficient. It is not surprising, therefore, that in most if not all preferred whole-body movements, many of the muscles involved in controlling the various joints are initially stretched before being allowed to shorten. For example, jumping movements generally begin with a downward movement followed by upward movement (figure 7.30). Similarly, in throwing actions, the arm is usually swung backward before being accelerated forward (see figure 1.16).

In such actions the initial movement in the opposite direction to that of the final movement is usually referred to as a countermovement. A countermovement involves two phases. In the first phase, the body (as in a vertical jump) or body segments (such as the arm in a throwing action) develop speed of movement in the opposite direction to that of the final movement. Before the final movement can be initiated, the movement of the body or body segments in the opposite direction must be arrested. Consequently, in the second phase of a countermovement the muscles contract to arrest the movement of the body or body segments; in doing so the muscles are forcibly stretched and contract eccentrically. The eccentric phase is usually immediately followed by a concentric phase

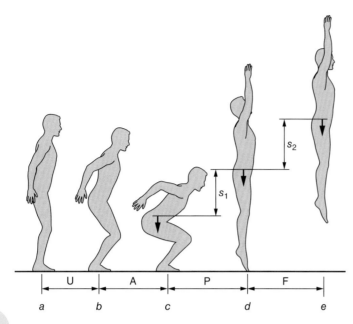

U   unweighting phase

A   absorption phase (or weighting phase)

P   propulsion phase

F   flight phase from takeoff to the top of the flight

s_1 vertical displacement of center of gravity during propulsion phase

s_2 vertical displacement of center of gravity from takeoff to top of flight

FIGURE 7.30   Sequence of movement in a countermovement vertical jump *(a-e)* and a squat jump from a stationary position *(c-e)*. *(a)* Standing upright at rest. *(b)* Instant in countermovement when downward velocity of the body is at its maximum. *(c)* Instant at which the vertical velocity of the body is zero (maximum displacement downward). *(d)* Takeoff. *(e)* Top of flight.

to produce the final movement. The pattern of eccentric contraction followed without a pause by concentric contraction is referred to as the **stretch-shorten cycle** (Komi 2003). Most whole-body actions, including relatively slow walking, use the stretch-shorten cycle. However, use of the stretch-shorten cycle is more obvious in movements involving a backswing. For example, in a golf swing, the backswing is arrested by eccentric contraction of the same muscles that then contract concentrically to produce the downswing. The eccentric phase of the backswing is usually accentuated by starting to turn the trunk to face the flag before the backswing has been completed.

Key Term

stretch-shorten cycle   A sequence of muscle activity involving an eccentric contraction followed immediately by a concentric contraction.

Storage and Use of Elastic Strain Energy

A force does mechanical work when it moves its point of application in the direction of the force; that is, work done is the product of the magnitude of the force and the distance moved by the point of application of the force. Consequently, when a muscle contracts isometrically, it expends energy (metabolic energy expenditure by the muscle cells to maintain the contraction) but it does no work because the length of the musculotendinous unit does not change. When a muscle contracts concentrically, it expends energy in creating tension and does work by pulling its skeletal attachments closer together. The amount of work done by a muscle in a concentric contraction is the product of the muscle force and the distance over which the musculotendinous unit shortens. When a muscle contracts eccentrically it expends energy in creating tension, but in contrast to a concentric contraction, work is done on the musculotendinous unit; it is lengthened by the external load. The amount of work done on the musculotendinous unit by the external load is the product of the muscle force and the distance over which the musculotendinous unit is lengthened. When work is done on a musculotendinous unit—when energy is expended in stretching a musculotendinous unit—the energy is absorbed by the musculotendinous unit in the form of strain energy (like the energy stored in a drawn bow, which is subsequently used to propel the arrow). The extent to which the strain energy can be used to move the body, rather than being dissipated as heat within the muscle, will depend on the speed of the changeover from eccentric to concentric activity within the stretch-shorten cycle.

For any given task, movements using the stretch-shorten cycle may be more effective (in terms of performance) and more efficient (in terms of use of energy) than movements relying on concentric contractions (Anderson and Pandy 1993, Komi 2003). For example, most people are able to jump higher using a countermovement jump (CMJ) (figure 7.30, a-e) than a squat jump (figure 7.30, c-e). In performing a CMJ a person starts from a standing position and then performs a downward countermovement immediately followed by a maximum effort upward jump. The leg extensor muscles contract eccentrically to arrest the countermovement and then contract concentrically to drive the body upward into the jump. Performing a squat jump requires a maximum effort jump from a stationary squat position. The leg extensor muscles contract isometrically in the stationary squat position and then contract concentrically to drive the body upward into the jump. Assuming the same level of squat is achieved in both the CMJ and the squat jump, the difference in performance—height jumped—will be largely due to differences in the storage and use of elastic strain energy in the connective tissues of the leg extensor muscles and the amount of force produced by the muscle components of the leg extensor muscles.

KEY POINT

For any given task, movements using the stretch-shorten cycle may be more effective (in terms of performance) and more efficient (in terms of energy use) than movements that rely on concentric contractions.

During the eccentric phase of the countermovement in a CMJ, the connective tissue components (series elastic and parallel elastic) are forcibly stretched, and therefore elastic strain energy builds up in them like the energy stored in a stretched spring. The amount of elastic strain energy absorbed by the connective tissues is determined by the stretching force and the amount of stretch (tension strain). However, the magnitude of the stretching force largely determines the amount of stretch; the higher the force, the greater the stretch, and the greater the amount of elastic strain energy absorbed. The magnitude of the stretching force largely depends on the velocity of stretch—the rate at which stretching occurs.

The higher the velocity of stretch, the greater the amount of elastic strain energy absorbed. However, a high-velocity stretch of the connective tissues can only be achieved if the force produced by the muscle components is high; a low to moderate force results in lengthening of the muscle components with little or no stretch on the connective tissues. With regard to amount of stretch, the muscle components can only produce a high force within the central region of the length–tension range; consequently, the amount of stretch should be low to moderate. In general, the amount of elastic strain energy absorbed in the connective tissues of the leg extensor muscles during the countermovement of a CMJ will be maximized by a low to moderate amount of stretch combined with a high velocity of stretch.

In the squat position of a squat jump, the leg extensor muscles will contract isometrically and their connective tissue components will be in a stretched position. Consequently, the connective tissues will have a certain amount of elastic strain energy (dependent on the amount of static stretch), but it is likely to be less than that stored during the countermovement of a CMJ (dependent on the amount of dynamic stretch). In both the squat jump and CMJ (and any other movement involving absorption of elastic strain energy by connective tissue components), the elastic strain energy can contribute to movement in the form of force (additional to that produced by the muscle components) exerted during the recoil of the connective tissue components from their stretched position.

It should be possible to absorb more elastic strain energy in the dynamic countermovement phase of a CMJ than in the static squat phase of a squat jump. However, the ability to use the additional strain energy depends on the delay between the eccentric and concentric phases of the movement. Some of the additional strain energy is inevitably converted to heat and therefore is not available to contribute to movement. The amount of elastic strain energy converted to heat depends on the speed of the changeover from eccentric to concentric contraction. Generally, the faster the changeover, the smaller the proportion of strain energy converted to heat and consequently the greater the proportion available to contribute to movement (Gregor 1993, Komi 2003).

KEY POINT

The amount of elastic strain energy absorbed in the connective tissues of muscles during a countermovement will be maximized by a low to moderate amount of stretch combined with a high velocity of stretch.

Sometimes the elastic strain energy that is converted to heat is referred to as lost energy. However, this is misleading because energy can never be completely lost; it can only be converted from one form of energy to another. For example, the muscles convert energy in the form of chemical substances (such as adenosine triphosphate) into mechanical energy in the form of movement (kinetic energy), which itself may be converted into another form of mechanical energy (strain energy). With regard to elastic strain energy in musculotendinous units, the proportion of this energy converted to heat is only lost in the sense that it is not available to contribute to movement.

The concentric phase of the squat jump is preceded by isometric contraction of the leg extensor muscles, whereas the concentric phase of the CMJ is preceded by eccentric contraction. The elicitation of the stretch reflex during the eccentric phase of the CMJ is likely to enhance performance over the squat jump in two ways:

1. The high force during the eccentric phase (greater than isometric force) is likely to increase the possibility of a high velocity of stretching of the connective tissues and consequently increase the amount of elastic strain energy absorbed in the connective tissues compared with the squat jump.

2. The force generated during the concentric phase of the CMJ is likely to be greater than that generated in the concentric phase of the squat jump.

Stretch Load and Jump Performance in Drop Jumping

A person's ability to use the stretch-shorten cycle in jumping activities depends on her ability to tolerate stretch loading, that is, the force exerted in the musculotendinous units undergoing eccentric action. As stretch load and velocity of stretch increase, performance increases up to a particular optimum velocity of stretch and then decreases with further increases in velocity of stretch. For example, it has been shown that in drop jumping—when subjects are asked to drop onto the floor from various heights and then immediately perform a maximum vertical jump—performance (jump height) increases as drop height increases up to approximately 50 to 60 cm and then decreases as drop height increases further (Komi and Bosco 1978). As drop height increases, stretch load (reflected in the magnitude of the ground reaction force) during the eccentric phase of ground contact tends to increase. However, each person has a drop height D that corresponds to a maximum tolerable stretch load (a maximum stretch load at which maximal effort in the drop jump can be maintained). This drop height may be associated with maximal performance resulting from the beneficial effects of the stretch-shorten cycle. At drop heights greater than D, the person is unable to tolerate the stretch load associated with maximal effort and may reduce the stretch load by increasing the amount of hip, knee, and ankle flexion during the eccentric phase of landing. This increases the magnitude of stretch of the leg extensors. The increased magnitude of stretch reduces the velocity of stretch of the leg extensors and thus reduces the potential benefits of increased muscle force and elastic strain energy, resulting in decreased performance. Not surprisingly, jump training based on drop jumping has been shown to increase performance and the size of the stretch load that can be tolerated (Komi 2003).

Using the Stretch-Shorten Cycle in Movements Without a Countermovement

In movements such as the CMJ, it is only possible to use the stretch-shorten cycle by incorporating a countermovement. In some whole-body movements it may be possible to use the stretch-shorten cycle to accelerate the relevant body parts in the intended direction during the eccentric phase as well as during the concentric phase. In such actions there is no countermovement, but there is a premovement that results in eccentric contraction of the relevant muscles. For example, in the final stage of throwing a javelin, the javelin is pulled by the throwing arm from a position behind the trunk to a release point in front of the trunk (figure 7.31). If, at the start of this pulling action, the (right-handed) thrower can thrust his right shoulder forward fast enough (figure 7.31, c and d), the inertia of the arm and javelin results in eccentric contraction of the shoulder flexors and abductors and elbow flexors, which are (in association with muscles in the legs and trunk) responsible for the pulling action. The arm and javelin are accelerated forward though the shoulder and elbow muscles contract eccentrically. The

eccentric phase of the elbow flexors is relatively short, and the elbow flexes and then extends prior to release of the javelin. In contrast, the eccentric phase of the shoulder flexors and abductors may last for most of the pulling action with only a short concentric phase just prior to release (see figure 7.31, e and f). Ideally, the thrower should thrust his right shoulder forward for as long as possible during the pulling action, because this will maximize the force exerted on the javelin by prolonging the eccentric (high-force) phases of the shoulder and elbow muscles and increasing the force exerted in the concentric phases of the muscles by decreasing their speed of shortening.

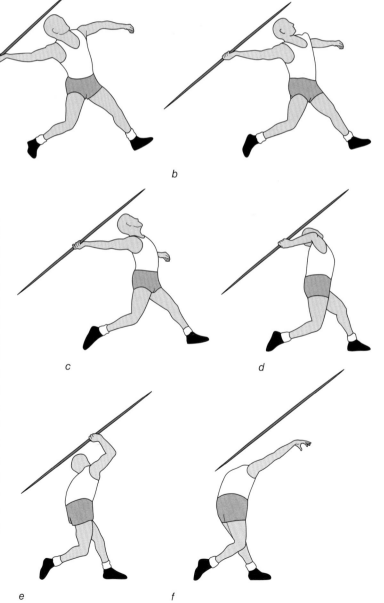

a

b

c

d

e

f

FIGURE 7.31 The delivery phase in throwing a javelin.

Summary

Muscles vary considerably in the type, arrangement, and innervation ratio of muscle fibers, and these variations are reflected in the force and excursion capabilities of the individual muscles. Most body movements involve simultaneous activity in a number of muscles, with each muscle performing a particular role such that the functional characteristics of the muscles complement each other and maximize the effectiveness and efficiency of the movement. The neuromuscular system also has a role in kinesthetic sense. This chapter concludes part I. Part II describes the effects of loading on the structure and function of musculoskeletal components.

Review Questions

1. Differentiate between the cerebrospinal nervous system and the central nervous system.
2. Differentiate between resting potential and action potential.
3. Describe the general organization of nerve tissue in the spinal cord and spinal nerves.
4. Describe the sequence of events within the nervous system in a typical voluntary movement.
5. Describe the various categories of pennate and nonpennate muscles.
6. Describe the possible causes and effects of compartment pressure syndrome.
7. Describe the sliding filament theory of muscular contraction.
8. Differentiate between kinesthetic sense and proprioception.
9. Differentiate between active and passive insufficiency in skeletal muscle.
10. Describe the fundamental relationships between muscle architecture and muscle function.
11. Describe the roles of prime mover, synergist, and neutralizer in skeletal muscle.
12. Describe the stretch-shorten cycle of muscular contraction in relation to a countermovement jump.
13. Explain how the stretch-shorten cycle can be used in a movement that does not have a countermovement.

PART II
Musculoskeletal Response and Adaptation to Loading

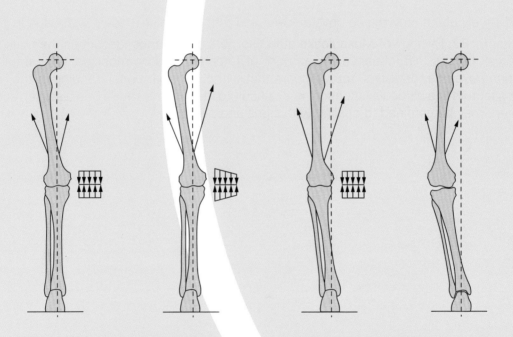

Your genes determine your general pattern of growth and development, but environmental influences, in particular your nutrition and physical activity, largely determine the timing, rate, extent, and type of growth and development that occur so that your body constantly adapts its structure to maximize function—this process is called structural adaptation. Your musculoskeletal system constantly adapts to changes in loading that result from changes in your level of physical activity. However, the capacity of your musculoskeletal system to adapt to changes in loading decreases with age. For example, a hip abductor–adductor imbalance prior to maturity will result in adaptation of the femur and tibia to maintain normal knee joint function. After maturity, the same imbalance is likely to result in osteoarthritis of the knee joint. See page 334 for a complete description of this figure.

Part II describes the immediate and long-term effects of loading on the external form and internal architecture of the musculoskeletal system.

- Chapter 8, Elementary Biomechanics, develops knowledge and understanding of elementary biomechanical concepts and principles that will enable you to understand how forces in muscles and joints vary in relation to body position and body movement and, in particular, how a change in function, such as an increase in the intensity of training, results in changes in structure.

- Chapter 9, Forces in Muscles and Joints, describes the effect of the open-chain arrangement of the skeleton on the forces exerted in muscles and joints. You will learn that the leverage of muscles is usually low, such that muscle forces and joint reaction forces are high relative to the weights of the body segments that the muscles control.

- Chapter 10, Mechanical Characteristics of Musculoskeletal Components, describes how the muscles, bones, and joints respond to the forces exerted on the body, in particular the force of body weight and impact forces that arise, for example, from contact with the ground when running. You will learn that the musculoskeletal system tends to minimize the risk of injury through shock absorption and minimize energy expenditure through energy conservation.

- Chapter 11, Structural Adaptation of the Musculoskeletal System, describes how the muscles, bones, and joints adapt their structure over time to withstand the forces exerted on them. You will learn that insufficient loading, arising from insufficient physical activity, is likely to adversely affect growth and development and that excessive loading will result in injury.

- Chapter 12, Etiology of Musculoskeletal Disorders and Injuries, describes the main groups of risk factors that influence the amount of force exerted on muscles, bones, and joints during everyday activities. You will learn that without a clear understanding of the relevant risk factors in a particular situation, it is difficult to identify the cause of a disorder, and treatment may be confined to dealing with signs and symptoms.

Elementary Biomechanics

To understand the relationship between structure and function of the musculoskeletal system and, in particular, the way that a change in function (due to a change in loading) results in a change in structure (structural adaptation), it is necessary to understand elementary biomechanics, that is, the relationship between external loading (in particular, body weight and ground reaction force) and internal loading (muscle forces and joint reaction forces). The purpose of this chapter is to develop knowledge and understanding of elementary biomechanical concepts and principles.

8

Mechanics and Biomechanics

Forces tend to affect bodies in two ways (Watkins 2007):

1. Forces tend to deform bodies, that is, change the shape of the bodies by stretching, squashing, bending, or twisting. For example, squeezing a tube of toothpaste changes the shape of the tube.

2. Forces determine the movement of bodies; that is, the forces acting on a body determine whether it moves or remains at rest and determine its speed and direction of movement if it does move.

Mechanics is the study of the forces that act on bodies and the effects of the forces on the size, shape, structure, and movement of the bodies. The effect that a force or a combination of forces has on a body, that is, the amount of deformation and change of movement that occurs, depends on the size of the force in relation to the mass of the body and the mechanical properties of the body.

The **mass of a body** is the product of its volume and its density. The **volume of a body** is the amount of space that the mass occupies, and its **density** is the concentration of matter (atoms and molecules) in the mass, that is, the amount of mass per unit volume. The greater the concentration of mass, the greater the density. For example, the density of iron is greater than that of wood and the density of wood is greater than that of polystyrene. Similarly, with regard to the structure of the human body, bone is more dense than muscle and muscle is more dense than fat.

The mass of a body is a measure of its **inertia,** which is its reluctance to start moving if it is at rest and its reluctance to change its speed or direction if it is already moving. The larger the mass, the greater the inertia and, consequently, the larger the force that will be needed to move the mass or change the way it is moving. For example, the inertia of a stationary soccer ball (a small mass) is small in comparison to that of a heavy barbell (a large mass), so much more force will be required to move the barbell than to move the ball.

Whereas the effect of a force on the movement of a body is largely determined by its mass, the amount of deformation that occurs is largely determined by its mechanical properties, in particular, its stiffness (the resistance of the body to deformation) and strength (the amount of force required to break the body). For a given amount of force, the higher the stiffness and the greater the strength of a body, the smaller the deformation that will occur.

Biomechanics is the study of the forces that act on and within living organisms and the effect of the forces on the size, shape, structure, and movement of the organisms (Watkins 2007). In relation to humans, biomechanics is the study of the relationship between the external forces (body weight and the forces that arise from physical contact with the environment) and internal forces (active forces generated by muscles and passive forces exerted on connective tissues) that act on the body and the effect of these forces on the size, shape, structure, and movement of the body.

Key Terms

mechanics   The study of the forces that act on bodies and the effects of the forces on the size, shape, structure, and movement of the bodies.

mass of a body   The amount of matter (atoms and molecules) that comprises a body (mass = volume × density).

volume of a body   The amount of space that a body occupies.

density of a body   The amount of mass per unit volume (the concentration of matter).

inertia   The reluctance of an object to start moving if it is at rest and its reluctance to change its speed and/or direction if it is already moving.

biomechanics   The study of the forces that act on and within living organisms and the effect of the forces on the size, shape, structure, and movement of the organisms.

Center of Gravity

The human body consists of a number of segments linked by joints. Each segment contributes to the body's total weight (figure 8.1*a*). Movement of the body segments relative to each other alters the weight distribution of the body. However, in any particular body posture the body behaves (in terms of the effect of body weight on the movement of the body) as if the total weight of the body is concentrated at a single point called the **center of gravity** (also referred to as center of mass) (figure 8.1*b*). Body weight acts vertically downward from the center of gravity along a line called the line of action of body weight. The concept of center of gravity applies to all bodies, animate and inanimate.

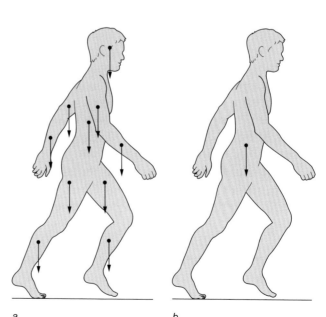

FIGURE 8.1   *(a)* Centers of gravity of body segments and lines of action of the weights of the segments. *(b)* Center of gravity of the whole body and line of action of body weight.

a   b

The center of gravity of an object is the point at which the whole weight of the object can be considered to act.

The position of an object's center of gravity depends on the distribution of the weight of the object. For a regular-shaped object with uniform density such as a cube, oblong, or sphere, the center of gravity is located at the object's geometric center. However, if the object has an irregular shape or nonuniform density, like the human body, the position of the center of gravity will reflect the mass distribution and it may be inside or outside the body. For example, when a person is standing upright, the center of gravity is located inside the body at about the level of the navel (figure 8.2a). However, flexing the trunk from the upright position in figure 8.2a will move the center of gravity forward and downward relative to the hips such that it will be located outside the body when the trunk is fully flexed (figure 8.2b). Movements that involve continuous change in the orientation of the body segments to each other, such as walking and running, result in continuous change in the position of the center of gravity.

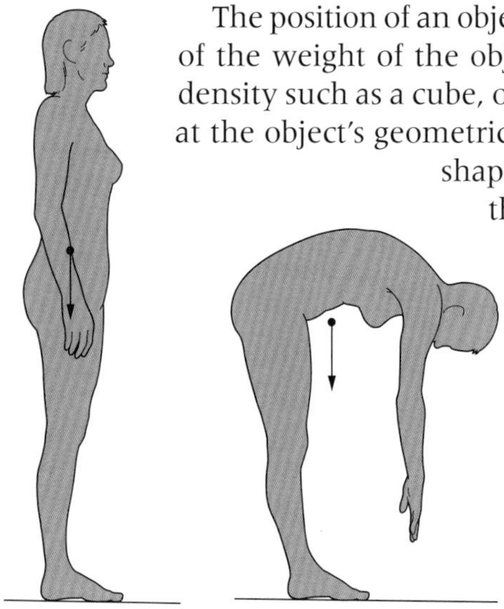

FIGURE 8.2 Position of the whole-body center of gravity when (a) standing upright and (b) bending forward.

Stability

Figure 8.3 shows a cube-shaped block of wood resting on a horizontal surface. The center of gravity of the block of wood is located at its geometric center, and the line of action of the weight of the wood intersects the base of support ABCD on which it is resting (figure 8.3, a and b). If the block of wood is tilted over on any of the edges of the base of support, AB, BC, CD, or AD, it will return to its original position provided that at release, the line of action of its weight intersects the plane of the original base of support ABCD. This situation is shown in figure 8.3c with respect to the edge BC. However, if at release the line of action of the block's weight does not intersect the original base of support, the block of wood will fall onto one of its other faces, as shown in figure 8.3, d and e. With respect to a particular base of support, an object is stable when the line of action of its weight intersects the plane of the base of support and unstable when it does not. Consequently, the block of wood in figure 8.3 is stable with respect to the base of support ABCD in the positions shown in parts b and c and unstable with respect to the base of support ABCD in the position shown in part d.

With respect to a particular base of support, an object is stable when the line of action of its weight intersects the plane of the base of support and unstable when it does not.

With regard to human movement, the terms *stability* and *balance* are often used synonymously. Maintaining **stability** of the human body is a fairly complex, albeit largely unconscious process (Roberts 1995). When one is standing upright, the line of action of body weight intersects the base of support formed by the area beneath and between the feet (figure 8.4, *a* and *b*). The size of the base of support can be increased by moving the feet farther apart. For example, moving one foot to the side increases side-to-side stability (figure 8.4*c*) and moving one foot in front of the other increases anteroposterior stability (figure 8.4*d*).

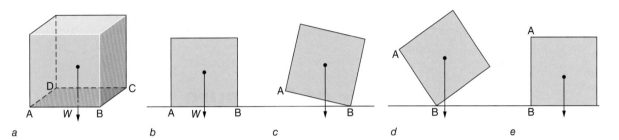

FIGURE 8.3 The line of action of the weight of a cube in relation to its base of support.

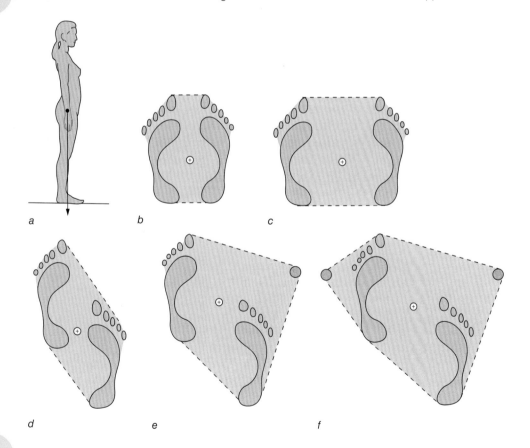

FIGURE 8.4 The line of action of body weight in relation to the base of support. *(a, b)* Standing upright. *(c)* Standing upright with feet apart, side-by-side. *(d)* Standing upright with the left foot in front of the right foot. *(e)* Standing with the aid of a walking stick in the right hand. *(f)* Standing with the aid of crutches or two walking sticks. The symbol ⊕ denotes the point of intersection of the line of action of body weight with the base of support. The dotted lines indicate the limits of the base of support in each position.

In general, the lower the center of gravity and the larger the area of the base of support, the greater the stability. The recumbent position (lying down) is the most stable position of the human body because it is the position in which the area of the base of support is greatest and the height of the center of gravity lowest. The degree of muscular effort needed to maintain stability tends to decrease as the area of the base of support increases. For example, it is usually easier, in terms of muscular effort, to maintain stability when standing on both feet than when standing on one foot. Similarly, it is usually less tiring to sit than to stand and less tiring to lie down than to sit. A person recovering from a leg injury may use crutches or a walking stick or cane to relieve the load on the injured limb. The use of crutches or a walking stick also increases the area of the base of support and makes it easier for the user to maintain stability (figure 8.4, *e* and *f*).

Center of Pressure

The ground reaction force is distributed across the whole of the area of contact between the feet and the floor. Figure 8.5*a* shows the contact area when one is standing barefoot on both feet; the contact area is much smaller than the area of the base of support. Figure 8.5*b* shows the contact area when a person is standing on one foot, and figure 8.5*c* shows the contact area when the person is standing on one foot with the heel raised off the ground. In figure 8.5, *b* and *c*, the contact area is very similar to the base of support. Whereas the ground reaction force is distributed across the whole of the contact area, the effect is as if the ground reaction force acts at a single point, which is referred to as the **center of pressure** (just as the whole weight of the body appears to act at the whole-body center of gravity in terms of the effect of body weight on the movement of the body). If we could stand upright perfectly still, the center of pressure would coincide with the point of intersection of the line of action of body weight with the base of support. However, it is very difficult to stand perfectly still; this is reflected in a certain amount of sway (forward–backward and sideways movement of the center of gravity). The balance mechanisms are constantly active to control sway and maintain upright posture.

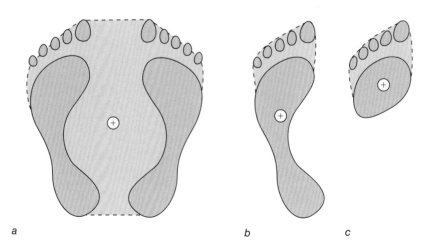

a   b   c

FIGURE 8.5 Center of pressure (⊕) when one is *(a)* standing upright on both feet, *(b)* standing on the left foot, and *(c)* standing on the left foot with the heel off the floor. The dotted lines and associated areas of the feet indicate the base of support in each position.

Key Terms

center of gravity The point at which the whole weight of an object can be considered to act.

stability The ability of an object to return to a balanced position with the same base of support after being disturbed.

center of pressure The point at which the ground reaction force can be considered to act.

Vector and Scalar Quantities

All quantities in the physical and life sciences can be categorized as either scalar or vector quantities. Quantities that can be completely specified by their magnitude (size) are called **scalar quantities.** These include volume, area, time, temperature, mass, distance, and speed. Quantities that require specification in both magnitude and direction are called **vector quantities.** These include displacement (distance in a given direction), velocity (speed in a given direction), acceleration (rate of change of velocity), and force. In a vector diagram, each vector is represented by an arrow; the length of the arrow, with respect to an appropriate scale, corresponds to the magnitude of the vector, and the orientation of the arrow, with respect to an appropriate reference axis (usually horizontal or vertical), indicates the direction.

> **KEY POINT**
>
> All quantities in the physical and life sciences can be categorized as either scalar or vector quantities. Quantities that can be completely specified by their magnitude (size) are called scalar quantities. Quantities that require specification in both magnitude and direction are called vector quantities.

Force Vectors and Resultant Force

When one is standing upright, three forces act on the human body: body weight W acting at the center of gravity of the body, the ground reaction force F_L on the left foot, and the ground reaction force F_R on the right foot (figure 8.6a). Figure 8.6b shows the corresponding vector diagram; the resultant ground reaction force (resultant of F_L and F_R) is equal in magnitude but opposite in direction to W, that is, the resultant force acting on the body is zero. In figure 8.6c, the force F_1 is the resultant ground reaction force (resultant of F_L and F_R), and figure 8.6d shows the corresponding vector diagram. To start walking or running (or move horizontally by any other type of movement such as jumping or hopping), the body must push or pull against something to provide the necessary resultant force to move it in the required direction. In walking and running, forward movement is achieved by pushing obliquely downward and backward against the ground. Provided that the foot does not slip, the leg thrust results in a ground reaction force directed obliquely upward and forward; this is F_2 in figure 8.6e. The resultant force R of W and F_2 moves the body forward while it maintains an upright posture; that is, the center of gravity of the body is accelerated in the direction of R. The corresponding vector diagram is shown in figure 8.6f. The vector diagram in figure 8.6f illustrates the **vector chain** method of determining the resultant of a number of vectors; the component vectors (W and F_2) are linked together in a chain (in any order) and the resultant vector runs from the

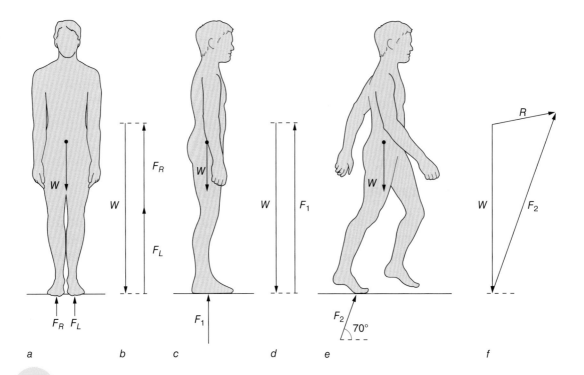

FIGURE 8.6 *(a, b)* The forces acting on the body when one is standing upright (front view) and the corresponding vector diagram. *(c, d)* The forces acting on the body when one is standing upright (side view) and the corresponding vector diagram. *(e, f)* The forces acting on the body just after the person starts to walk (side view) and the corresponding vector diagram. F_R = force on the right foot; F_L = force on the left foot; W = 70 kgf, F_1 = 70 kgf; F_2 = 80 kgf at 70° to the horizontal; R = 28 kgf at 11° to the horizontal.

start of the first component vector to the end of the last component vector. In figure 8.6, *b* and *d,* all of the forces are vertical forces. Consequently, for clarity, the upward and downward force vectors are presented in parallel (rather than overlapping) in the vector diagrams.

Key Terms

scalar quantity   A quantity that can be completely specified by its magnitude.

vector quantity   A quantity that must be specified in both magnitude and direction.

vector chain   The method of determining the resultant of a number of component vectors by joining the component vectors together in a chain.

Trigonometry of a Right-Angled Triangle

Whereas the vector chain method of determining the resultant of a number of component vectors is very useful, it is often more practical to use trigonometry, especially when there are a large number of vectors. Trigonometry is the branch of mathematics that deals with the relationships between the lengths of the sides and the sizes of the angles in a triangle. Figure 8.7 shows a right-angled triangle in which one angle (between sides *a* and *b)* is 90°. The angles between sides *a* and *c* and sides *b* and *c* are α and θ, respectively.

In a right-angled triangle, the two angles less than 90° (α and θ in figure 8.7) can be specified by the ratio between the lengths of any two sides of the triangle. The three most common ratios are sine, cosine, and tangent, and they are defined in relation to the particular angle under consideration. For example, in relation to angle θ in figure

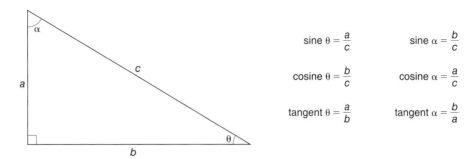

FIGURE 8.7   Definition of sine, cosine, and tangent in a right-angled triangle.

8.7, side a is referred to as the opposite side, side b is referred to as the adjacent side, and side c is the hypotenuse, the side of the triangle opposite the right angle.

The sine of θ is defined as the ratio of the opposite side to the hypotenuse:

$$\text{sine } \theta = \frac{\text{opposite side}}{\text{hypotenuse}} = \frac{a}{c}$$

The cosine of θ is defined as the ratio of the adjacent side to the hypotenuse:

$$\text{cosine } \theta = \frac{\text{adjacent side}}{\text{hypotenuse}} = \frac{b}{c}$$

The tangent of θ is defined as the ratio of the opposite side to the adjacent side:

$$\text{tangent } \theta = \frac{\text{opposite side}}{\text{adjacent side}} = \frac{a}{b}$$

Most electronic calculators provide a range of trigonometric ratios including sine (sin), cosine (cos), and tangent (tan). Alternatively, tables of sine, cosine, and tangent (for angles between 0° and 90°) can be obtained online (e.g., NASA 2008) or in publications such as that by Castle (1969). The lengths of sides and sizes of angles in right-angled triangles can be calculated using sine, cosine, and tangent functions provided that two sides or one side and one other angle are known. With reference to figure 8.7, for example, if c = 10 cm and θ = 30°, the lengths of sides a and b and the size of angle α can be determined as follows.

1. Calculation of the length of side a:

$$\frac{a}{c} = \sin \theta$$

$$a = c \cdot \sin \theta \tag{8.1}$$

In equation 8.1 and other equations throughout the book in which variables are denoted by letters or symbols, a center dot ($\cdot$) between variables indicates multiplication. When the variables being multiplied together are replaced by numbers and units, multiplication is denoted by a multiplication symbol ($\times$) as in equation 8.2.

$$a = c \cdot \sin 30°$$

From sine tables, sin 30° = 0.5 (i.e., the ratio of the length of side a to the length of side c is 0.5). Because c = 10 cm and sin 30° = 0.5, it follows that

$$a = 10 \text{ cm} \times 0.5 \tag{8.2}$$

$$a = 5 \text{ cm}$$

251

2. Calculation of the length of side *b:*

$$\frac{b}{c} = \cos\theta$$

$$b = c \cdot \cos\theta$$

$$b = c \cdot \cos 30°$$

From cosine tables, $\cos 30° = 0.866$ (i.e., the ratio of the length of side *b* to the length of side *c* is 0.866). Because $c = 10$ cm and $\cos 30° = 0.866$, it follows that

$$b = 10 \text{ cm} \times 0.866$$

$$b = 8.66 \text{ cm}$$

3. Calculation of angle α:

   Angle α can be determined a number of ways:

   - The sum of the three angles in any triangle (with or without a right angle) is 180°. Because the sum of θ and the right angle is 120°, it follows that $\alpha = 180° - 120° = 60°$.

   - The lengths of all three sides of the triangle are known: $a = 5$ cm, $b = 8.66$ cm, and $c = 10$ cm. Consequently, α can be determined by calculating the sine, cosine, or tangent of the angle.

In each case, the final step of converting the number into an angle can be determined from sine, cosine, and tangent tables.

$$\sin\alpha = \frac{b}{c} = \frac{8.66 \text{ cm}}{10 \text{ cm}} = 0.866. \text{ Therefore } \alpha = 60°.$$

$$\cos\alpha = \frac{a}{c} = \frac{5 \text{ cm}}{10 \text{ cm}} = 0.5. \text{ Therefore } \alpha = 60°.$$

$$\tan\alpha = \frac{b}{a} = \frac{8.66 \text{ cm}}{5 \text{ cm}} = 1.732. \text{ Therefore } \alpha = 60°.$$

Pythagoras' Theorem

Pythagoras, a Greek mathematician (572-497 BC), showed that in a right-angled triangle the square of the hypotenuse is equal to the sum of the squares of the other two sides. Therefore, with respect to figure 8.7,

$$c^2 = a^2 + b^2$$

$$c = \sqrt{(a^2 + b^2)}$$

This can be demonstrated with the data from the preceding example, where $a = 5$ cm, $b = 8.66$ cm, and $c = 10$ cm: $a^2 = 25$, $b^2 = 75$, and $c^2 = 100$.

Resolution of a Vector Into Component Vectors

Just as the resultant of any number of component vectors can be determined, any single vector can be replaced by any number of component vectors that have the same effect as the single vector. The process of replacing a vector by two or more component vectors is referred to as the **resolution of a vector.** When analyzing human movement, we frequently resolve displacement, velocity, and force vectors into vertical and horizontal components using trigonometry. The example of the man in figure 8.6*f*, replicated in figure 8.8, is used next to illustrate how the resolution of forces by trigonometry is used to determine the resultant force acting on the man. There are three steps in the process.

1. *Resolve all of the forces into their vertical and horizontal components.*

   Figure 8.8*a* shows the forces acting on the man, body weight *W* and the force *F* on the right foot. *W* = 70 kgf and *F* = 80 kgf at 70° to the horizontal. In figure 8.8*b*, *F* has been replaced by its vertical F_V and horizontal F_H components. From figure 8.8*b*,

$$F_V / F = \sin 70°$$
$$F_V = F \cdot \sin 70° = 80 \text{ kgf} \times 0.9397 = 75.17 \text{ kgf}$$
$$F_H / F = \cos 70°$$
$$F_H = F \cdot \cos 70° = 80 \text{ kgf} \times 0.3420 = 27.36 \text{ kgf}$$

2. *Calculate the vertical component R_V and horizontal component R_H of the resultant force R acting on the man.*

   If we use the convention that positive forces act upward and to the right and negative forces act downward and to the left, it follows from figure 8.8*b* that

$$R_V = F_V - W$$
$$R_V = 75.17 \text{ kgf} - 70 \text{ kgf} = 5.17 \text{ kgf}$$
$$R_H = F_H = 27.36 \text{ kgf}$$

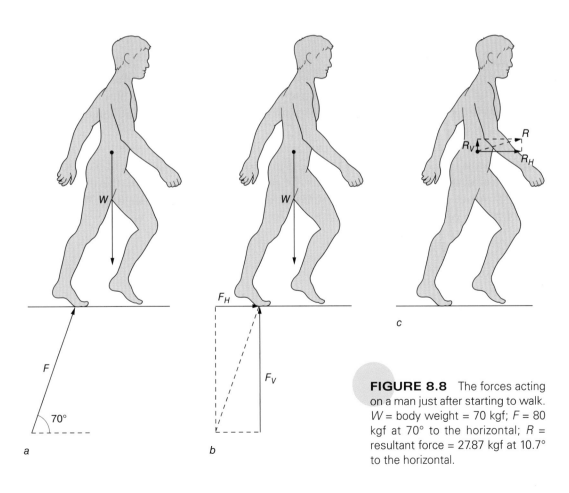

FIGURE 8.8 The forces acting on a man just after starting to walk. *W* = body weight = 70 kgf; *F* = 80 kgf at 70° to the horizontal; *R* = resultant force = 27.87 kgf at 10.7° to the horizontal.

3. *Calculate R and the angle θ of R with respect to the horizontal.*

   R, R_V, and R_H are shown in figure 8.8c. From Pythagoras' theorem,

   $$R^2 = R_V^2 + R_H^2$$
   $$R^2 = 26.73 + 748.57 = 775.30$$
   $$R = 27.84 \text{ kgf}$$

   $$\tan\theta = \frac{R_V}{R_H} = 0.1889$$

   $$\theta = 10.7°$$

As expected, R and $θ$ are the same (allowing for error in drawing the vector chain) as in the vector chain solution in figure 8.6f.

Key Term

resolution of a vector   The process of replacing a vector by two or more component vectors.

KEY POINT

Any single vector can be replaced by any number of component vectors that have the same effect as the single vector. The process of replacing a vector by two or more component vectors is referred to as the resolution of a vector.

Moment of a Force

Consider a rectangular block of wood resting on a table, as shown in figure 8.9a. The center of gravity of the block of wood is located at its geometric center, and the line of action of its weight intersects the base of support ABCD. If the block is tilted on one of the edges of the base of support, such as the edge BC, as in figure 8.9b, the weight of the block W will tend to rotate the block about the supporting edge back to its original resting position in figure 8.9a. The tendency to restore the block to its original position is the result of the moment (or turning moment) of W about the axis of rotation BC. The magnitude of the moment of W about the axis BC is equal to the product of W and the perpendicular distance d between the axis BC and the line of action of W (figure 8.9b):

moment of W about axis AB = $W \cdot d$ (*W* multiplied by *d*).

If W = 2 kgf and d = 0.1 m, then

moment of W about axis AB = 2 kgf × 0.1 m = 0.2 kgf · m.

In general, when a force F acting on an object rotates or tends to rotate the object about some specified axis, the moment of F is defined as the product of F and the perpendicular distance d between the axis of rotation and the line of action of F, that is, moment of F = $F \cdot d$. The axis of rotation is often referred to as the **fulcrum,** and the perpendicular distance between the line of action of the force and the axis of rotation is usually referred to as the **moment arm** of the force. The **moment of a force** is sometimes referred to as torque. For a given moment of force, the greater the force, the smaller the moment arm of the force and vice versa. For example, when one is trying to push open a heavy door, much less force will be required if the force is applied to the side of the door farthest away from the hinges (i.e., a large moment arm) than if the force is applied to the door close to the hinges (i.e., a small moment arm) (figure 8.10).

KEY POINT

For a given moment of force, the greater the force, the smaller the moment arm of the force and vice versa.

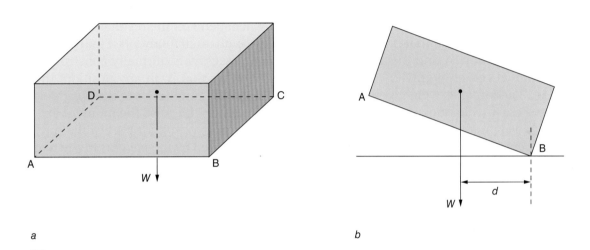

a

b

FIGURE 8.9 The turning moment of a force. *(a)* Block of wood at rest on base of support ABCD. *(b)* Turning moment $W \cdot d$ of the weight of the block W tending to restore to the block to its original resting position after being tilted on edge BC. d = moment arm of W about edge BC.

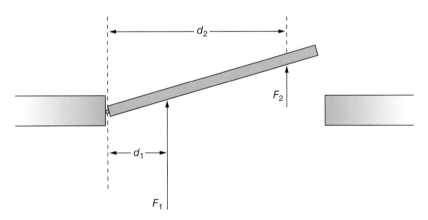

FIGURE 8.10 Effect of length of moment arm on magnitude of force needed to produce a particular moment of force to open a door. $F_1 \cdot d_1 = F_2 \cdot d_2$. If $d_2 = 3d_1$, then $F_1 = 3F_2$.

fulcrum   Axis of rotation.

moment arm   The perpendicular distance between the line of action of a force and the fulcrum.

moment of a force   The product of a force and the perpendicular distance between the line of action of the force and the axis of rotation. The moment of a force, also referred to as the turning moment of a force and, in engineering textbooks, as torque, is a measure of the tendency of the force to rotate an object about a particular axis of rotation.

Clockwise and Anticlockwise Moments

When an object is acted upon by two or more forces that tend to rotate the object, the actual amount and speed of rotation that occur will depend on the resultant moment acting on the object, that is, the resultant of all the individual moments. Figure 8.11a shows two boys A and B sitting on a seesaw (S). A seesaw is normally constructed so that its center of gravity coincides with the fulcrum; that is, in any position, the line of action of its weight will pass through the fulcrum and therefore the weight of the seesaw will not exert a turning moment on the seesaw because the moment arm of its weight about the fulcrum will be zero. Consequently, in figure 8.11a the only moments tending to rotate the seesaw will be those exerted by the weights of the two boys. The weight of A will exert an anticlockwise moment $W_A \cdot d_A$ and the weight of B will exert a clockwise moment $W_B \cdot d_B$. When $W_A \cdot d_A$ is greater than $W_B \cdot d_B$, there will be a resultant anticlockwise moment acting on the seesaw such that B will be lifted as A descends. When $W_A \cdot d_A$ is equal to $W_B \cdot d_B$, that is, when the clockwise moment is equal to the anticlockwise moment, the resultant moment acting on the seesaw will be zero and the seesaw will not rotate in either direction but will remain perfectly still in a balanced position. Consequently, if the weight of one of the boys is known, the weight of the other boy can be found by balancing the seesaw with one boy on each side of the fulcrum with both boys off the floor and then equating the clockwise and anticlockwise moments. For example, if W_A = 40 kgf and in the balanced position d_A = 1.5 m and d_B = 2.0 m, then by equating moments about the fulcrum,

$$\text{anticlockwise moments (ACM)} = \text{clockwise moments (CM)}$$
$$40 \text{ kgf} \times 1.5 \text{ m} = W_B \times 2.0 \text{ m}$$
$$W_B = \frac{40 \text{ kgf} \times 1.5 \text{ m}}{2.0 \text{ m}} = 30 \text{ kgf}$$

KEY POINT

When an object is acted upon by two or more forces that tend to rotate the object, the amount and speed of rotation that occurs will depend on the resultant moment.

Equilibrium

In the situation illustrated in figure 8.11a, the moments of the weights of the two boys acting about the fulcrum of the seesaw are equal and opposite such that the resultant moment acting on the seesaw is zero and the seesaw is at rest. As the seesaw is at rest, the resultant force on the seesaw is also zero. Consequently, the resultant downward force on the seesaw, that is, the weights of the seesaw and the two boys, must be counteracted by one or more forces whose resultant is equal and opposite to the weights of the seesaw and the two boys. In this case the counteracting force is a single force R exerted by the seesaw support through the fulcrum. Figure 8.11b shows a **free-body diagram** of the seesaw (a diagram showing all of the external forces acting on the seesaw) in the situation illustrated in figure 8.11a. As $W_A = 40$ kgf, $W_B = 30$ kgf, and $W_S = 20$ kgf, then $R = W_A + W_B + W_S = 90$ kgf.

When the resultant force and resultant moment acting on an object are both zero, the object is in **equilibrium.** These two equilibrium conditions are the basis for calculating forces acting on various parts of a multisegment system, such as muscle forces and joint reaction forces in the human body (see chapter 9).

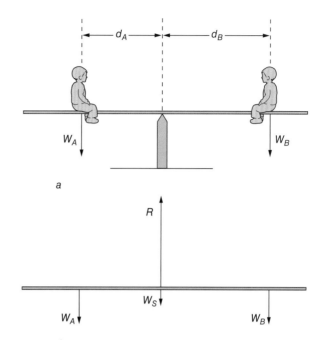

FIGURE 8.11 (a) Two boys sitting on a seesaw. (b) Free-body diagram of the seesaw. W_A = weight of boy A; W_B = weight of boy B; d_A = moment arm of W_A; d_B = moment arm of W_B; W_S = weight of seesaw; R = force exerted by the seesaw support on the seesaw = $W_A + W_B + W_S$.

Key Terms

free-body diagram   A diagram showing all of the external forces acting on an object.

equilibrium   When the resultant force and resultant moment acting on an object are both zero.

Levers

Whereas weight forces are always vertical, other external forces acting on a body are likely to be oblique. For example, consider using a claw hammer to pull out a nail from a piece of wood, as shown in figure 8.12a. In this example, the line of contact between the claw of the hammer and the surface of the wood constitutes the fulcrum. E is the force exerted on the handle of the hammer, and R is the resistance of the nail. Figure 8.12b shows a force-moment arm diagram; the forces E and R and their moment arms d_E and d_R are shown in relation to the fulcrum to more clearly illustrate the turning effects of the forces. The nail will be pulled out if the anticlockwise moment exerted by E is greater than the clockwise moment exerted by R, that is, if $E \cdot d_E > R \cdot d_R$.

In this situation the hammer is being used as a **lever,** a rigid or quasi-rigid object that can be made to rotate about a fulcrum to exert a force on another object. As in the example of the hammer, a lever encounters a resistance force R in response to an effort force E. The simplest form of lever, which is actually the simplest form of machine (i.e., a powered mechanism designed to apply force; Dempster 1965), is exemplified by a crowbar as shown in figure 8.13. In this case, the power is supplied by the person using the crowbar. The ratio of d_E to d_R is referred to as the leverage of the lever; the greater the leverage, the smaller the effort force E required to overcome the moment of the resistance force R.

The bones of the skeleton are essentially levers, and each joint constitutes a fulcrum. The muscles pull on the bones to control the movement of the joints. The resistance to movement exerted by a body segment is in the form of the segment's weight and any other external loads attached to the segment. The leverage of most of the muscles is very low (<1.0) such that muscle forces and the associated joint reaction forces tend to be much larger that the weights of the body segments that they control (see chapter 9).

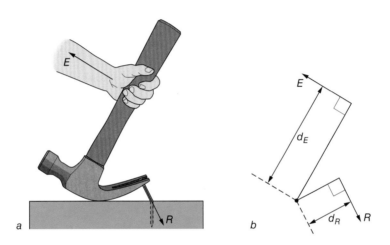

FIGURE 8.12 *(a)* Pulling a nail out of a piece of wood using a claw hammer. *(b)* The corresponding force-moment arm diagram. E = force exerted on the handle of the hammer; d_E = moment arm of E; R = resistance force exerted by the nail; d_R = moment arm of R.

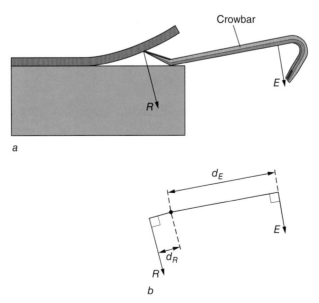

FIGURE 8.13 *(a)* Use of a crowbar to open a box. *(b)* The corresponding force-moment arm diagram. E = force exerted on the crowbar by the person using it; d_E = moment arm of E; R = resistance force exerted on the crowbar by the lid of the box; d_R = moment arm of R.

Key Term

lever A rigid or quasi-rigid object that can be made to rotate about a fulcrum to exert a force on another object.

KEY POINT

The bones of the skeleton act as levers and each joint constitutes a fulcrum.

Summary

The movements of inanimate objects and living organisms are governed by the same mechanical principles. Knowledge of elementary biomechanical concepts and principles is necessary to understand the relationship between the external forces that act on the body and the internal forces that the body exerts to counteract them. In particular, this chapter has described the concepts of stability (line of action of body weight, base of support, center of pressure), vectors (resultant force), determination of resultant force by graphical (vector chain) and numerical (trigonometry) methods, the turning effect of a force (turning moment, moment arm, lever, free-body diagram, reaction force), and equilibrium. The next chapter describes the effect of the open-chain arrangement of the skeleton on forces exerted in muscles and joints.

Review Questions

1. Differentiate between center of gravity and center of pressure.
2. Differentiate between scalar and vector quantities.
3. Define the trigonometric ratios sine, cosine, and tangent with respect to a right-angled triangle.
4. If the angle θ in figure 8.7 is 40° and the length of side c is 20 cm, calculate the length of sides a and b given that cos 40° = 0.766 and sin 40° = 0.643.
5. Differentiate stability and equilibrium.

Forces in Muscles and Joints

The musculoskeletal system exerts internal forces (muscle forces that result in joint reaction forces) to counteract external forces acting on the body. The leverage of most muscles is low such that muscle forces and joint reaction forces are high relative to the external loads. However, muscles tend to work together, which spreads the load on the muscles and reduces bending stress on bones. Using examples of the forces acting on the head (about the atlantooccipital joint), pelvis (about the hip joint), lumbar region (about the intervertebral joints), and forearm (about the elbow joint), this chapter examines the effect of changes in the leverage of the weight of body segments on the magnitude of muscle forces and joint reaction forces.

OBJECTIVES

After reading this chapter, you should be able to do the following:

1. Describe the effects of leverage of external loads on muscle forces and joint reaction forces.

2. Describe the effect of using a walking stick, or cane, on hip abductor muscle force and hip joint reaction force.

3. Describe the effect of intratruncal pressure (pressure inside the thorax and abdomen) on low back extensor muscle force and intervertebral joint reaction force.

4. Describe the effect of joint angle on the swing and stabilization components of a muscle.

5. Describe the contribution of the prime movers and synergists to elbow flexion (swing component) and elbow joint reaction force (stabilization component).

Selective Recruitment of Motor Units to Match Functional Requirements

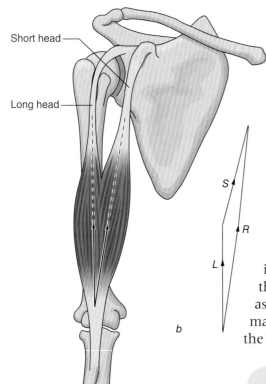

Short head

Long head

S

R

L

b

a

Figure 9.1*a* shows the right biceps brachii muscle, which has two heads of origin. The short head arises from the coracoid process, and the long head arises from the supraglenoid tubercle (tubercle on the superior border of the glenoid fossa). The tendon of the long head passes through the bicipital groove of the humerus. In the upper half of the muscle, the muscle fibers in the long head portion are separate from those in the short head portion, but the two groups of fibers merge in the lower half of the muscle. The muscle is attached by a single tendon to the radial tuberosity. When the whole muscle is stimulated to contract, the line of action of the force exerted by the fibers in the long head portion is slightly different from that in the short head portion because of the separation in the upper half of the muscle (figure 9.1*a*). However, the net effect of the two forces is a single resultant force as shown by the vector chain method in figure 9.1*b*. The magnitude and direction of the resultant force produced by the biceps brachii in a particular movement depend on the

FIGURE 9.1 Resultant force exerted by the biceps brachii. *(a)* Anterior aspect of the right biceps brachii. *(b)* Vector chain representation of the resultant *R* of the force *L* exerted by the long head and the force *S* exerted by the short head.

component forces produced by the short head and long head portions of the muscle. By varying the component forces (by selective recruitment of appropriate motor units) and, therefore, the resultant force, the human body can match the action of the biceps brachii (in association with other muscles) to the specific requirements of each particular movement that involves the biceps brachii.

The capacity to vary the magnitude and direction of force is characteristic of most muscles and reflects the tendency of muscles to work together to produce movements. The extent of this capacity depends considerably on the size, shape, and number of attachments of the muscle or musculotendinous unit to the skeletal system. In general, the greater the size, breadth, and number of attachments of the muscle, the greater the capacity of that muscle to vary the magnitude and direction of force produced (in muscles of similar muscle mass).

KEY POINT

The magnitude and direction of the resultant force produced by a muscle in a particular movement depend on the component forces produced by the various portions of the muscle. By selective recruitment of appropriate motor units, the component forces and the resultant force can be matched to the specific requirements of each particular movement.

Forces Acting on the Head in Upright Postures

Figure 9.2*a* shows the position of the head in normal upright standing. In this position, the line of action of the weight of the head passes in front of the vertebral column and exerts a clockwise moment that tends to rotate the head forward and downward about a mediolateral axis through the fulcrum, which is the joint between the occipital bone and the atlas. The tendency of the weight W of the head to rotate the head forward and downward is counteracted by the neck extensor muscles. Figure 9.2*b* shows a free-body diagram of the head where F is the force exerted by the neck extensor muscles and J is the force exerted at the fulcrum, that is, the **joint reaction force** exerted by the atlas on the occipital bone. Figure 9.2*c* shows the corresponding force-moment arm diagram. In this example, it is assumed that F acts vertically downward. The weight of the head of an adult man is approximately 6.94% of total body weight (see table 9.1).

Consequently, if total body weight is 70 kgf, the weight of the head is approximately 4.86 kgf. The moment arms of W (d_W) and F (d_F) are in the region of 2 cm and 7 cm, respectively (An et al 1984). Because the head is in equilibrium, the resultant moment acting on the head is zero and the resultant force acting on the head is zero. Consequently, the forces F and J can be determined by equating moments and then equating forces as follows:

1. Equating moments:

$$(W \cdot d_W) - (F \cdot d_F) = 0$$
$$W \cdot d_W = F \cdot d_F$$
$$F = \frac{W \cdot d_W}{d_F}$$

where $W = 4.86$ kgf, $d_W = 2$ cm, and $d_F = 7$ cm

$$F = \frac{4.86 \text{ kgf} \times 2 \text{ cm}}{7 \text{ cm}} = 1.39 \text{ kgf}$$

2. Equating forces: Because F and W are vertical forces, J must also be vertical, that is,

$$J = F + W$$
$$J = 1.39 \text{ kgf} + 4.86 \text{ kgf} = 6.25 \text{ kgf}$$

Thus, to counteract W, an external force, the musculoskeletal system has to exert two internal forces, F and J. Force F is an active force, a muscle force, and J is a passive force, a joint reaction force. This illustrates the relationship between the internal and external forces that act on the body; the musculoskeletal system exerts internal forces to counteract or overcome the external forces.

Key Term

joint reaction force   The force exerted by opposed articular surfaces on each other; opposed articular surfaces exert equal and opposite joint reaction forces on each other.

KEY POINT

In counteracting or overcoming external forces, the musculoskeletal system exerts two kinds of internal forces: active forces (muscle forces) and passive forces (joint reaction forces and forces in passive joint support structures).

In the previous example, all of the forces acting on the head are vertical forces and thus the vector chain of the forces is a straight line. The downward and upward forces have been separated in figure 9.2d to demonstrate that when the resultant force acting on an object is zero, the vector chain of the component forces starts and ends at the same point. In most musculoskeletal lever systems, the internal and external forces are not usually parallel. For example, figure 9.3a shows the head position of a person writing at a desk. The head and trunk are tilted forward such that the moment arm of the weight of the head about the atlantooccipital joint is greater than when the head is in the upright position (see figure 9.2), that is, about 4.0 cm compared with 2.0 cm. The moment arm of the force F exerted by the neck extensor muscles is about the same as when the head is in the upright

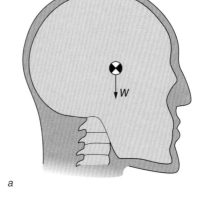
a

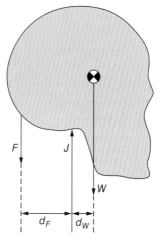
b

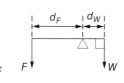
c

d

FIGURE 9.2   Forces acting on the head during normal upright standing. *(a)* Location of the center of gravity of the head. *(b)* Free-body diagram of the head. *(c)* Force-moment arm diagram corresponding to b. *(d)* Vector chain of the forces acting on the head. W = weight of the head; F = force exerted by the neck extensors; d_W = moment arm of W; d_F = moment arm of F; J = joint reaction force exerted by the atlas on the occipital bone.

Table 9.1 Segmental Masses and Mass Center Loci for Young Adult Women and Men

| | Segment end points | | Mass[a] (%) | | CG locus[b] (%) | |
Segment	Proximal	Distal	Women	Men	Women	Men
Head	Top of head	C1	6.68	6.94	58.94	59.76
Trunk	C7	Mid-hip	42.57	43.46	41.51	44.86
Upper arm	Shoulder	Elbow	2.55	2.71	57.54	57.72
Forearm	Elbow	Wrist	1.38	1.62	45.59	45.74
Hand	Wrist	Head of third metacarpal	0.56	0.61	74.74	79.00
Thigh	Hip	Knee	14.78	14.16	36.12	40.95
Lower leg	Knee	Ankle	4.81	4.33	44.16	44.59
Foot	Heel	Toe tip	1.29	1.37	40.14	44.15

C1 = first cervical vertebra; C7 = seventh cervical vertebra.

[a]Segment masses as a percentage of total body mass; [b]segment mass center loci expressed as a percentage of segment length as measured from the proximal endpoint.

From *Journal of Biomechanics*, Vol. 29, P. de Leva, "Adjustments to Zatsiorsky-Seluyanov's segment inertia parameters," pp. 1223-1230, Copyright 1996, with permission from Elsevier.

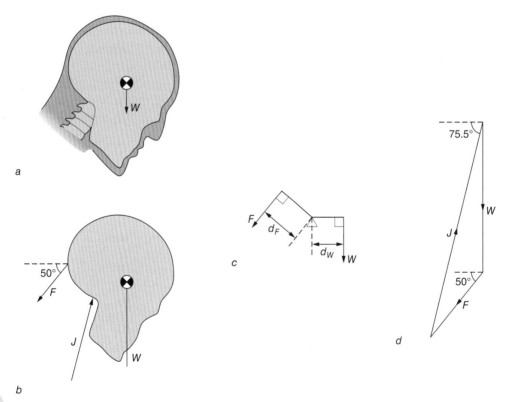

FIGURE 9.3 Forces acting on the head while a person writes at a desk. W = weight of the head; F = force exerted by the neck extensors; d_W = moment arm of W; d_F = moment arm of F; J = joint reaction force exerted by the atlas on the occipital bone.

position, that is, about 7.0 cm, but its line of action will be at approximately 50° to the horizontal. Figure 9.3*b* shows a free-body diagram of the head, and figure 9.3*c* shows the corresponding force-moment arm diagram. By taking moments about the fulcrum

$$(W \cdot d_W) - (F \cdot d_F) = 0$$

$$W \cdot d_W = F \cdot d_F$$

$$F = \frac{W \cdot d_W}{d_F}$$

where W = 4.86 kgf, d_W = 4 cm, and d_F = 7 cm

$$F = \frac{4.86 \, \text{kgf} \times 4 \, \text{cm}}{7 \, \text{cm}} = 2.78 \, \text{kgf}$$

The joint reaction force J can be found two ways: constructing the vector chain and equating the forces. Using the vector chain method, as demonstrated earlier in figure 8.6f, we find that J has a magnitude of approximately 7.2 kgf and acts at an angle of approximately 75.5° to the horizontal (figure 9.3d). The vector chain method of determining the resultant of a number of component forces is useful and fairly quick. However, the accuracy with which the resultant is determined depends on the accuracy of the scale diagram; the greater the number of component vectors, the more difficult it becomes to determine the resultant accurately.

The resultant force can be determined accurately by equating the forces acting on the object. Figure 9.4a shows a free-body diagram of the head (same as figure 9.3b). Figure 9.4b shows the force F resolved into its horizontal and vertical components, and figure 9.4c shows the force J resolved into its horizontal and vertical components. Figure 9.4d shows the forces acting on the head in terms of their horizontal and vertical components. Because the head is in equilibrium, the resultant force on the head is zero. Consequently, the resultant of the horizontal forces is zero and the resultant of the vertical forces is zero. Using the convention that forces acting to the right or upward are positive and that forces acting to the left or downward (with respect to figure 9.4) are negative, we find that

Horizontal forces:

$$J_H - F_H = 0$$

where J_H = horizontal component of J and F_H = horizontal component of F. Therefore,

$$J_H = F_H$$
$$J_H = F \cdot \cos 50°$$
$$J_H = 2.78 \, \text{kgf} \times 0.6428$$
$$J_H = 1.79 \, \text{kgf}$$

Vertical forces:

$$J_V - F_V - W = 0$$
$$J_V = F_V + W$$
$$J_V = F \cdot \sin 50° + W$$
$$J_V = (2.78 \, \text{kgf} \times 0.76604) + 4.86 \, \text{kgf}$$
$$J_V = 2.13 \, \text{kgf} + 4.86 \, \text{kgf}$$
$$J_V = 6.99 \, \text{kgf}$$

Because J_H and J_V are at right angles to each other, the magnitude of their resultant J can be found by applying Pythagoras' theorem, and the direction of J can be found by calculating the cosine of the angle θ that it makes to the horizontal from J_H and J (figure 9.4e).

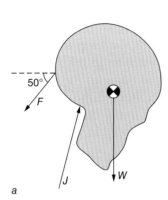

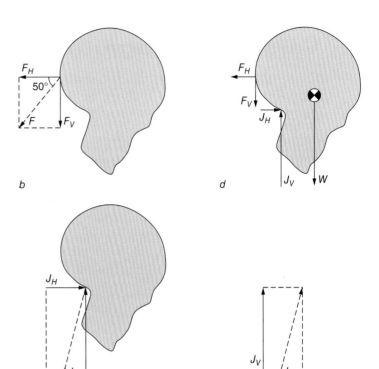

FIGURE 9.4 Determination of joint reaction force by equating forces. W = weight of the head; F = force exerted by the neck extensors; F_H = horizontal component of F; F_V = vertical component of F; J = joint reaction force exerted by the atlas on the occipital bone; J_H = horizontal component of J; J_V = vertical component of J.

Magnitude of J, by Pythagoras' theorem:

$$J^2 = J_V^2 + J_H^2$$
$$J^2 = 6.99^2 \text{ kgf}^2 + 1.79^2 \text{ kgf}^2$$
$$J^2 = 48.86 \text{ kgf}^2 + 3.20 \text{ kgf}^2$$
$$J^2 = 52.06 \text{ kgf}^2$$
$$J = 7.21 \text{ kgf}$$

Direction of J:

$$\cos \theta = \frac{J_H}{J}$$
$$\cos \theta = \frac{1.79 \text{ kgf}}{7.21 \text{ kgf}}$$
$$\cos \theta = 0.2482$$
$$\theta = 75.6°$$

In this example, the vector chain method and the calculation method (equating the forces) produced almost identical results, indicating, in this case, the accuracy of the vector chain method.

Moment of External Forces
Versus the Magnitude of Internal Forces

The examples of the forces acting on the head in the upright (see figure 9.2) and writing positions (see figure 9.3) show the general effect of increases in the moment of external forces on the magnitude of the internal forces that counteract them; increasing the moment of external forces results in an increase in muscle forces, which in turn increases joint reaction forces. This is illustrated in figure 9.5, which shows a 6 m beam of weight W balanced in three positions on a knife-edge support. In figure 9.5a, the beam is balanced with the line of action of its weight passing through the knife-edge support; that is, the moment of the weight of the beam about the fulcrum is zero. In this situation the reaction force R is sufficient to counteract W and maintain equilibrium (figure 9.5b). In figure 9.5c, the beam has been displaced 1 m to the right such that W exerts a clockwise turning moment on the beam of $W \times 1$ m. The beam is held in equilibrium by the reaction force R_1 and a force F_1 exerted by a tie that attaches the left end of the beam to the base of support. If it is assumed

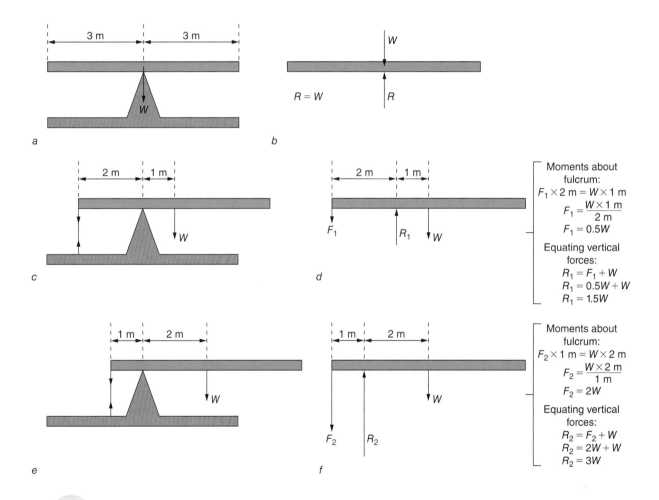

FIGURE 9.5 Effect of increasing the moment arm of the weight W of a beam balanced on a knife edge on the restraining forces needed to maintain equilibrium. *(a)* Beam balanced with the line of action of W through the base of support. *(b)* Free-body diagram of *(a)*. *(c)* Beam balanced with the aid of a restraining force F_1. *(d)* Free-body diagram of *(c)*. *(e)* Beam balanced with the aid of a restraining force F_2. *(f)* Free-body diagram of *(e)*.

that F_1 acts vertically, R_1 will also act vertically. In this situation, F_1 and R_1 are found to be $0.5W$ and $1.5W$, respectively (figure 9.5*d*). In figure 9.5*e*, the beam has been displaced 2 m to the right with respect to its original position such that W exerts a clockwise turning moment on the beam of $W \times 2$ m. The beam is held in equilibrium by the reaction force R_2 and a force F_2 exerted by the tie; it is assumed that F_2 and consequently R_2 act vertically. In this situation, F_2 and R_2 are found to be $2W$ and $3W$, respectively (figure 9.5*f*).

> **KEY POINT**
>
> Increasing the moment of external forces results in an increase in the magnitude of muscle forces, which increases in the magnitude of joint reaction forces.

Forces About the Hip Joint in One-Leg Stance

The situation depicted in figure 9.5*f* is similar to that of single-leg support during standing or walking (figure 9.6, *a-c*). In this position the pelvis tends to rotate about the hip joint under the action of the weight of the body and the hip abductor muscles. Figure 9.6*c* shows a free-body diagram of the pelvis in this situation, and figure 9.6*d* shows the corresponding force-moment arm diagram. W is the weight of the body less the weight of the grounded leg. For a man of total body weight of 73.62 kgf, W is approximately 59 kgf (80.14% of total body weight; see table 9.1). W exerts a clockwise moment on the pelvis. The hip abductor muscles maintain the pelvis in a level position by exerting an equal and opposite moment on the pelvis. The line of action of the force A exerted by the hip abductors is approximately 80° with respect to the horizontal. Force J is the joint reaction force, that is, the force exerted by the head of the femur on the pelvis via the acetabulum. When the pelvis is in equilibrium, the resultant of W, A, and J is zero. The moment arms of A (d_A) and W (d_W) with respect to the hip joint axis of rotation are approximately 6 cm and 11 cm, respectively. By taking moments about the fulcrum

$$(W \cdot d_W) - (A \cdot d_A) = 0$$

$$W \cdot d_W = A \cdot d_A$$

$$A = \frac{W \cdot d_W}{d_A}$$

where $W = 59$ kgf, $d_W = 11$ cm, and $d_A = 6$ cm.

$$A = \frac{59 \text{ kgf} \times 11 \text{ cm}}{6 \text{ cm}} = 108 \text{ kgf}$$

The vector chain determination of J is shown in figure 9.6*e*; $J = 166$ kgf at an angle of 83° to the horizontal. The determination of J by equating the forces is shown in figure 9.6, *f* and *g*; $J = 166.4$ kgf at an angle of 83.5° to the horizontal. In this example, the vector chain and calculation methods of determining J give almost identical results. The results show that the joint reaction force in one-leg stance is in the region of 2.4 times body weight. The force S—the force exerted by the pelvis on the head of the femur—is equal and opposite to that of J (see figure 9.6*b*).

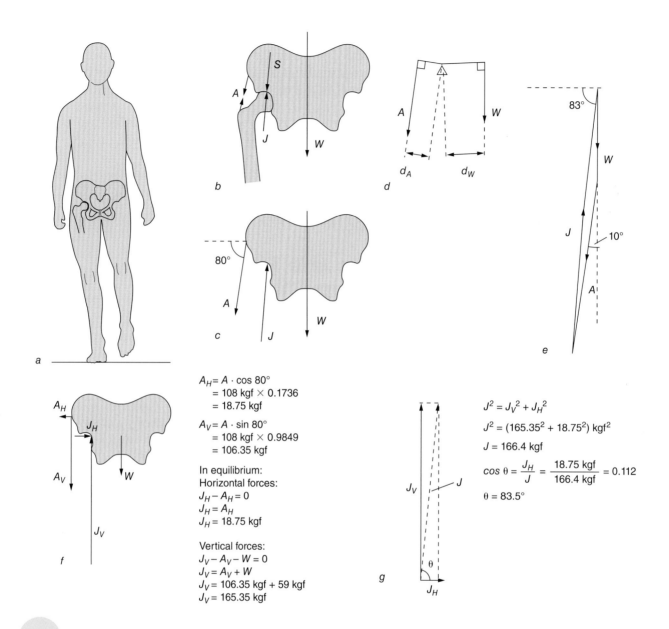

$A_H = A \cdot \cos 80°$
$= 108 \text{ kgf} \times 0.1736$
$= 18.75 \text{ kgf}$

$A_V = A \cdot \sin 80°$
$= 108 \text{ kgf} \times 0.9849$
$= 106.35 \text{ kgf}$

In equilibrium:
Horizontal forces:
$J_H - A_H = 0$
$J_H = A_H$
$J_H = 18.75 \text{ kgf}$

Vertical forces:
$J_V - A_V - W = 0$
$J_V = A_V + W$
$J_V = 106.35 \text{ kgf} + 59 \text{ kgf}$
$J_V = 165.35 \text{ kgf}$

$J^2 = J_V^2 + J_H^2$
$J^2 = (165.35^2 + 18.75^2) \text{ kgf}^2$
$J = 166.4 \text{ kgf}$
$\cos \theta = \dfrac{J_H}{J} = \dfrac{18.75 \text{ kgf}}{166.4 \text{ kgf}} = 0.112$
$\theta = 83.5°$

FIGURE 9.6 Effect of one-leg stance on hip abductor muscle force and hip joint reaction force. *(a)* One-leg stance. *(b)* Forces on the pelvis and hip joint in one-leg stance. *(c)* Free-body diagram of the pelvis. *(d)* Force-moment arm diagram of *(c)*. *(e)* Vector chain determination of the hip joint reaction force. *(f)* Forces on the pelvis resolved into their horizontal and vertical components. *(g)* Direction of the joint reaction force. A = force exerted by the hip abductors; J = hip joint reaction force exerted by the head of the femur on the acetabulum; S = hip joint reaction force exerted by the acetabulum on the head of the femur; W = weight of the body less the weight of the grounded leg.

When someone is recovering from a serious leg injury, such as a broken femur, it is desirable to reduce the load on the leg during weight-bearing activities such as standing and walking by using crutches and walking sticks or canes. Figure 9.7*a* shows a man walking with the aid of a cane in his left hand. During the right-leg single-support phase of the walking cycle, the cane helps to support the weight of the body and reduce the load on the right leg. The cane acts as an extension of the left arm, enabling the left arm to support the weight of the body by pushing against the floor. In this example the left arm can be considered to be a lateral extension of the pelvis such that the free-body diagram of the pelvis can be represented as in figure 9.7b. The corresponding force-moment arm diagram is shown in figure 9.7c.

When the man in this example is walking without a cane, as in figure 9.6a, the moment of W is counteracted by the moment of A on its own. However, when the man is walking with a cane, the moment of W is counteracted by the combined moment of A and the force C exerted by the cane on the man's left hand. When the pelvis is in equilibrium, the resultant of W, A, C, and J will be zero. For a man of total body weight 70 kgf, C is about 16 kgf with a moment arm of approximately 35 cm

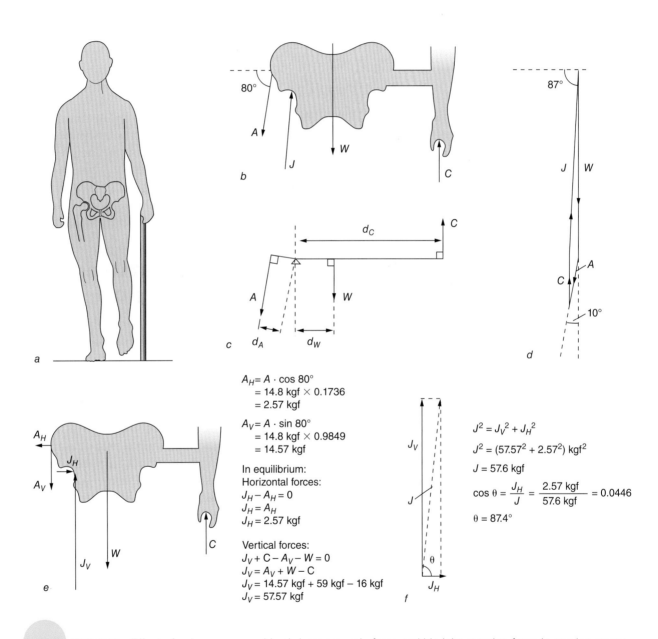

$A_H = A \cdot \cos 80°$
$\quad = 14.8 \text{ kgf} \times 0.1736$
$\quad = 2.57 \text{ kgf}$

$A_V = A \cdot \sin 80°$
$\quad = 14.8 \text{ kgf} \times 0.9849$
$\quad = 14.57 \text{ kgf}$

In equilibrium:
Horizontal forces:
$J_H - A_H = 0$
$J_H = A_H$
$J_H = 2.57 \text{ kgf}$

Vertical forces:
$J_V + C - A_V - W = 0$
$J_V = A_V + W - C$
$J_V = 14.57 \text{ kgf} + 59 \text{ kgf} - 16 \text{ kgf}$
$J_V = 57.57 \text{ kgf}$

$J^2 = J_V^2 + J_H^2$
$J^2 = (57.57^2 + 2.57^2) \text{ kgf}^2$
$J = 57.6 \text{ kgf}$

$\cos \theta = \dfrac{J_H}{J} = \dfrac{2.57 \text{ kgf}}{57.6 \text{ kgf}} = 0.0446$

$\theta = 87.4°$

FIGURE 9.7   Effect of using a cane on hip abductor muscle force and hip joint reaction force in one-leg stance. *(a)* One-leg stance with the aid of a cane. *(b)* Free-body diagram of the pelvis during one-leg stance with a cane. *(c)* Force-moment arm diagram corresponding to *(b)*. *(d)* Vector chain determination of the hip joint reaction force. *(e)* Forces on the pelvis resolved into their horizontal and vertical components. *(f)* Direction of the joint reaction force. A = force exerted by the hip abductors; J = hip joint reaction force exerted by the head of the femur on the acetabulum; W = weight of the body less the weight of the grounded leg; C = force exerted by the cane on the left hand.

with respect to the hip joint axis of rotation; it is assumed that C acts vertically (An et al 1984). By taking moments about the fulcrum

$$(W \cdot d_W) - (A \cdot d_A) - (C \cdot d_C) = 0$$

$$A \cdot d_A = (W \cdot d_W) - (C \cdot d_C)$$

$$A = \frac{(W \cdot d_W) - (C \cdot d_C)}{d_A}$$

where $W = 59$ kgf, $C = 16$ kgf, $d_W = 11$ cm, $d_A = 6$ cm, and $d_C = 35$ cm.

$$A = \frac{(59 \text{ kgf} \times 11 \text{ cm}) - (16 \text{ kgf} \times 35 \text{ cm})}{6 \text{ cm}}$$

$$A = 14.8 \text{ kgf}$$

The vector chain determination of J is shown in figure 9.7d; $J = 57.5$ kgf at an angle of 87° to the horizontal. The determination of J by equating the forces is shown in figure 9.7, e and f; $J = 57.6$ kgf at an angle of 87.4° to the horizontal. As in the previous example, the vector chain and calculation methods of determining J give almost identical results. The results indicate that using a cane to aid one-leg stance considerably reduces the force exerted in the hip abductor muscles (from 108 kgf to 14.8 kgf) and thus considerably reduces the hip joint reaction force (from 166 kgf to 57.5 kgf).

Effect of Squat and Stoop Postures on Forces in the Lumbar Region

When the trunk is inclined forward from the upright position, the weight of the upper body (head, neck, arms, and trunk) exerts a moment that tends to flex the trunk; the greater the degree of inclination of the trunk, the greater the trunk flexor moment exerted by upper-body weight. When the trunk is inclined forward, but in equilibrium, the trunk flexor moment exerted by upper-body weight is counteracted by the trunk extensor muscles (figure 9.8, a and b).

The moment arm of the trunk extensor muscles about a mediolateral axis through the center of each intervertebral joint is similar throughout the vertebral column. However, the moment arm of upper-body weight (the proportion of upper-body weight superior to a particular intervertebral joint) increases as the degree of forward inclination of the trunk increases. Consequently, the greater the degree of forward inclination of the trunk, the greater the force exerted by the trunk extensors to maintain equilibrium. Because the moment arm of the trunk extensors is usually much shorter than the moment arm of upper-body weight, the force exerted by the trunk extensors and the joint reaction forces exerted on the intervertebral joints are relatively large when maintaining the trunk in a forward inclined position. The proportion of upper-body weight above each intervertebral joint increases the lower the intervertebral joint is in the vertebral column; for example, the proportion of upper-body weight above the L4-L5 intervertebral joint is slightly less than that above the L5-S1 intervertebral joint. The lower the intervertebral joint in the column, the greater the proportion of upper-body weight above the joint and the greater the moment arm of upper-body weight about the joint. Consequently, for a given angle of trunk inclination, the lower the intervertebral joint in the column, the greater the trunk extensor muscle force and the greater intervertebral joint reaction force.

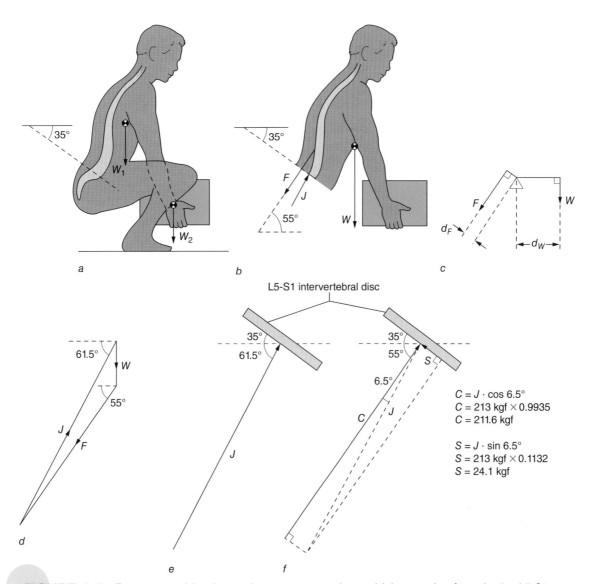

FIGURE 9.8 Force exerted by the trunk extensor muscles and joint reaction force in the L5-S1 intervertebral joint when one is lifting a load of 10 kgf from a squat position. *(a)* Orientation of the trunk and L5–S1 joint. *(b)* Free-body diagram of upper-body weight and load relative to the L5-S1 joint. *(c)* Force-moment arm diagram corresponding to *(b)*. *(d)* Vector chain determination of the L5-S1 joint reaction force. *(e)* Orientation of the joint reaction force to the plane of the L5-S1 joint. *(f)* Compression and shear components of the joint reaction force.

Squat Posture

Any load held or lifted in front of the body exerts a flexor moment on the trunk and increases trunk extensor muscle forces and intervertebral joint reaction forces. Figure 9.8*a* shows a person in a squat posture holding a load of 10 kgf just above the floor. Figure 9.8*b* shows a free-body diagram of the upper body and load, where W is the combined weight of the upper body (W_1) and load (W_2), F is the force exerted by the trunk extensor muscles across the L5-S1 intervertebral joint, and J is the L5-S1 intervertebral joint reaction force. Figure 9.8*c* shows the corresponding force-moment arm diagram. The weight W_1 of the upper body above the L5-S1 joint, that is, the head, neck, arms, and proportion of the trunk above the L5-S1 joint, is approximately 48% of total body weight (Morris et al 1961). Consequently, for a man of total body weight

70 kgf, W_1 is approximately 34 kgf. The moment arm of W is approximately 20 cm. The plane of the L5-S1 joint in the squat position is approximately 35° with respect to the horizontal. The line of action of F is approximately perpendicular to the plane of the L5-S1 joint, and the moment arm of F about the fulcrum (i.e., the middle of the L5-S1 joint) is approximately 5 cm. By taking moments about the fulcrum

$$(W \cdot d_W) - (F \cdot d_F) = 0$$

$$W \cdot d_W = F \cdot d_F$$

$$F = \frac{W \cdot d_W}{d_F}$$

where W = 44 kgf, d_W = 20 cm, and d_F = 5 cm.

$$F = \frac{44\ \text{kgf} \times 20\ \text{cm}}{5\ \text{cm}}$$

$$F = 176\ \text{kgf}$$

The vector chain determination of J is shown in figure 9.8d; J = 213 kgf at an angle of 61.5° to the horizontal. Consequently, for a person to hold a load of 10 kgf just above the floor in a squat position, the force exerted by the trunk extensors adjacent to the L5-S1 intervertebral joint is in the region of 2.5 times body weight and the L5-S1 joint reaction force is about 3 times body weight.

As shown in figure 9.8e, the line of action of J is oblique to the plane of the L5-S1 joint. Consequently, J has a compression component C (perpendicular to the plane of the joint) and a shear component S (parallel to the plane of the joint) (figure 9.8f). Because J (213 kgf) makes an angle of 83.5° with the plane of the joint, C is relatively large (212 kgf) and S is relatively small (24 kgf).

Stoop Posture

In figure 9.9a, the person is shown holding the 10 kgf just above the floor in a stoop posture. Figure 9.9b shows a free-body diagram of the upper body and load relative to the L5-S1 joint. Figure 9.9c shows the corresponding force-moment arm diagram. In this situation the moment arm of W is approximately 30 cm (compared with 20 cm in the squat posture). The plane of the L5-S1 joint in the stoop posture is approximately 70° with respect to the horizontal (compared with 35° in the squat posture). If the line of action of F is assumed to be perpendicular to the plane of the L5-S1 joint with a moment arm of 5 cm (as in the squat position), then by taking moments about the fulcrum

$$(W \cdot d_W) - (F \cdot d_F) = 0$$

$$W \cdot d_W = F \cdot d_F$$

$$F = \frac{W \cdot d_W}{d_F}$$

where W = 44 kgf, d_W = 30 cm, and d_F = 5 cm.

$$F = \frac{44\ \text{kgf} \times 30\ \text{cm}}{5\ \text{cm}}$$

$$F = 264\ \text{kgf}$$

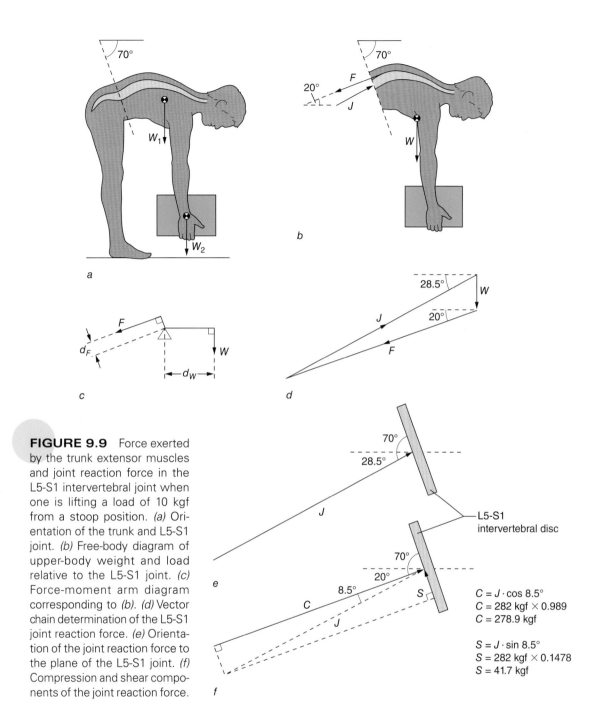

FIGURE 9.9 Force exerted by the trunk extensor muscles and joint reaction force in the L5-S1 intervertebral joint when one is lifting a load of 10 kgf from a stoop position. *(a)* Orientation of the trunk and L5-S1 joint. *(b)* Free-body diagram of upper-body weight and load relative to the L5-S1 joint. *(c)* Force-moment arm diagram corresponding to *(b)*. *(d)* Vector chain determination of the L5-S1 joint reaction force. *(e)* Orientation of the joint reaction force to the plane of the L5-S1 joint. *(f)* Compression and shear components of the joint reaction force.

$C = J \cdot \cos 8.5°$
$C = 282 \text{ kgf} \times 0.989$
$C = 278.9 \text{ kgf}$

$S = J \cdot \sin 8.5°$
$S = 282 \text{ kgf} \times 0.1478$
$S = 41.7 \text{ kgf}$

The vector chain determination of J is shown in figure 9.9*d*; $J = 282$ kgf at an angle of 28.5° to the horizontal. Consequently, for someone to hold a load of 10 kgf just above the floor in a stoop position, the force exerted by the trunk extensor muscles adjacent to the L5-S1 intervertebral joint will be about 3.8 times body weight (compared with 2.5 times body weight in the squat posture), and the L5-S1 joint reaction force will be about 4 times body weight (compared with 3 times body weight in the squat posture). As in the squat posture, the line of action of J is oblique to the plane of the L5-S1 joint (figure 9.9*e*). In this case the compression component is approximately 279 kgf (compared with 212 kgf in the squat posture) and the shear component is approximately 42 kgf (compared with 24 kgf in the squat posture) (figure 9.9*f* and table 9.2).

Table 9.2 Effect of Type of Posture (Squat, Stoop) and Intra-truncal Pressure (P) on Trunk Extensor Muscle Force and L5-S1 Joint Reaction Force

	Force (kgf)		Squat as a percentage of stoop (%)	(2) as a percentage of (1) (%)	
	Squat	Stoop		Squat	Stoop
Trunk extensor muscle force					
(1) Without P	176	264	67		
				77	77
(2) With P	135	203	66		
Joint reaction force J					
(1) Without P	213	282	75		
				73	68
(2) With P	155	192	81		
Compression component of J					
(1) Without P	212	279	76		
				72	67
(2) With P	153	188	81		
Shear component of J					
(1) Without P	24	42	60		
				100	95
(2) With P	24	40	60		

Effect of Intratruncal Pressure

In the preceding estimates of trunk extensor muscle forces and L5-S1 joint reaction forces, no account was taken of the effect of intratruncal pressure. As described in chapter 5, intratruncal pressure can reduce trunk extensor muscle forces and intervertebral joint reaction forces in postures involving forward inclination of the trunk by exerting a trunk extensor moment. It has been estimated that the trunk extensor moment exerted by intratruncal pressure is approximately 30% of the moment exerted by the trunk extensor muscles and that the moment arm of intratruncal pressure is approximately 11 cm (Morris et al 1961). Figure 9.10*a* shows the orientation of the trunk and the thoracic and abdominal cavities in the squat posture. Figure 9.10*b* shows a free-body diagram of the upper body and load relative to the L5-S1 joint, including the force P exerted by intratruncal pressure. Figure 9.10*c* shows the corresponding force-moment arm diagram. In this situation the trunk flexor moment exerted by W is counteracted by F and P. By taking moments about the fulcrum,

$$(W \cdot d_W) - (F \cdot d_F) - (P \cdot d_P) = 0$$
$$W \cdot d_W = (F \cdot d_F) + (P \cdot d_P) \tag{9.1}$$

If the moment exerted by P is approximately 30% of the moment exerted by F, then

$$P \cdot d_P = 0.3(F \cdot d_F) \tag{9.2}$$

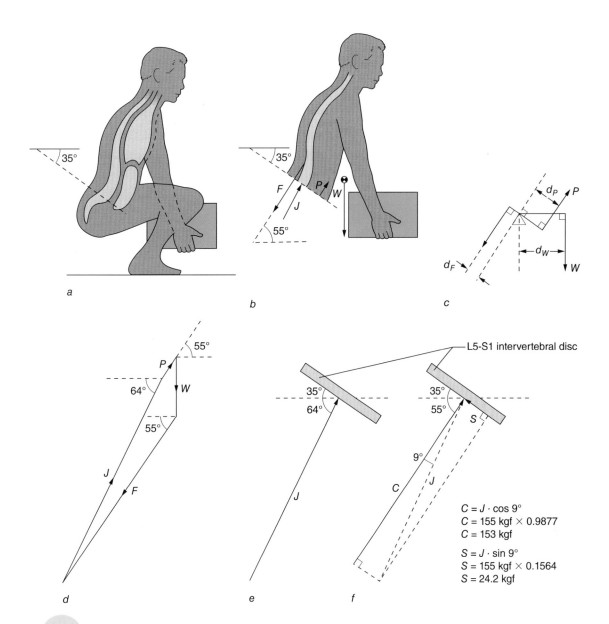

FIGURE 9.10 Effect of intratruncal pressure on the force exerted by the trunk extensor muscles and joint reaction force in the L5-S1 intervertebral joint when one is lifting a load of 10 kgf from a squat position. *(a)* Orientation of the trunk and L5-S1 joint. *(b)* Free-body diagram of upper-body weight and load relative to the L5-S1 joint. *(c)* Force-moment arm diagram corresponding to *(b)*. *(d)* Vector chain determination of the L5-S1 joint reaction force. *(e)* Orientation of the joint reaction force to the plane of the L5-S1 joint. *(f)* Compression and shear components of the joint reaction force.

Therefore, from equations 9.1 and 9.2,

$$W \cdot d_W = (F \cdot d_F) + 0.3(F \cdot d_F)$$
$$W \cdot d_W = 1.3(F \cdot d_F)$$

$$F = \frac{W \cdot d_W}{1.3 d_F}$$

where W = 44 kgf, d_W = 20 cm, and d_F = 5 cm.

$$F = \frac{44 \text{ kgf} \times 20 \text{ cm}}{1.3 \times 5 \text{ cm}}$$

$$F = 135.4 \text{ kgf}.$$

From equation 9.2,

$$P = \frac{0.3(F \cdot d_F)}{d_P}$$

where $F = 135.4$ kgf, $d_F = 5$ cm, and $d_P = 11$ cm.

$$P = \frac{0.3 \times 135.5 \text{ kgf} \times 5 \text{ cm}}{11 \text{ cm}}$$

$$P = 18.4 \text{ kgf}.$$

The vector chain determination of J is shown in figure 9.10d; $J = 155$ kgf at an angle of 64° to the horizontal. As in the previous examples (figures 9.8 and 9.9), J is oblique to the plane of the L5-S1 joint (figure 9.10e). Figure 9.10f shows the compression and shear components of J. In this example, intratruncal pressure reduces the trunk extensor muscle force, the L5-S1 joint reaction force, and the compression component of the joint reaction force by 23%, 27%, and 28%, respectively, in the squat posture (see table 9.2).

Figure 9.11a shows the orientation of the trunk and thoracic and abdominal cavities in the stoop posture. Figure 9.11b shows a free-body diagram of the upper body and load relative to the L5-S1 joint, including the force P exerted by intratruncal pressure. Figure 9.11c shows the corresponding force-moment arm diagram. By taking moments about the fulcrum,

$$(W \cdot d_W) - (F \cdot d_F) - (P \cdot d_P) = 0$$
$$W \cdot d_W = (F \cdot d_F) + (P \cdot d_P).$$

As previously discussed, if the moment exerted by P is approximately 30% of the moment exerted by F, then

$$F = \frac{W \cdot d_W}{1.3 d_F}$$

where $W = 44$ kgf, $d_W = 30$ cm, and $d_F = 5$ cm.

$$F = \frac{44 \text{ kgf} \times 30 \text{ cm}}{1.3 \times 5 \text{ cm}}$$

$$F = 203 \text{ kgf}.$$

Also

$$P = \frac{0.3(F \cdot d_F)}{d_P}$$

where $F = 203$ kgf, $d_F = 5$ cm, and $d_P = 11$ cm.

i.e.,
$$P = \frac{0.3 \times 203 \text{ kgf} \times 5 \text{ cm}}{11 \text{ cm}}$$

$$P = 27.7 \text{ kgf}.$$

The vector chain determination of J is shown in figure 9.11d; J = 192 kgf at an angle of 32° to the horizontal. As in the previous examples (figures 9.8-9.10), J is oblique to the plane of the L5-S1 joint (figure 9.11e). The compression and shear components are shown in figure 9.11f. In this example, intratruncal pressure reduces the trunk extensor muscle force, the L5-S1 joint reaction force, and the compression component of the joint reaction force by 23%, 32%, and 33%, respectively, in the stoop posture (see table 9.2). Case study 6 examines the role of intratruncal pressure during lifting tasks.

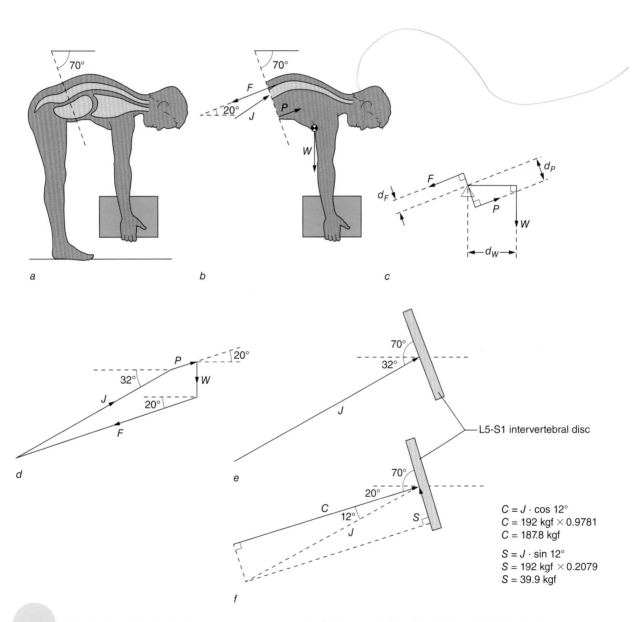

$C = J \cdot \cos 12°$
$C = 192 \text{ kgf} \times 0.9781$
$C = 187.8 \text{ kgf}$

$S = J \cdot \sin 12°$
$S = 192 \text{ kgf} \times 0.2079$
$S = 39.9 \text{ kgf}$

FIGURE 9.11 Effect of intratruncal pressure on the force exerted by the trunk extensor muscles and joint reaction force in the L5-S1 intervertebral joint when one is lifting a load of 10 kgf from a stoop position. (a) Orientation of the trunk and L5-S1 joint. (b) Free-body diagram of upper-body weight and load relative to the L5-S1 joint. (c) Force-moment arm diagram corresponding to (b). (d) Vector chain determination of the L5-S1 joint reaction force. (e) Orientation of the joint reaction force to the plane of the L5-S1 joint. (f) Compression and shear components of the joint reaction force.

CASE STUDY 6 **BREATHING PATTERNS AND LIFTING TECHNIQUE**

Hagins M, Lamberg EM. 2006. Natural breath control during lifting tasks: effect of load. *European Journal of Applied Physiology* 956:453-458.

Studies have shown that voluntary control of the breath influences intra-abdominal pressure (IAP) and that increases in IAP are related to an increase in lumbar stability. The aims of the study were to determine the naturally occurring breath patterns during lifting tasks in healthy adult subjects and the effects of different levels of load during lifting tasks on natural breath control patterns in healthy adult subjects. The subjects were 10 males and 10 females (age 25 ± 3.6 years; mass 68 ± 16.5 kg; height 173 ± 5.8 cm).

After adequate practice, each subject performed nine trials of a lifting–lowering task in which the subject was required to lift a loaded plastic milk crate (27 cm × 47 cm × 32 cm) from a lifting platform (11.4 cm above the floor), turn 90° to the right, and place the crate on a placement platform set at 30% of the subject's height. Trials 1-3, 4-6, and 7-9 were performed sequentially with loads of 5%, 10%, and 25% of body weight, respectively. During each set of three trials each subject wore a face mask (covering nose and mouth) attached to a pneumotachograph, which recorded the timing and magnitude of inspiration, expiration, and breath holding during the trials.

When the breath pattern was examined across all loads, there was a significant increase in inspired volume relative to normal breathing immediately prior to liftoff, and breath holding occurred most frequently during load placement. In addition, the volume inspired and retained during lifting was positively related to load. Other researchers have suggested that a failure of coordination between the respiratory and motor systems is a contributory factor in certain types of low back pain.

Application

The results of the present study suggest that breathing pattern influences lifting and lowering technique and that the greater the load, the greater the importance of stabilizing the trunk. This result is consistent with the development of intratruncal pressure to reduce trunk extensor muscle forces and intervertebral joint reaction forces when lifting and lowering. In association with the development of intratruncal pressure, good lifting and lowering technique is characterized by the following:

- A squat posture; flexed knees with the trunk held as upright as possible
- Maintenance of a hollow in the lumbar region (lumbar lordosis)
- Head up when lifting and lowering, because this will help to maintain a hollow in the lumbar region
- The load to be lifted kept as close to the body as possible (to minimize the trunk flexion moment)
- Feet apart to increase the base of support

Swing and Stabilization Components of Muscle Force

In postures other than very relaxed postures such as lying down or sitting in an arm chair, most of the muscles of the body are active to control the movements of the joints. In controlling joint movements, muscles exert two effects on joints: stabilization and rotation. Because joints need to be stabilized (i.e., joint congruence needs to be maintained), regardless of whether the joints are moving, stabilization of joints is a major function of the muscles. The contribution of a muscle to stabilization and rotation of a joint is determined by the stabilization component and the swing component, respectively, of the muscle force. The **stabilization component** is directed at the axis of rotation of the joint, that is, tending to compress the articular surfaces and maintain joint congruence. The **swing component** is at right angles to the stabilization component and exerts a turning moment about the joint. In movement of a particular joint, it is likely that all of the muscles involved contribute to both stabilization and swing, but the contributions of each muscle depend on the angle of the joint.

Figure 9.12 shows the orientation of the line of action of the biceps brachii at three elbow angles. In figure 9.12*a*, the elbow is close to full extension. In this position the stabilization component is much larger than the swing component, that is, more force is directed at the elbow joint than is flexing the joint. As the elbow flexes from the position in figure 9.12*a* toward the position in figure 9.12*b*, the stabilization component progressively decreases and the swing component progressively increases such that a point is reached, when the elbow angle is about 90°, where the stabilization component is zero (figure 9.12*b*), that is, no muscle force is directed at the elbow joint and all of the muscle force is flexing the joint. As the elbow flexes from the position in figure 9.12*b* toward the position in figure 9.12*c*, the component of muscle force in line with the axis of rotation is directed away from the joint and, as such, it is a subluxation component; that is, it tends to pull apart the articular surfaces of the elbow joint instead of pushing them together (as is the case between the positions in figure 9.12, *a* and *b*). Under normal circumstances this is not likely to be a problem because the stabilization component of the muscle is relatively small because of active insufficiency, and the other muscles that control elbow flexion, including the pronator teres, brachialis, and wrist flexors (figure 9.13), are likely to counteract the subluxation component of the biceps brachii.

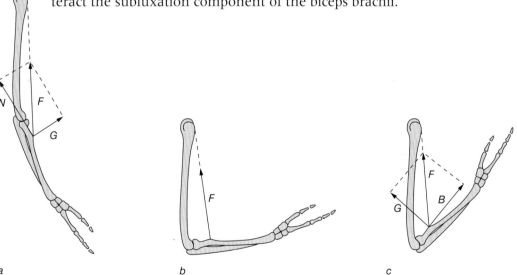

a   *b*   *c*

FIGURE 9.12 Swing *(G)*, stabilization *(N)*, and subluxation *(B)* components of the biceps brachii muscle force *(F)*. *(a)* As the elbow joint is close to full extension, the stabilization component *(N)* of *F* is larger than the swing component *(S)*. *(b)* The elbow joint angle is close to 90° such that there is no stabilization component of *F*. *(c)* As the elbow joint angle is less than 90°, the component of *F* in line with the elbow joint is a subluxation component *(B)*.

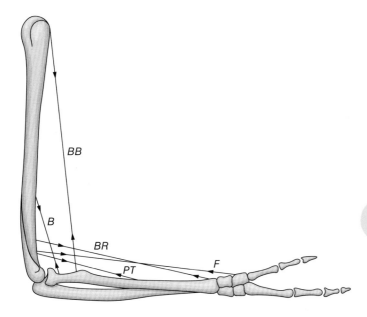

FIGURE 9.13 Lines of action of muscles that contribute to elbow flexion: *BB* = biceps brachii; *B* = brachialis; *BR* = brachioradialis; *PT* = pronator teres; *F* = wrist and finger flexors.

Key Terms

stabilization component The component of the force exerted by a muscle that is directed through the axis of rotation of a joint that the muscle crosses.

swing component The component of the force exerted by a muscle that is perpendicular to the stabilization component.

Prime Movers and Synergists in Elbow Flexion

The biceps brachii and brachialis are prime movers in elbow flexion, and they exert relatively large swing components and small stabilization components during most of the elbow flexion and extension range of motion; they also tend to exert subluxation components close to full flexion. Figure 9.13 shows the lines of action of the biceps brachii *(BB)* and brachialis *(B)* when the forearm is held horizontal with the upper arm vertical. The biceps brachii and brachialis are assisted in elbow flexion by a number of other muscles in the role of synergist. These muscles, which include the pronator teres *(PT)*, brachioradialis *(BR)*, and wrist and finger flexors *(F)*—flexor carpi ulnaris, flexor carpi radialis, and flexor digitorum sublimis—exert relatively small swing components and large stabilization components during most of the elbow flexion and extension range. Consequently, during elbow flexion, these muscles mainly function to stabilize the elbow joint while providing some assistance to the biceps brachii and brachialis in terms of swing. Figure 9.13 shows these muscles' lines of action.

Calculation of Muscle Forces

Figure 9.14, *a* and *b*, shows a free-body diagram of the forearm of an adult held in a horizontal position with the upper arm vertical and a load *(L)* of 2 kgf in the palm of the hand. For a man weighing 67 kgf, the weight of the lower arm and hand *(A)* is approximately 1.5 kgf. (2.23% of total body weight; see table 9.1). In this position, *A* and *L* exert clockwise moments on the forearm about the elbow. In equilibrium, these clockwise moments are counteracted by the combined counterclockwise moment exerted by the five muscles or muscle groups shown in figure 9.14*a*: brachialis *(B)*, biceps brachii *(BB)*, pronator teres *(PT)*, brachioradialis *(BR)*, and the wrist and finger flexors *(F)*. The wrist and finger flexors comprise the flexor carpi ulnaris, flexor carpi

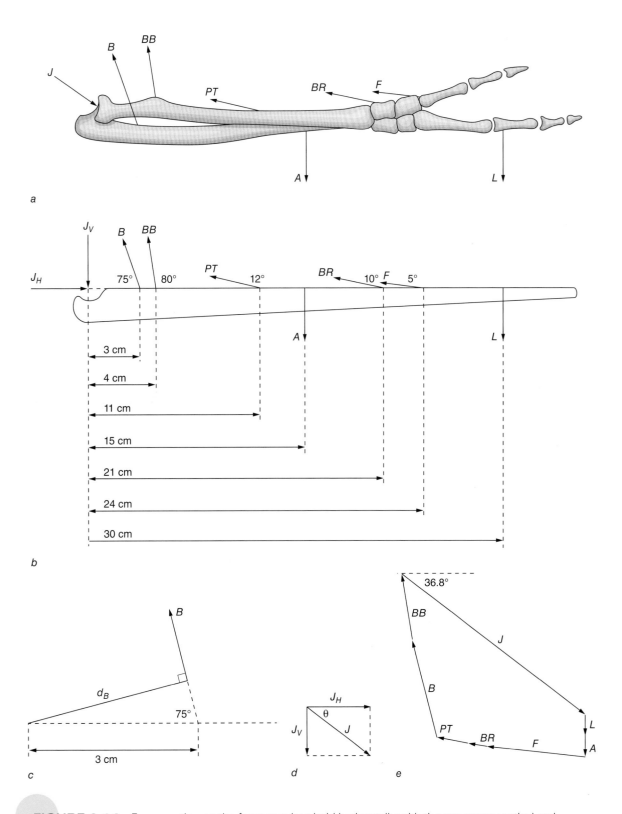

FIGURE 9.14 Forces acting on the forearm when held horizontally with the upper arm vertical and a load of 2 kgf in the palm of the hand. *(a)* Free-body diagram of the forearm and hand. *(b)* Simplified free-body diagram of the forearm and hand. *(c)* Determination of the moment arm of the force exerted by the brachialis. *(d)* Determination of the elbow joint reaction force from horizontal and vertical components. *(e)* Vector chain determination of the elbow joint reaction force.

radialis, and the portion of the flexor digitorum sublimis that crosses the elbow joint. The pronator teres represents that portion of the muscle that crosses the elbow joint. Figure 9.14*b* shows a simplified version of the free-body diagram in figure 9.14*a*. The simplified version assumes that the points of application of the muscle forces are in the same horizontal plane as the axis of elbow flexion and extension, that is, the fulcrum.

The calculation of the moment arm of each muscle is illustrated with reference to the brachialis. Figure 9.14*c* shows the line of action of the brachialis and its moment arm d_B. Because the line of action of the brachialis makes an angle of 75° with the horizontal plane through the fulcrum, it follows that

$$\frac{d_B}{3\text{ cm}} = \sin 75°$$
$$d_B = \sin 75° \times 3 \text{ cm}$$
$$d_B = 0.9659 \times 3 \text{ cm}$$
$$d_B = 2.9 \text{ cm}$$

The moment arms of the other muscles can be calculated in the same way. By taking moments about the fulcrum,

$$A \cdot d_A + L \cdot d_L = B \cdot d_B + BB \cdot d_{BB} + PT \cdot d_{PT} + BR \cdot d_{BR} + F \cdot d_F$$

where A = 1.5 kgf; L = 2 kgf, B = force exerted by the brachialis, BB = force exerted by the biceps brachii, PT = force exerted by the pronator teres, BR = force exerted by the brachioradialis, F = force exerted by the wrist and finger flexors, d_A = moment arm of A = 15 cm; d_L = moment arm of L = 30 cm; d_B = moment arm of B = 2.9 cm; d_{BB} = moment arm of BB = 3.94 cm; d_{PT} = moment arm of PT = 2.29 cm; d_{BR} = moment arm of BR = 3.65 cm; d_F = moment arm of F = 2.09 cm. Thus,

$$(1.5 \text{ kgf} \times 15 \text{ cm}) + (2 \text{ kgf} \times 30 \text{ cm}) = (B \times 2.9 \text{ cm}) + (BB \times 3.94 \text{ cm})$$
$$+ (PT \times 2.29 \text{ cm}) + (BR \times 3.65 \text{ cm}) + (F \times 2.09 \text{ cm})$$

$$22.5 \text{ kgf} \cdot \text{cm} + 60 \text{ kgf} \cdot \text{cm} = (B \times 2.9 \text{ cm}) + (BB \times 3.94 \text{ cm})$$
$$+ (PT \times 2.29 \text{ cm}) + (BR \times 3.65 \text{ cm}) + (F \times 2.09 \text{ cm})$$

$$82.5 \text{ kgf} \cdot \text{cm} = (B \times 2.9 \text{ cm}) + (BB \times 3.94 \text{ cm})$$
$$+ (PT \times 2.29 \text{ cm}) + (BR \times 3.65 \text{ cm}) + (F \times 2.09 \text{ cm}) \qquad (9.3)$$

Table 9.3 Muscle Forces

Muscle	Muscle force relative to brachioradialis (BR)	Muscle force (kgf)
Brachioradialis (BR)	BR	1.93
Pronator teres (PT)	1.3 BR	2.51
Biceps brachii (BB)	3.1 BR	5.98
Brachialis (B)	4.7 BR	9.07
Wrist and finger flexors (F)	4.9 BR	9.46

To calculate the muscle forces, it is necessary to estimate the force produced by each muscle relative to the other muscles. These estimates of relative force are made on the basis of the physiological cross-sectional areas of the muscles (An et al 1981; Lieber et al 1990). Consequently, in relation to the force BR exerted by the brachioradialis, the relative magnitudes of the muscle forces are as shown in table 9.3.

By substitution of the relative muscle forces in equation 9.3,

$$82.5 \text{ kgf} \cdot \text{cm} = (4.7BR \times 2.9 \text{ cm}) + (3.1BR \times 3.94 \text{ cm})$$
$$+ (1.3BR \times 2.29 \text{ cm}) + (BR \times 3.65 \text{ cm}) + (4.9BR \times 2.09 \text{ cm})$$

$$82.5 \text{ kgf} \cdot \text{cm} = (13.63BR \cdot \text{cm}) + (12.21BR \cdot \text{cm})$$
$$+ (2.98BR \cdot \text{cm}) + (3.65BR \cdot \text{cm}) + (10.24BR \cdot \text{cm})$$

$$82.5 \text{ kgf} \cdot \text{cm} = 42.71BR \cdot \text{cm}$$
$$BR = \frac{82.5 \text{ kgf} \cdot \text{cm}}{42.71 \text{ cm}}$$
$$BR = 1.93 \text{ kgf}$$

Consequently, table 9.3 shows the forces exerted by the muscles.

Calculation of Elbow Joint Reaction Force

Because the forearm and hand are in equilibrium, the resultant force acting on the forearm and hand is zero. Consequently, the sum of the horizontal forces is zero, and the sum of the vertical forces is zero. With respect to figure 9.14*b*, the horizontal and vertical forces can be calculated as follows:

Horizontal forces:

$$J_H - B \cdot \cos 75° - BB \cdot \cos 80° - PT \cdot \cos 12° - BR \cdot \cos 10° - F \cdot \cos 5° = 0$$
$$J_H = B \cdot \cos 75° + BB \cdot \cos 80° + PT \cdot \cos 12° + BR \cdot \cos 10° + F \cdot \cos 5°$$
$$J_H = 2.35 \text{ kgf} + 1.04 \text{ kgf} + 2.45 \text{ kgf} + 1.90 \text{ kgf} + 9.42 \text{ kgf}$$
$$J_H = 17.16 \text{ kgf}$$

Vertical forces:

$$B \cdot \sin 75° + BB \cdot \sin 80° + PT \cdot \sin 12° + BR \cdot \sin 10° + F \cdot \sin 5° - A - L - J_V = 0$$
$$J_V = B \cdot \sin 75° + BB \cdot \sin 80° + PT \cdot \sin 12° + BR \cdot \sin 10° + F \cdot \sin 5° - A - L$$
$$J_V = 8.76 \text{ kgf} + 5.89 \text{ kgf} + 0.52 \text{ kgf} + 0.33 \text{ kgf} + 0.82 \text{ kgf} - 1.5 \text{ kgf} - 2 \text{ kgf}$$
$$J_V = 12.82 \text{ kgf}$$

Magnitude of elbow joint reaction force *J*:
By Pythagoras' theorem

$$J^2 = J_V^2 + J_H^2$$
$$J^2 = (12.82 \text{ kgf})^2 + (17.16 \text{ kgf})^2$$
$$J^2 = (164.35 + 294.45) \text{ kgf}^2$$
$$J^2 = 458.81 \text{ kgf}^2$$
$$J = 21.42 \text{ kgf}$$

Direction of *J* (figure 9.14*d*):

$$\cos \theta = \frac{J_H}{J} = \frac{17.16 \text{ kgf}}{21.42 \text{ kgf}} = 0.8011$$

$$\theta = 36.8° \text{ with respect to the horizontal}$$

Figure 9.14*e* shows the vector chain determination of *J*.

Contribution of Prime Movers and Synergists to Swing and Stabilization

The contribution of the prime movers and synergists to swing is given by their contributions to the total swing moment. As shown in table 9.4, the prime movers and

Table 9.4 Contribution of Prime Movers and Synergists to Total Elbow Flexor Swing Moment

Group	Muscle	Moment (kgf · cm)	Group total (kgf · cm)	Group total (%)
Prime mover	Brachialis	26.28		
			49.84	60.5
Prime mover	Biceps brachii	23.56		
Synergist	Pronator teres	5.74		
Synergist	Brachioradialis	7.04	32.57	39.5
Synergist	Wrist and finger flexors	19.79		
	Total swing moment	82.41	82.41	100

Table 9.5 Contribution of Prime Movers and Synergists to Total Elbow Flexor Joint Stabilization Component

Group	Muscle	Stabilization component (kgf)	Group total (kgf)	Group total (%)
Prime mover	Brachialis	2.35		
			3.39	19.7
Prime mover	Biceps brachii	1.04		
Synergist	Pronator teres	2.45		
Synergist	Brachioradialis	1.90	13.77	80.3
Synergist	Wrist and finger flexors	9.42		
	Total stabilization moment	17.16	17.16	100

synergists contribute approximately 60% and 40%, respectively, to the total swing moment. The contributions of the prime movers and synergists to joint stabilization are given by their contributions to the total stabilization component, which, in this example, are equal and opposite to J_H. As shown in table 9.5, the prime movers and synergists contribute approximately 20% and 80%, respectively, to the total stabilization component.

Contribution of Synergists to Reducing Bending Stress on the Forearm and Hand

Although an object may be in equilibrium, it may also be subject to bending or torsional stress depending on the orientation of the forces acting on it. In the elbow example, the synergists not only contribute to swing and joint stabilization but also reduce bending stress on the forearm and hand. In the absence of the synergists, the clockwise moment CM exerted by A and L about the radial tuberosity (site of attachment of the biceps brachii) would bend the arm downward. With respect to figure 9.14b, the moment arm of A about the point of application of BB is 11 cm (15 cm – 4 cm). The moment arm of L about the point of application of BB is 26 cm (30 cm – 4 cm). Therefore, it follows that

$$CM = (A \times 11 \text{ cm}) + (L \times 26 \text{ cm})$$
$$CM = 16.5 \text{ kgf} \cdot \text{cm} + 52 \text{ kgf} \cdot \text{cm}$$
$$CM = 68.5 \text{ kgf} \cdot \text{cm}.$$

However, the synergists counteract the bending moment by exerting a counterclockwise moment AM (about the radial tuberosity), that is,

$$AM = (PT \cdot \sin 12° \times 7 \text{ cm}) + (BR \cdot \sin 10° \times 17 \text{ cm}) + (F \cdot \sin 5° \times 20 \text{ cm})$$
$$AM = 3.65 \text{ kgf} \cdot \text{cm} + 5.69 \cdot \text{kgf} \cdot \text{cm} + 16.49 \text{ kgf} \cdot \text{cm}$$
$$AM = 25.83 \text{ kgf} \cdot \text{cm}.$$

Consequently, in this example, the synergists reduce the bending moment on the forearm and hand by approximately 38%. It is likely that the inclusion of more muscles in the free-body diagram and more accurate data regarding muscle forces and their moment arms would reduce the bending load even further.

Summary

Changes in the leverage of external loads affect the magnitude of internal forces required to counteract the external loads. Increasing the leverage of external loads increases the magnitude of muscle forces, which in turn increases the magnitude of joint reaction forces. Furthermore, because of the low leverage of most muscles, the muscle forces and joint reaction forces are much larger than the external loads. Muscles work together to counteract external loads; this distributes the load over many muscles, reduces the force exerted by each muscle, and reduces bending stress on bones. Under normal circumstances the musculoskeletal components adapt their size, shape, and structure to more readily withstand the loads exerted on them. The response and adaptation of the musculoskeletal system to loading are covered in the remaining chapters, starting with consideration of the mechanical characteristics of musculoskeletal components in the next chapter.

Review Questions

1. Explain the relationship between the moment of external forces and the magnitude of muscle forces and joint reaction forces.

2. With regard to muscle forces, differentiate stabilization, subluxation, and swing components.

3. By equating the horizontal and vertical components of force, determine the magnitude and direction of the joint reaction force in each of the examples given in figures 9.8, 9.9, 9.10, and 9.11.

Mechanical Characteristics of Musculoskeletal Components

All materials deform in response to loading in a manner that reflects their mechanical characteristics, including, for example, their stiffness and their strength. The components of the musculoskeletal system (muscles, tendons, ligaments, cartilage, bone) have different mechanical characteristics, but in combination they have two main functions: absorbing shock and minimizing energy expenditure. The musculoskeletal system as a whole, often in combination with protective clothing, such as running shoes, or equipment, such as soft mats, acts as a shock absorber to protect the body from potentially harmful external loads, in particular, the ground reaction forces that act on the feet in weight-bearing activities such as walking, running, and landing. Most of the musculoskeletal components are elastic to a certain extent and, in combination with the stretch-shorten cycle of muscle actions (as described in chapter 9), are able to function as springs that can significantly reduce energy expenditure. Using examples of shock absorption and energy storage and release in walking and running, this chapter describes the mechanical characteristics of the musculoskeletal components and the response of the musculoskeletal system to loading.

10

OBJECTIVES

After reading this chapter, you should be able to do the following:

1. Differentiate between a load–deformation curve and a stress–strain curve.
2. Describe the general stress–strain behavior of the musculoskeletal components.
3. Differentiate between passive and active loading.
4. Describe the causes of shock and vibration.
5. Describe the shock absorption mechanisms that operate within the human body.
6. Describe the effects of impact loading on synovial joints.

Stress–Strain Relationships in Solids

As described in chapter 1, there are three types of load: tension, compression, and shear. When subjected to a load, all materials deform to a certain extent. For example, resting your wrists on a gel-filled computer wrist rest exerts two compression loads (mainly the weight of your left forearm and left hand and the weight of your right forearm and right hand) on the wrist rest. In response to these compression loads,

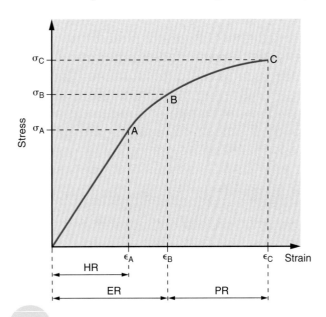

FIGURE 10.1 Generalized stress–strain curve. ER = elastic range; PR = plastic range; HR = Hookean range; A = proportional limit; B = yield point (elastic limit); C = failure point (complete rupture).

the wrist rest is deformed (compressed) under each wrist. The amount of compression of the rest under each wrist depends on the compression load (the greater the load, the greater the deformation) and the mechanical characteristics of the materials that comprise the wrist rest (the softer the materials, the greater the deformation). A material's mechanical characteristics are reflected in its **load–deformation curve,** that is, the change in deformation that occurs in response to a change in load. Most solid materials, including cartilage and bone, deform differently in response to different types of load. Consequently, for most solid materials, the load–deformation curves for tension load, compression load, and shear load are different.

To compare the mechanical characteristics of materials, it is usual to express load and deformation in terms of stress (load per unit cross-sectional area) and strain (deformation as a proportion of the dimensions of the material when unloaded), respectively. Stress is denoted by the Greek letter sigma (σ), and strain is denoted by epsilon (ϵ). Figure 10.1 shows a generalized **stress–strain curve.** In response to tension loading, a material tends to lengthen in the direction of the tension load. Consequently, tension strain refers to increase in length of the material as a proportion of the length of the material when it is not loaded.

Similarly, in response to compression loading, a material tends to shorten in the direction of the compression load—compression strain refers to decrease in length of the material as a proportion of the length of the material when it is not loaded. Note that stress is measured in units of force per unit cross-sectional area, such as N/cm², but strain has no units because it is the ratio of two lengths.

KEY POINT

All materials deform in response to loading. The load–deformation relationship of a material reflects its mechanical characteristics. To compare the mechanical characteristics of materials, load is expressed in terms of stress and deformation in terms of strain.

Most materials are elastic to a certain extent; that is, they deform in response to loading and then restore their original dimensions when the load is removed. In figure 10.1 the point B represents the **yield point** or elastic limit—the point at which the material starts to tear or fracture. The strain range between zero strain and the strain at B (ϵ_B) is the **elastic range;** that is, provided that the strain on the material is within the elastic range, the material will not be damaged and will remain perfectly elastic. However, if the strain on the material exceeds the elastic range, the material will be deformed plastically, that is, in the case of an inanimate object, permanently damaged to a degree corresponding to the amount of strain. This material will lose some of its elasticity and will not return to its original dimensions when the load is removed. If the amount of strain is allowed to increase progressively beyond the yield point, the material will eventually fail completely; this is the **failure point** and is represented by point C in figure 10.1. The strain range between ϵ_B and ϵ_C is the **plastic range**. The stress at C (σ_C) is the **ultimate stress** or strength of the material.

Units of Force

The Système International d'Unites, usually referred to as the SI system or the metric system, is the most widely used system of units in commerce and scientific communication. In the SI system, the unit of force is the newton (N). In accordance with Newton's second law of motion ($F = m \cdot a$, where a is the acceleration experienced by a mass m when acted upon by a force F), a newton is defined as the force acting on a mass of 1 kg that accelerates it at 1 m/s², that is,

$$1 \text{ N} = 1 \text{ kg} \times 1 \text{ m/s}^2 \text{ (i.e., 1 N} = 1 \text{ kg} \cdot \text{m/s}^2)$$

The weight W of an object is the product of its mass m and the acceleration caused by gravity g, that is,

$$W = m \cdot g$$

Consequently, because $g = 9.81$ m/s², the weight of a mass of 1 kg, referred to as 1 kgf (kilogram force), is 9.81 N, that is,

$$1 \text{ kgf} = 1 \text{ kg} \times 9.81 \text{ m/s}^2 = 9.81 \text{ N}$$

The kilogram force is referred to as a gravitational unit of force. Most weighing machines in everyday use, such as kitchen scales and bathroom scales, are graduated in kgf or lbf (pounds force). Thus, the weight of a man of mass 70 kg can be expressed as 70 kgf or 686.7 N, that is,

$$70 \text{ kgf} = 70 \text{ kg} \times 9.81 \text{ m/s}^2 = 686.7 \text{ N}$$

In calculations using SI units, the unit of force is the newton, and all forces, including weights, must be expressed in newtons.

Key Terms

load–deformation curve The relationship between load and deformation for a material.

stress–strain curve The relationship between stress and strain for a material.

yield point The upper limit of the elastic range; also referred to as elastic limit.

elastic range The strain range within which a material remains perfectly elastic.

failure point The point at which a material fails.

plastic range The strain range between the yield point and the failure point.

ultimate stress The stress experienced by a material immediately prior to failure; also referred to as the strength of the material.

Stiffness and Compliance

In most materials, all or part of the stress–strain curve within the elastic range is linear—the increase in strain is directly proportional to the increase in stress. Any material that behaves this way is said to obey Hooke's law (after Robert Hooke, 1635-1703, who first described this characteristic of elastic behavior). Consequently, the linear region of the stress–strain curve is referred to as the **Hookean region.** The Hookean range is that part of the elastic range that corresponds to the Hookean region of the curve (figure 10.1). The upper limit of the Hookean region is the **proportional limit.** In some materials the proportional limit and the yield point coincide, but in most materials the proportional limit occurs before the yield point. The gradient of the stress–strain curve in the Hookean region reflects the **stiffness** of the material, that is, the resistance of the material to deformation. The greater the stress required to produce a given amount of strain, the stiffer the material.

The gradient of the stress–strain curve in the Hookean region is referred to as **Young's modulus** of elasticity (or Young's modulus or elastic modulus) for the material. Young's modulus (after Thomas Young, 1773-1829) indicates the amount of stress needed to produce 100% strain. Most materials fail long before

100% strain, but Young's modulus provides a standard measure of stiffness for comparing materials. Young's modulus is denoted by uppercase epsilon ϵ. With reference to figure 10.1, if $\sigma_A = 6{,}000$ N/cm^2, and $\epsilon_A = 0.015$ (1.5% strain), then

$$E = \frac{\sigma_A}{\epsilon_A}$$

$$E = \frac{6{,}000 \text{ N/cm}^2}{0.015} = 400{,}000 \text{ N/cm}^2 = 400 \text{ kN/cm}^2$$

(1 kN = 1 kilonewton = 1,000 N)

The larger the Young's modulus, the stiffer the material. Slight changes in the composition of a material can affect its stiffness (and other mechanical characteristics). For example, different kinds of steel have different levels of stiffness. Similarly, the stiffness of musculoskeletal components depends considerably on a person's age, nutrition status, and physical activity level. Furthermore, the mechanical characteristics of similar musculoskeletal components vary with location in the body. For example, the stiffness of compact bone in the femur is different from that in the tibia of the same individual (Burstein and Wright 1994). Given the variation in the magnitude of the mechanical properties of different types of the same material, there are no standard reference data. Table 10.1 shows some data reported in the literature concerning the stiffness of certain materials in tension.

> **KEY POINT**
>
> Slight changes in the composition of a material can affect its stiffness (and other mechanical characteristics). The stiffness of musculoskeletal components depends considerably on a person's age, nutrition status, and physical activity level.

Figure 10.2 shows generalized stress–strain curves for bone and ligament. It is clear that bone is stiffer and stronger than ligament. However, bone tends to fail suddenly, whereas ligament exhibits progressive, albeit rapid failure following the proportional limit. Furthermore, in contrast to bone, where the stress–strain curve is linear throughout the elastic range,

Table 10.1 Young's Modulus for Some Materials Under Tension (kN/cm²)

Material		Mean	Range
1. Normal compact bone (sixth decade)	Femur	1,700	
	Tibia	2,000	
2. Normal compact bone	Femur	1,569	1,269-1,943
3. Osteoporotic compact bone	Femur	1,155	397-1,834
4. Patellar ligament		40	
5. Elastin		0.06	
6. Tool steel		19,000	
7. Stainless steel		17,000	
8. Glass		7,000	
9. Oak		1,000	
10. Lightly vulcanized rubber		0.14	

1 and 4 reported by Burstein and Wright 1994; 2 and 3 reported by Dickenson et al. 1981; 5, 9, and 10 reported by Alexander 1968; 6, 7, and 8 reported by Frost 1967.

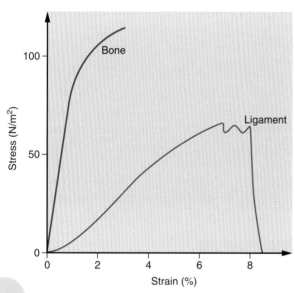

FIGURE 10.2 Generalized stress–strain curves for bone and ligament in tension.

the stress–strain curve of ligament is nonlinear throughout the elastic range. The first part of the stress–strain curve of ligament is referred to as the toe region, where a relatively small increase in stress results in a relatively large increase in strain during the first 1% of strain. The toe region reflects the straightening out of the collagen fibers, which, under resting conditions, have a wavy arrangement. Tendons, muscle–tendon units, and cartilage respond to loading in a similar manner to ligament.

In real-life situations, many materials, like bone, are loaded simultaneously in tension, compression, and shear. For example, when one is standing upright, the fifth lumbar vertebra experiences shear load at both the left and the right pars interarticularis, compression load on the body of the vertebra, and tension loads (resulting from attached ligaments and tendons) on the transverse processes and spinous process (see figure 5.7c). Young's modulus can be calculated for each type of loading. When the Young's modulus of a particular material is the same in all three forms of loading, the material is said to exhibit **strain isotropy.** When the Young's modulus is different for the three forms of loading, the material is said to exhibit **strain anisotropy.** Most materials with a physical grain—like wood, bone, tendon, ligament, and cartilage—exhibit strain anisotropy (Frost 1967). For example, bone is stiffer in compression than in tension, and stiffer in tension than in shear.

Compliance is the reciprocal of stiffness; that is, increasing the stiffness of a material decreases its compliance and vice versa. The greater the strain produced by a given amount of stress, the more compliant the material. The yield point of the load–deformation (or stress–strain) curve of a material is usually signified by a marked increase in compliance (decrease in stiffness).

Key Terms

Hookean region The linear region of a stress–strain curve prior to the yield point.

proportional limit The upper limit of the Hookean region.

stiffness The resistance of a material to deformation; stress per unit strain.

Young's modulus The gradient of the stress–strain curve in the Hookean region.

strain isotropy Property of a material where the Young's modulus of a material is the same in tension, compression, and shear.

strain anisotropy Property of a material where the Young's modulus of a material is different in tension, compression, and shear.

compliance The reciprocal of stiffness; strain per unit stress.

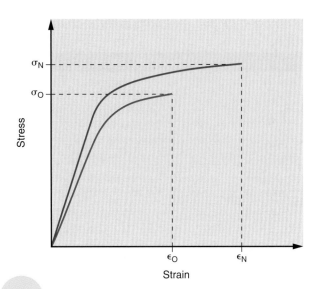

FIGURE 10.3 The effects of osteoporosis on the stress–strain characteristics of compact bone. N =normal bone; O = osteoporotic bone.

Toughness, Fragility, and Brittleness

For any given piece of material, the area under the load–deformation curve represents the strain energy absorbed by the material. The **toughness** of a material is defined as the amount of strain energy the material absorbs prior to failure; the greater the amount of energy absorbed, the tougher the material. For the purpose of comparing the toughness of various materials, the **toughness modulus** is defined as the strain energy absorbed per unit volume of the material prior to failure. The toughness modulus of a material is represented by the area under the stress–strain curve of the material.

Figure 10.3 shows typical stress–strain curves for normal compact bone and osteoporotic compact bone. The area under the osteoporotic curve is much smaller than that under the normal curve; that is, normal bone is much tougher than osteoporotic bone. **Fragility** refers to a low level of toughness—a fragile material absorbs relatively little strain energy prior to failure. It follows that osteoporotic bone is more fragile than normal bone. Figure 10.3 shows that the greater toughness of normal bone is the result of greater strength (ultimate stress) and greater strain to failure compared with osteoporotic bone. A brittle material fails after relatively little strain; the lower the strain to failure, the greater the **brittleness** of the material. It follows that osteoporotic bone is more brittle than normal bone.

Key Terms

toughness The amount of strain energy a material can absorb prior to failure; the greater the energy absorption, the tougher the material.

toughness modulus Strain energy absorbed to failure per unit volume of material.

fragility The quality of being easily broken or destroyed. A fragile material has a low toughness modulus, that is, it absorbs a low level of strain energy prior to failure.

brittleness The amount of strain to failure; the lower the strain to failure, the more brittle the material.

Work, Strain Energy, and Kinetic Energy

As described in chapter 7, a force does **work** when it moves its point of application in the direction of the force, and the amount of work done is defined as the product of the force and the distance moved by the point of application of the force. For example, in drawing a bow, the archer does work on the bow by pulling on the arrow, which in turn pulls on the bowstring (figure 10.4*a*). Figure 10.4*b* shows the corresponding load–deformation curve of the force exerted on the bow and the corresponding deformation of the bow. As the bow is drawn, the work done on the bow, represented by the area between the curve and the deformation axis, is stored in the bow which deforms like a spring. The work done on the bow is stored in the bow as **strain energy,** that is, energy that can be used to do work on the arrow (propel it forward via the bowstring) when the bowstring is released; the greater the deformation of the bow, the greater the amount of strain energy that is stored in it. Strain energy is a form of **mechanical energy,** that is, energy that can do work. The amount of work done by the archer on the bow, that is, the amount of strain energy stored in the bow, is $F \cdot d$, where F is the average force exerted on the bowstring and d is the distance that the bow is drawn back. When the arrow is released, the bow recoils and the strain energy stored in the bow is transformed into work on the arrow via the bowstring. The arrow separates from the bowstring with kinetic energy equivalent to the strain energy stored in the drawn bow. **Kinetic energy,** another form of mechanical energy, is the energy possessed by a body as a result of its speed of movement. An object of mass m and speed of movement v has kinetic energy equal to $m \cdot v^2 / 2$. Consequently, in the bow and arrow example, $F \cdot d = m \cdot v^2 / 2$, where F is the average force exerted on the arrow by the bowstring, d is the distance over which F is applied, m is the mass of arrow, and v is the release speed of the arrow.

Many materials store strain energy in response to loading, for example, a stretched elastic band, a trampoline, a springboard in diving, a beat board in vaulting, a pole in pole vaulting, and a musculotendinous unit in an eccentric action. Strain energy is a form of **potential energy,** that is, stored energy that

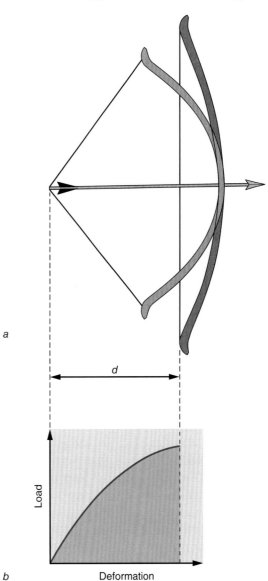

FIGURE 10.4 Storage of strain energy in a bow. *(a)* Deformation of a bow through a distance *d*. *(b)* Load–deformation curve of the force exerted on the bow and the corresponding deformation of the bow. The work done on the bow, that is, the strain energy stored in the bow, is represented by the shaded area under the load–deformation curve.

can be used to do work, given appropriate conditions.

In the SI system, the unit of work is the **joule** (J) (after James Joule, 1818-1889). One joule is the amount of work done by a force of 1 N when it moves its point of application a distance of 1 m in the direction of the force, that is, 1 J = 1 N × 1 m. The units of work, kinetic energy, and moment of force consist of the same base units, $kg \cdot m^2/s^2$. To distinguish these quantities, the unit for work and kinetic energy is the joule (J) and the unit for moment of force is the newton meter (N·m).

Human movement is brought about by coordinated actions of the skeletal muscles. The muscles do work in moving the body segments. In doing work, the muscles transform chemical energy stored in the muscles into kinetic energy of the moving body segments. The stored energy in the muscles is potential energy in the form of complex chemical substances (chiefly adenosine triphosphate and creatine phosphate). When a muscle contracts isometrically (does not change in length), it expends energy but does no work. When a muscle contracts concentrically (shortens), it expends energy in doing positive work, that is, it pulls its skeletal attachments closer together. When a muscle contracts eccentrically (lengthens) it expends energy, but in contrast to a concentric contraction, work is done on the muscle, that is, it is forcibly stretched; this type of work by muscles is called negative work. As a musculotendinous unit lengthens in an eccentric action, the elastic components of the musculotendinous unit absorb energy in the form of strain energy. For example, when a gymnast is landing from a jump or vault, the kinetic energy of the body is transformed into strain energy in the support surface and strain energy in the musculotendinous units that control the hip, knee, and ankle joints by eccentric action of these units. When the purpose of the landing is to bring the body to rest, the strain energy in the musculotendinous units is rapidly dissipated in the form of heat in the musculotendinous units and subsequently in the rest of the body and the surrounding air. However, if the landing is immediately followed by a rebound, some of the strain energy

in the musculotendinous units can be recycled in the subsequent movement in the form of work (additional to that produced by concentric action of the muscles) resulting from recoil of the elastic components of the musculotendinous units. As described in chapter 7, the use of this strain energy depends largely on the speed of changeover from eccentric to concentric muscle contraction. Generally, the faster the changeover, the smaller the proportion of strain energy dissipated as heat and consequently the greater the proportion available to contribute to the subsequent movement.

Key Terms

work The product of a force and the distance moved by the point of application of the force in the direction of the force.

mechanical energy Energy that can do work.

strain energy The energy absorbed by an object when it is deformed.

kinetic energy The energy possessed by an object as a result of its speed of movement.

potential energy Stored energy that, given appropriate conditions, can be used to do work.

joule The amount of work done by a force of 1 N when it moves its point of application a distance of 1 m in the direction of the force; 1 J = 1 N × 1 m.

KEY POINT

The muscles do work in moving the body segments. In doing work, the muscles transform chemical energy stored in the muscles into kinetic energy of the moving body segments.

Gravitational Potential Energy

If an object is held above ground level and then released, it will fall to the ground because of the force of its own weight. The work done on the object by the force of its own weight W when it falls a distance h is given by $W \cdot h$.

Consequently, when an object of weight W is held a distance h above the ground, it possesses gravitational potential energy equivalent to $W \cdot h$, which can be transformed into kinetic energy if it is allowed to fall. **Gravitational potential energy** is usually expressed as $m \cdot g \cdot h$ (where $W = m \cdot g$, m is the mass of the object, and g is the acceleration due to gravity).

Figure 10.5a shows a rubber ball held at rest at a height h_1 above the floor where the floor is the reference level ($h = 0$) for the measurement of gravitational potential energy. While it is held at rest, the ball has no kinetic energy, but its gravitational potential energy will be equal to $m \cdot g \cdot h_1$. If the ball is allowed to fall, its gravitational potential energy will be transformed into kinetic energy, that is, its gravitational potential energy will decrease and its kinetic energy will increase. When the ball hits the floor, its gravitational potential energy will be zero and its kinetic energy will be equal to $m \cdot v^2 / 2$, where v is the velocity of the ball at impact, that is,

$$m \cdot g \cdot h_1 = m \cdot v^2 / 2$$
$$v = \surd(2g \cdot h_1)$$

If $h_1 = 1$ m, then $v = \surd(2 \times 9.81$ m/s$^2 \times 1$ m) $= 4.43$ m/s.

Key Term

gravitational potential energy   The energy possessed by an object as a result of its height above a reference position, usually ground level.

Hysteresis, Resilience, and Damping

In the preceding example, the ball strikes the floor with kinetic energy equivalent to the gravitational potential energy it possessed at release. During contact with the floor the ball will undergo a loading phase in which it is compressed and the kinetic energy of the ball is transformed into strain energy in the compressed ball. Following the loading phase, the

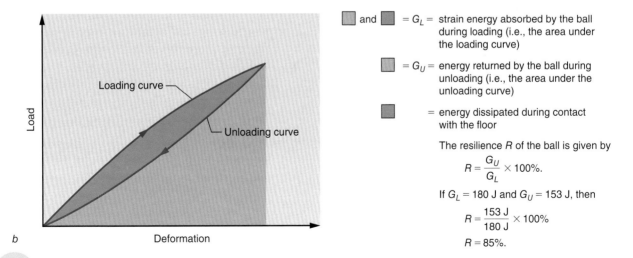

h_1 Drop height

h_2 Bounce height

□ and ■ $= G_L =$ strain energy absorbed by the ball during loading (i.e., the area under the loading curve)

□ $= G_U =$ energy returned by the ball during unloading (i.e., the area under the unloading curve)

■ $=$ energy dissipated during contact with the floor

The resilience R of the ball is given by

$$R = \frac{G_U}{G_L} \times 100\%.$$

If $G_L = 180$ J and $G_U = 153$ J, then

$$R = \frac{153 \text{ J}}{180 \text{ J}} \times 100\%$$

$$R = 85\%.$$

FIGURE 10.5   Load–deformation characteristics of a bouncing rubber ball.

ball undergoes an unloading phase in which it recoils and the strain energy is released as kinetic energy in the form of the upward bounce of the ball. However, the ball will not bounce as high as the point from which it was dropped. This situation is shown in figure 10.5*a*, where h_1 is the drop height and h_2 is the bounce height. As the ball is at rest at A and B, some of the energy of the ball is dissipated during contact with the floor in the form of, for example, heat and sound. The amount of energy dissipated is reflected in the load deformation curves of the ball during loading and unloading (figure 10.5*b*).

The amount of strain energy absorbed by the ball during loading, the area under the loading curve, is greater than the amount of energy returned during unloading, the area under the unloading curve. The loop described by the loading and unloading curves is the **hysteresis loop** (from the Greek word *husteros* meaning "later" or "delayed"). The area of the hysteresis loop represents the energy dissipated. The extent of hysteresis in a material is reflected in the **resilience** of the material, which is defined as the amount of energy returned during unloading as a percentage of the amount of energy absorbed during loading. All materials exhibit hysteresis to a certain extent; no material is 100% resilient.

Figure 10.6*a* shows the load–deformation characteristics of highly resilient material such as ligament and tendon, and figure 10.6*b* shows the load–deformation characteristics of low-resilience material, such as some forms of vinyl acetate foam. **Damping** refers to a low level of resilience; a damping material returns very little energy during unloading compared with the amount of energy that it absorbs during loading. For protection during transportation, fragile goods are usually packed in materials with good damping properties. Similarly, in walking and running, shock-absorbing soles and insoles in shoes are used to protect the body from high-impact loads at heel strike. In human movement, shock absorption refers to the dissipation of the work done on the body as a result of a collision with the environment in a way that prevents high impact loads. Shock absorption systems are low-resilience energy absorption systems.

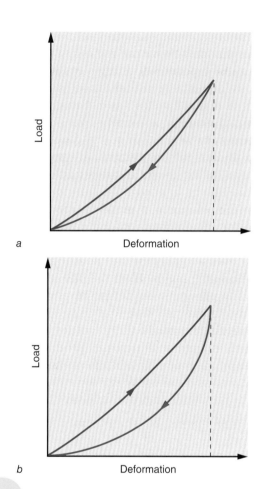

a   Deformation

b   Deformation

FIGURE 10.6 Load–deformation characteristics of materials with *(a)* high resilience and low damping and *(b)* low resilience and high damping.

Key Terms

hysteresis loop The loop on a load–deformation curve defined by the loading and unloading phases; the loop defines the amount of strain energy dissipated between the end of the loading phase and the end of the unloading phase.

resilience The amount of energy returned during unloading as a percentage of the amount of energy absorbed during loading.

damping A low level of resilience; the lower the resilience, the greater the damping.

KEY POINT

The human body converts chemical energy into three forms of mechanical energy: kinetic energy, gravitational potential energy, strain energy.

Viscosity and Viscoelasticity

As described previously, stiffness is a measure of the resistance of a solid material to deformation by a load. The viscosity of a liquid or semiliquid substance is a measure of the resistance of the substance to shear deformation in response to a shear load. Figure 10.7 shows a flat, square piece of wood separated from a larger flat surface by a layer of liquid or semiliquid substance. If the area of the piece of wood is A and a horizontal force F is applied to the piece of wood, then the shear stress on the substance is given by F / A. The movement of the wood is resisted by the viscosity of the substance; the greater the viscosity, the greater the resistance. The viscosity V of the substance is defined as follows:

$$V = \frac{\text{shear stress}}{\text{shear strain rate}}$$

Consequently, for a given level of shear stress, the lower the shear strain rate, the higher the viscosity of the substance and vice versa (Alexander 1968).

In contrast to an elastic material in which the response to loading (and unloading) is immediate, the response of a viscous substance to loading is time dependent; deformation occurs gradually. Most biological materials, including all the musculoskeletal components, have elastic and viscous properties—they behave viscoelastically. As described in chapter 3, a viscoelastic material deforms gradually in response to loading and gradually restores its original dimensions when unloaded.

Mechanical Model of Viscoelasticity

In mechanical models of viscoelastic materials, the elastic elements are usually represented by springs and the viscous elements by Newton's model of a hydraulic piston (a "dashpot") (Taylor et al 1990; figure 10.8). Figure 10.8c shows a mechanical model of a viscoelastic material based on a piston and three springs; one of the springs is in series with the piston and the other two springs are in parallel with the piston. In response to tension loading, the model gradually lengthens in the direction of T at a rate that depends on the stiffness of the springs and the viscosity of the piston. When

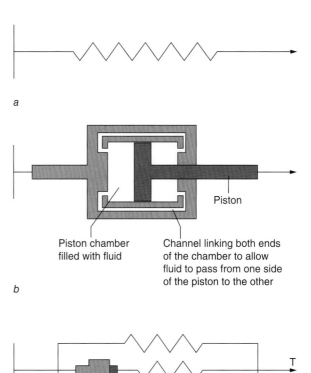

FIGURE 10.8 A mechanical model of viscoelastic behavior. *(a)* A spring represents elastic behavior. *(b)* A piston represents viscoelastic behavior. *(c)* Model of a viscoelastic material that incorporates a piston and three springs with one of the springs in series with the piston and the other springs in parallel with the piston.

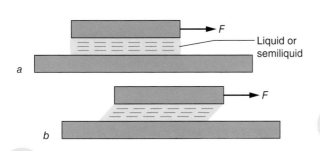

FIGURE 10.7 Response of a liquid or semiliquid to shear stress. *(a)* At the start of application of *F*. *(b)* Some time after the start of application of *F*.

the load is removed, the model gradually restores its original length at a rate depending on the force exerted by the springs and the viscosity of the piston.

Properties of Viscoelastic Materials

The response of a viscoelastic material to loading is always time dependent. However, the actual response depends on the type of load. When a viscoelastic material is subjected to a constant load (lower than yield stress), the material deforms asymptotically with time—it gradually deforms at a progressively decreasing rate until a point at which further deformation ceases. This property of viscoelastic materials is called **creep.** If the load is then removed, the material gradually restores its original dimensions. For example, if a viscoelastic material is subjected to a constant tension load (lower than yield stress), the material gradually lengthens until a point is reached where lengthening ceases (figure 10.9*a*). If the load is then removed, the material gradually restores its original dimensions.

If a viscoelastic material is deformed (within its elastic range) and then held in the deformed position, the stress experienced by the material decreases asymptotically with time until a point at which no further decrease in stress occurs. This property of viscoelastic materials is called **stress relaxation.** If the load is then removed, the material gradually restores its original dimensions. For example, if a viscoelastic material is stretched within its elastic range and then held at a constant length, the tension stress experienced by the material decreases with time until a point where no further decrease in tension stress occurs (figure 10.9*b*). If the load is then removed, the material gradually restores its original dimensions.

In addition to demonstrating creep and stress relaxation, viscoelastic materials also exhibit **strain rate dependency**—the mechanical characteristics of a viscoelastic material depend on the rate of strain. The higher the rate of strain, the smaller the degree to which creep and stress relaxation can occur during deformation, which in turn increases the stiffness, strength (ultimate stress), and toughness of

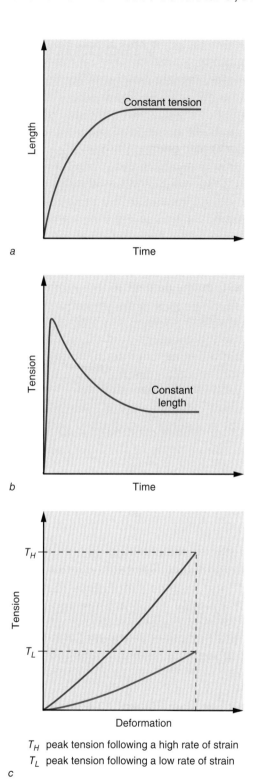

FIGURE 10.9 Properties of viscoelastic materials. *(a)* Creep: length–time curve for a viscoelastic material subjected to a constant tension load. *(b)* Stress relaxation: tension–time curve for a viscoelastic material stretched and then held at a constant length. *(c)* Strain rate dependency: effect of strain rate on stiffness, peak tension, and energy absorption in a viscoelastic material (at the same degree of deformation).

T_H peak tension following a high rate of strain
T_L peak tension following a low rate of strain

the material (Garrett et al 1987, Taylor et al 1990) (figure 10.9c). In general, the higher the strain rate, the greater the stiffness, strength, and toughness of the material. The increased toughness of a viscoelastic material in response to an increased strain rate enables the material to absorb more energy. With regard to the musculoskeletal system, the increased toughness of the components, especially musculotendinous units and bone, in response to increased strain rates can enable the body to dissipate large amounts of energy as heat. However, if the large amounts of stored energy associated with high strain rates are dissipated in the form of failure, then the damage to the musculoskeletal system can be considerable. For example, when a bone fails in response to a low rate of strain, the amount of energy dissipated is relatively low and the bone can fracture in the form of a clean break with relatively little damage to the surrounding soft tissues. However, if the bone fails in response to a high rate of strain, a large amount of energy is dissipated and the bone can shatter into many small pieces, resulting in considerable damage to the bone and sur-rounding soft tissues. Bones can be subjected to very high rates of strain, for example, in high-speed collisions.

Key Terms

creep   The property of viscoelastic materials to deform asymptotically in response to a constant load.

stress relaxation   The property of viscoelastic materials to reduce asymptotically the level of stress experienced in response to a constant level of deformation.

strain rate dependency   The change in the mechanical characteristics of a viscoelastic material in response to changes in strain rate.

KEY POINT

Biological materials, including all the musculoskeletal components, behave viscoelastically. Viscoelastic materials exhibit creep, stress relaxation, and strain rate dependency.

Active and Passive Loading

Figure 10.10a shows a woman standing upright. Her weight W is transmitted to the floor via her feet. She experiences a force R at her feet—the ground reaction force—equal and opposite to W. In this situation the ground reaction force is a vertical force. To move forward, the woman pushes downward and backward against the floor. Provided her foot does not slip, she experiences a ground reaction force equal and opposite to the push she exerts against the floor (figure 10.10b). In this case the ground reaction force has a vertical component R_V counteracting W and a horizontal component R_H enabling her to move forward. R_H does not move the body forward directly; it prevents the foot from slipping backward so that the leg can extend against a fixed point.

The impulse of a force is the product of the force and the duration of the force; the greater the impulse, the greater the change in speed. Consequently, in a sprint start, the greater the magnitude of R_H, the greater the horizontal speed of the body off the blocks is likely to be (figure 10.10c). Similarly, to jump upward it is necessary to push down hard on the floor; in response to the push against the floor the jumper experiences an equal and opposite ground reaction force, which enables her to thrust her body upward. Her upward speed at takeoff is determined by the impulse of the vertical component of the ground reaction force.

In the preceding examples the magnitude and direction of the ground reaction force are determined by muscular activity under the person's conscious control. In these circumstances, when the ground reaction force or any other external load (apart from body weight) is completely controlled by conscious muscular activity, the load is called an **active load.** Because they are under conscious control, active loads are unlikely to be harmful under normal circumstances. In everyday situations, the muscles respond to changes in external loading to ensure that the body is not subjected to harmful loads.

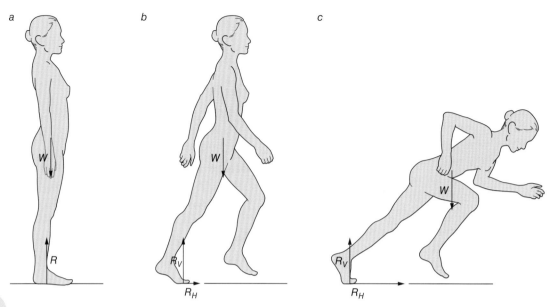

FIGURE 10.10 Horizontal (R_H) and vertical (R_V) components of the ground reaction force *(R)* in various activities. *(a)* Standing. *(b)* Push-off in walking. *(c)* Driving away from the blocks in a sprint start.

However, it takes a finite time for muscles to fully respond (in terms of appropriate changes in the magnitude and direction of muscle forces) to changes in external loading; this time lag is referred to as **muscle latency.** Muscle latency varies between approximately 30 and 75 ms in adults (Nigg et al 1984, Watt and Jones 1971). Consequently, muscles cannot fully respond to changes in external loading that occur in less than the latency period of the muscles. In these circumstances, the body is forced to respond passively (by passive deformation) to the external load; this type of load is a **passive load.** The body is unable to control passive loads and is vulnerable to injury from high passive loads.

The body is subjected to passive loading during, for example, the initial phase of contact of the foot with the ground during walking and running. Approximately 80% of runners and joggers are heel strikers; they contact the ground initially with the heel (Kerr et al 1983). The remaining 20% of runners and joggers contact the ground initially with either the middle of the foot (midfoot strikers) or the front part of the foot (forefoot strikers).

Figure 10.11 shows the variation with time of the vertical component of the ground reaction force of a heel striker running at approximately 4.5 m/s (6 min per mile). The period of ground contact (contact time) can be divided into component phases. In terms of changes in the gravitational potential energy and kinetic energy of the body, contact time can be divided into an absorption phase and a propulsion phase. During the absorption phase—55% to 60% of the contact time (Luethi and Stacoff 1987)—some of the body's gravitational potential energy and kinetic energy is transformed into strain energy in the form of deformation of the floor, shoe, sock, and entire body (especially the musculoskeletal system). Lowering of the body's center of gravity (due to leg flexion) decreases the body's gravitational potential energy. The decrease in the velocity of the center of gravity (due to the negative impulse, or braking effect, of the ground reaction force) decreases the body's kinetic energy (figure

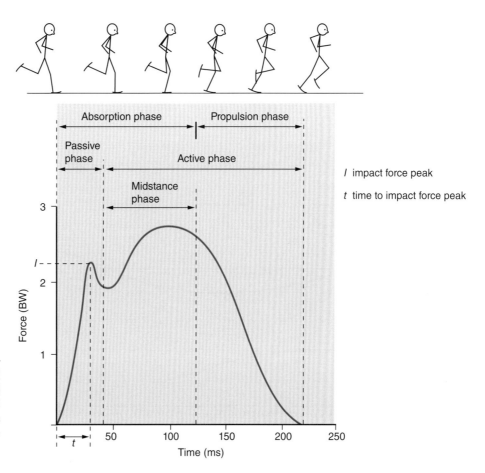

FIGURE 10.11 Force–time curve of the vertical component of the ground reaction force of a heel striker running at 4.5 m/s in running shoes. BW = body weight.

10.12). For an average running velocity of 4.5 m/s, the decrease in gravitational potential and kinetic energy during the absorption phase is about 12% of the average level of gravitational and kinetic energy (based on Alexander 1987).

Figure 10.12 illustrates the change in velocity of the center of gravity of a subject running at an average speed of 4.5 m/s. In figure 10.12*a*, the high point of the center of gravity, the runner is in midflight between the ground contact phase of the left foot and the next ground contact phase of the right foot. Forward velocity is close to maximum (decreased slightly by air resistance following toe-off of the left foot to the position shown in figure 10.12*a*). Vertical velocity is close to zero as the center of gravity is close to its high point. Figure 10.12*b* shows heel strike of the right foot. Forward velocity at heel strike is lower than at figure 10.12*a* because of air resistance. Forward velocity continues to decrease between figure 10.12, *b* and *c*, because of the braking effect of the absorption phase. Vertical velocity at heel

strike is maximal downward and decreases to zero between figure 10.12, *b* and *c*, because of the braking effect of the absorption phase. Figure 10.12*c* shows the end of the absorption phase of the right foot. Forward velocity is at its lowest and vertical velocity is zero. Figure 10.12*d* illustrates the end of the propulsion phase of the right foot. Between figure 10.12, *c* and *d*, the propulsion phase, forward velocity increases to maximum and vertical velocity upward increases to maximum. In figure 10.12*e*, the runner is in midflight between the ground contact phase of the right foot and the next ground contact phase of the left foot. Forward velocity is close to maximum (decreased slightly by air resistance between figure 10.12, *d* and *e*) and vertical velocity is close to zero as the center of gravity is close to its high point.

During the propulsion phase, concentric contraction of the leg extensor muscles, assisted by the release of strain energy in the floor, shoe, sock, and musculoskeletal system, increases the body's gravitational potential

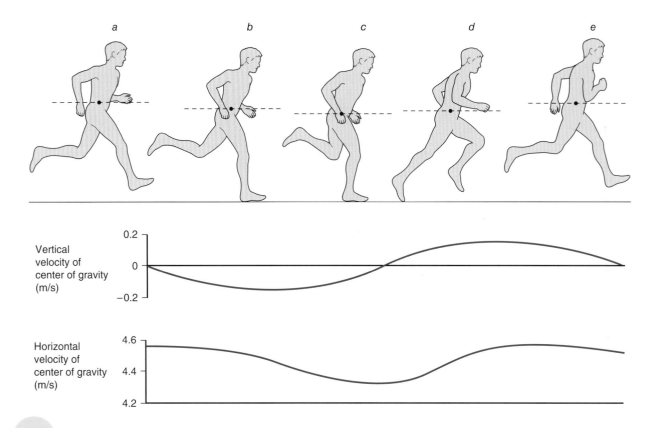

FIGURE 10.12 Change in velocity of the center of gravity of a subject running at an average speed of 4.5 m/s. *(a)* High point of the center of gravity. The runner is in midflight between the ground contact phase of the left foot and the next ground contact phase of the right foot. Forward velocity is close to maximum, and vertical velocity is close to zero. *(b)* Heel strike of the right foot. Forward velocity is lower than at *(a)* because of air resistance, and vertical velocity is maximal downward. *(c)* End of the absorption phase of the right foot. Forward velocity is at its lowest and vertical velocity is zero. *(d)* End of propulsion phase of the right foot. Forward velocity is maximum, and vertical velocity is maximal upward. *(e)* The runner is in midflight. Forward velocity is close to maximum, and vertical velocity is close to zero.

energy and kinetic energy. Raising of the center of gravity (due to leg extension) increases the body's gravitational potential energy. The increase in the velocity of the center of gravity (due to the positive impulse of the ground reaction force) increases the body's kinetic energy (figure 10.12). The ground reaction force is an active load in the propulsion phase. However, in the absorption phase the ground reaction force is initially a passive load and then an active load (see figure 10.11). The passive phase of the absorption phase—the period of muscle latency following heel strike—roughly corresponds to the period between heel strike and foot flat (the loading of the forefoot). The active phase of the absorption phase roughly corresponds to the period between flat foot and

heel-off—also called midstance phase (Luethi and Stacoff 1987). The propulsion phase corresponds to the period between heel-off and toe-off. Because the ground reaction force during midstance and propulsion is an active load, this phase of ground contact is called the active phase.

Key Terms

active load   Any external load (other than body weight) completely controlled by conscious muscular activity.

muscle latency   The time muscles take to fully respond to changes in external loading.

passive load   Any change in external load that occurs within the latency period of the muscles.

Impact and Shock

In running, a high rate of loading immediately following heel strike, which culminates in a peak force, characterizes the ground reaction force–time curve (vertical component) during the passive phase of ground contact. The force then declines slightly before rising again at the start of the active phase (see figure 10.11). The slope of the force–time curve reflects the rate of loading; the steeper the slope, the higher the rate of loading. The higher the rate of loading, the higher the rate of strain on the system. In this context the system refers to the collection of materials subjected to the passive load—the support surface, shoe, sock, and human body.

In mechanics, **impact** refers to a collision between two objects, typically of short duration (Goldsmith 1960). Because the heel of a runner collides with the support surface at heel strike, the peak force during the passive phase is called the **impact force peak** (Nigg et al 1995). The impact force peak reflects the amount of strain on the system; the higher the impact force peak, the greater the amount of strain. At the time of impact force peak, the rate and amount of strain experienced by the system are highest in the part of the system at and around the impact site and least in the parts of the system farthest from the impact site. Following impact force peak, the parts of the system closest to the impact site recoil and press on adjacent parts of the system, subjecting them to a rate and amount of strain similar to that experienced around the impact site at impact. The adjacent parts of the system subsequently recoil such that the impact is transmitted to other parts of the system in the form of a shock wave (or stress wave). The amplitude (amount of strain experienced by the system) and frequency (rate of strain experienced by the system) of the shock wave decline with distance from the impact site because of damping (see later section on shock absorption).

When the initial shock wave passes through a particular part of the system, it is reflected toward as well as away from the impact site. Consequently, shortly after impact the system experiences a large number of shock waves that cause the system to vibrate. The vibration declines and eventually ceases as the energy contained in the shock waves dissipates as heat. In walking and running, heel strike shock waves radiate upward throughout the body, with the amplitude and frequency decreasing with increased distance from the impact site.

Shock is a transitory state in which the equilibrium of the system is disrupted, resulting in a nonuniform distribution of stress (and strain) in the system (Nigg et al 1995). Shock produces **shock waves**—spatial propagation of mechanical discontinuity in a system (Nigg et al 1995). Impact loading causes shock, and the degree of shock produced by an impact depends on the rate of loading and the peak force during the impact; the greater the rate of loading and the greater the peak force, the greater the degree of shock produced and thus the greater the amplitude and frequency of vibration.

When a person is running at about 4.5 m/s, ground contact time is approximately 200 ms and the contributions of the passive, midstance, and propulsion phases to ground contact time are approximately 20%, 40%, and 40%, respectively (see figure 10.11) (Cavanagh and Lafortune 1980; Dickinson et al 1985; Luethi and Stacoff 1987). The body's center of gravity—reflecting the movement of the body as a whole—decelerates throughout the absorption phase; that is, transformation of gravitational potential energy and kinetic energy occurs throughout the absorption phase. However, the deceleration of individual body segments varies with distance from the point of impact; the closer the segment to the point of impact, the greater its deceleration, and the shorter the period over which its gravitational potential energy and kinetic energy are transformed. Consequently, the foot and lower leg experience high levels of deceleration because most of the gravitational potential energy and kinetic energy of these segments transformed during the absorption phase is transformed between heel strike and impact force peak—a relatively small portion of the absorption phase. The higher the deceleration of the segment, the greater the rate and amount of loading experienced by the segment. In running, the impact force peak and rate of loading prior to impact force peak reflect the deceleration of the foot and lower leg (Bobbert et al 1991).

Power of Impact

The rate of loading and peak loading in an impact depend on the **power of the impact**—the rate at which gravitational potential energy and kinetic energy in the system transform into other forms of energy such as strain energy during the impact. The power of the impact is given by E/t, where E is the amount of energy transformed and t is the time to impact force peak (see figure 10.11). The higher the power of the impact, the greater the rate of loading and impact force peak and the greater the degree of shock produced.

Key Terms

impact   A collision between two objects typically of short duration.

impact force peak   The peak force during the passive phase of impact of the body with another body or object.

shock   A transitory state in which the equilibrium of the system is disrupted, resulting in a nonuniform distribution of stress (and strain) in the system.

shock wave   Spatial propagation of mechanical discontinuity in a system as a result of shock.

power of the impact   The rate at which energy is transformed into other forms of energy in an impact.

KEY POINT

The body is subjected to passive loading following heel strike in walking and running; the greater the power of the impact, the greater the degree of shock produced.

The amount of energy transformed in an impact depends considerably on the velocities of the colliding objects. The greater the relative velocity of the objects at impact—the velocity of the objects relative to each other—the greater the amount of energy transformed. For example, if a train moving at 50 km/hr runs into the back of another train moving in the same direction at 40 km/hr, the relative velocity of the trains at impact is 10 km/hr. However, if the same trains were moving toward each other, then their relative velocity at impact would be 90 km/hr. Because the kinetic energy of the trains is proportional to the square of the relative velocity, the amount of energy transformed during a collision at a relative velocity of 90 km/hr would be approximately 80 times greater than in a collision at a relative velocity of 10 km/hr.

The time to impact force peak depends on the stiffness of the system; the higher the stiffness, the shorter the time and therefore the higher the power of the impact. A system's stiffness depends on the stiffness of its materials and the arrangement of the materials relative to the line of action of the impact load. Figure 10.13 shows the composition of the system subjected to impact loading during heel strike in running. The system's components are the support surface, the shoe (including insole and sock), and the foot and ankle (representing the body as a whole). The foot and ankle consist of a combination of fairly stiff material (bone) and fairly compliant material (soft tissues including skin, adipose tissue, musculotendinous units, ligaments, and cartilage).

KEY POINT

The stiffness of a system depends on the stiffness of the materials that comprise the system and the arrangement of the materials in relation to the line of action of the impact load.

At heel strike the calcaneus is protected naturally by a fairly thick layer of adipose tissue called the heel pad. The 2 cm thick heel pad is underneath the calcaneus (Steinbach and Russell 1964). It has been shown that the heel pad and overlying skin are deformed (compressed) by approximately 1 cm at heel strike when one is running at 3.6 m/s (Cavanagh et al 1984). The deformation of the skin and heel pad cushions the impact load, and thus

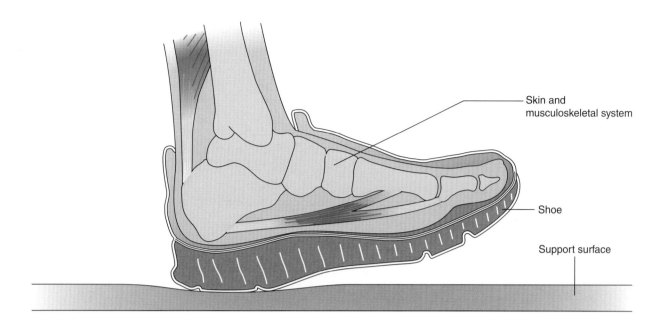

FIGURE 10.13   Components of the system subjected to impact loading during heel strike in running.

the time to impact force peak is greater than what would occur if these structures were not present. The increase in time to impact force peak decreases the power of the impact and consequently decreases the rate of loading and impact force peak, which in turn reduce the level of shock produced.

Effect of Shoes and Surfaces on Time to Impact Force Peak

The compliant soles of running shoes and compliant surfaces tend to increase time to impact force peak in the same way as the heel pad. Figure 10.14 shows the ground reaction force–time curve (vertical component) for the same subject running at approximately 4 m/s barefoot (10.14*a)* and in running shoes (10.14*b)* (Dickinson et al 1985). The duration of ground contact time is similar in both conditions, as is the force–time curve in the active phase, but the force–time curve during the passive phase is different. The impact force peak in the barefoot condition is approximately 3 BW (body weight) compared with approximately 1.9 BW with shoes. However, the most noticeable differences between the two conditions are the time to impact force peak and the rate of loading prior to impact force peak. The

time to impact force peak when the runner is barefoot is approximately 6 ms compared with 24 ms when the runner is wearing shoes. The large difference in time to impact force peak is mainly responsible for the large difference in rate of loading: 500 BW/s barefoot and 79 BW/s with shoes. The rate of loading in the barefoot condition can result in a high level of shock, whereas the rate of loading with shoes is likely to result in relatively little shock.

KEY POINT

Deformation of the skin and heel pad at heel strike cushions the impact load and decreases the power of the impact. The compliant soles of running shoes and compliant surfaces have a similar effect.

Figure 10.15 shows the ground reaction force–time curves for the same person running at 5 m/s in two types of running shoes, one with a hard (low compliance) sole and one with a soft (moderate compliance) sole (Nigg et al 1981). The hard-soled shoe results in a higher impact force peak, a higher rate of loading, and consequently a higher level of shock than the soft-soled shoe. Whereas soft-soled shoes produce less shock than hard-soled

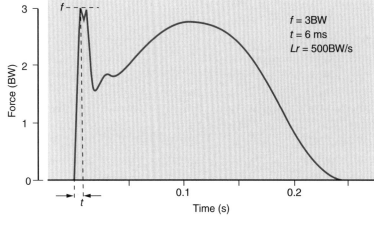

f = impact force peak

t = time to impact force peak

Lr = rate of loading = $\dfrac{f}{t}$

$f = 3BW$
$t = 6$ ms
$Lr = 500BW/s$

a

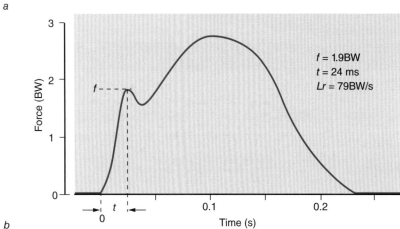

$f = 1.9BW$
$t = 24$ ms
$Lr = 79BW/s$

b

FIGURE 10.14   Force–time curve of the vertical component of the ground reaction force of a heel striker running at 4 m/s. *(a)* Barefoot. *(b)* While wearing running shoes. BW = body weight.

Adapted from *Journal of Biomechanics,* Vol. 18, J. Dickinson, S. Cook, and T. Leinhardt, "The measurement of shock waves following heel strike while running," p. 417, Copyright 1985, with permission from Elsevier Science.

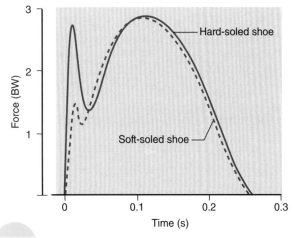

FIGURE 10.15   Vertical component of the ground reaction force while running at 5 m/s in hard- and soft-soled shoes. BW = body weight.

Adapted from B.M. Nigg, J. Denoth, and P.A. Neukomm, 1981, Quantifying the load on the human body: Problems and some possible solutions, In *Biomechanics, Vol. VIIB,* edited by A. Morecki et al. (Baltimore: University Park Press), 89. By permission of B.M. Nigg.

shoes, there is a limit to how soft a shoe can be if it is to be effective in increasing time to impact force peak. If the shoe material is too soft, it can bottom out quickly, such that the time to impact force peak is largely determined by other parts of the system.

Segmental Alignment and Time to Impact Force Peak

At foot strike the alignment of the upper and lower leg and foot in relation to the line of action of the ground reaction force affects the leg's stiffness. The closer the line of action of the ground reaction force to the hip, knee, and ankle joints, the more likely the leg will respond to the impact load like a rod subjected to a compression load at one end in line with the rod. Consequently, the system is stiff;

most of the gravitational potential energy and kinetic energy transformed during the impact occurs in the form of strain energy in bones and joints such that the system experiences a high level of shock. This situation is typical of running heel strike, where the line of action of the ground reaction force passes close to the joint centers of the ankle and knee (figure 10.16*a*).

The greater the moment arm of the ground reaction force about the hip, knee, and ankle joints at foot strike, the more likely the leg will respond to the impact load like a spring, resulting in flexion of the hip, knee, and ankle. This type of action is typical of a forefoot-striking runner, where the line of action of the ground reaction force dorsiflexes the ankle and flexes the knee (figure 10.16*b*). The stiffness of the spring depends considerably on the degree of

activation of the muscles controlling the joints. Whereas muscle latency limits the speed with which a muscle can fully respond to a change in external loading, the muscles can still influence the time to impact force peak by setting the rotational stiffness of the joints prior to foot strike. The greater the degree of tension in the muscles, the greater the degree of rotational stiffness of the joints. A certain amount of rotational stiffness of the joints is necessary to prevent the spring from bottoming out too quickly. In general, the more the leg behaves like a spring at foot strike, the longer the time to impact force peak. Furthermore, forced lengthening of the musculotendinous units results in their absorbing some of the gravitational potential energy and kinetic energy transformed during the impact as strain energy. The more energy the musculotendinous units absorb, the lower the stress (and strain) on other components of the musculoskeletal system, especially bones and joints. Striking the ground with the midfoot or forefoot is generally effective in preventing a high rate of loading following foot strike, such that most midfoot- and forefoot-strikers do not exhibit a discernible impact force peak (Cavanagh et al 1984) (figure 10.16*c*).

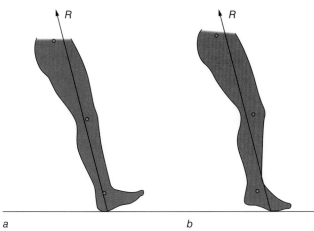

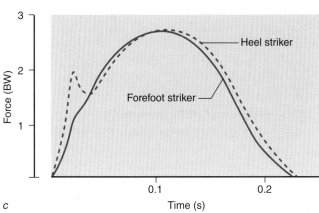

KEY POINT

In heel-strike running, the leg is likely to be fairly stiff at impact, tending to result in a high level of shock. In contrast, the stiffness of the leg in forefoot-strike running is likely to be relatively low, tending to result in little if any shock.

FIGURE 10.16 Alignment of the upper leg, lower leg, and foot in relation to the line of action of the ground reaction force *(R)* at *(a)* heel strike and *(b)* forefoot strike. *(c)* Ground reaction force–time curves (vertical component) for a heel striker and a forefoot striker. BW = body weight.

The shape of the heel of a running shoe can affect the center of pressure (point of application of the ground reaction force) and, as such, the line of action of the ground reaction force in relation to the hip, knee, and ankle joints. Figure 10.17 shows the effect of a flared heel on the line of action of the ground reaction force about the ankle at heel strike. The flared heel increases the moment arm of the ground reaction force about the ankle and tends to promote springlike activity at the ankle. Flared heels have been shown to decrease the magnitude of impact force peak in heel strikers

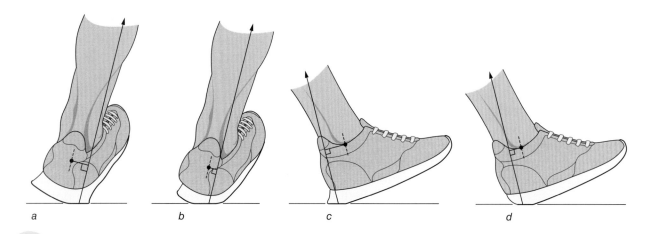

FIGURE 10.17 Effect of flared heels on the line of action of the ground reaction force. *(a, c)* Flared heels. *(b, d)* Nonflared heels.

(Nigg and Bahlsen 1988; Nigg and Morlock 1987). However, because flared soles increase the moment arm of the ground reaction force about the ankle, they also increase the load on the supinators (the muscles controlling the amount and rate of pronation (figure 10.17, *a* and *b)* and the dorsiflexors (the muscles that control the amount and rate of plantar flexion (figure 10.17, *c* and *d)*.

Shock Absorption

Shock absorption refers to the damping of vibrations generated in a system. With regard to the human body, all of the tissues are viscoelastic to a certain extent and dissipate a certain amount of energy as heat by hysteresis during each cycle of vibration. Consequently, the amplitude of vibration progressively decreases and vibration usually ceases fairly quickly. The speed with which vibrations are absorbed and the extent to which vibrations are transmitted to other regions of the body depend on the degree of damping afforded by relative motion between bone and soft tissues and joints. Bones are much stiffer than the soft tissues that surround them. Because of the difference in stiffness, the frequency of vibration of bone is greater than that of the surrounding soft tissues following impact loading. Consequently, the bone and soft tissues move relative to each other, and some of the energy contained in the vibration of the bone is absorbed as strain energy by the soft tissues and subsequently dissipated as heat. Because

hysteresis in the soft tissues is greater than in bone, the vibrational energy of bone dissipates fairly quickly.

Key Term

shock absorption The damping of vibrations generated in a system.

> **KEY POINT**
>
> All of the tissues of the body are viscoelastic to a certain extent and thus dissipate a certain amount of energy as heat by hysteresis during each cycle of vibration.

Just as bones are much stiffer than the soft tissues surrounding them, bones are stiffer than the joints linking the bones together. Synovial and cartilaginous joints deform in response to loading and in so doing absorb energy, including vibrational energy of the bones. In symphysis joints, such as the intervertebral joints and pubic symphysis, the amount of fibrocartilage is large compared with the amount of articular (hyaline) cartilage in synovial joints. Consequently, symphysis joints absorb shock better than synovial joints. Figure 10.18 shows a section through a typical synovial joint. In response to loading, both epiphyses deform to a certain extent. Cancellous bone is stiffer than subchondral bone, which is stiffer than articular cartilage. Consequently, articular cartilage and subchondral bone absorb shock better than an equivalent amount of cancellous bone. However, in a typical synovial joint, the

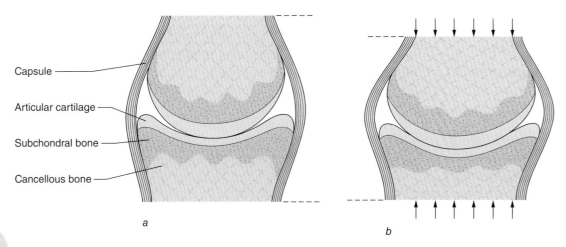

FIGURE 10.18 Response of a synovial joint to compression loading. *(a)* Unloaded. *(b)* Response to compression loading.

volume of cancellous bone is greater than the volume of articular cartilage and subchondral bone, such that in healthy synovial joints the cancellous bone can be the major contributor to shock absorption (Radin and Paul 1971, Hoshino and Wallace 1987).

Effect of Impact Loading on Synovial Joints

With regard to synovial joints, the greater the power of the impact, the greater the likelihood of microfractures to trabeculae in cancellous bone. Healing of fractured trabeculae increases the stiffness of the cancellous bone and consequently decreases its shock-absorbing capacity (Radin and Paul 1971, Simon et al 1972, Radin et al 1973). Repeated microfracture and healing of cancellous bone progressively increase stiffness and progressively decrease shock-absorbing capacity. As the stiffness of cancellous bone increases, the subchondral bone and articular cartilage are subjected to increased strain (rate and amount), thereby increasing the likelihood of microtrauma—crushing or tearing of the matrix in subchondral bone and articular cartilage. Such microtrauma decreases the shock-absorbing capacity of the subchondral bone and articular cartilage. Not surprisingly, it has been shown that the shock-absorbing capacity of healthy joints is greater than that of degenerated joints (Voloshin and Wosk 1982). Progressive damage to articular cartilage can culminate in osteoarthritis (figure 10.19).

KEY POINT

Repeated microfracture and healing of cancellous bone result in a progressive increase in stiffness, which increases the strain on subchondral bone and articular cartilage.

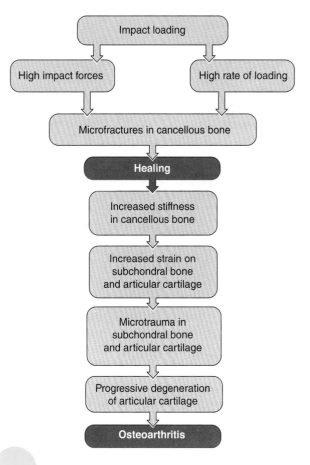

FIGURE 10.19 Development of osteoarthritis.

Viscoelastic Inserts and Shock Absorption

Heel strike during walking is the most common source of impact loading on the human body, and repeated exposure to high levels of shock resulting from walking heel strike appears to be a major cause of joint degeneration (Radin et al 1973, Wosk and Voloshin 1985). Viscoelastic inserts in shoes have been shown to reduce the amplitude of heel-strike-induced shock waves and can compensate to a certain extent for the decreased shock-absorbing capacity of degenerated joints (Voloshin and Wosk 1981, Johnson 1988). Voloshin and Wosk (1981) showed that prolonged use (18 months) of viscoelastic heel inserts can significantly reduce clinical symptoms, especially pain, in adults suffering from chronic degenerative joint conditions such as osteoarthritis and intervertebral disc damage.

In another study, the same researchers investigated the effect of prolonged use of viscoelastic heel inserts on a group of 254 males and 128 females, ages 14 to 75 years, suffering from chronic low back pain (Wosk and Voloshin 1985). The subjects used a 4-point scale to rate the effect of the treatment at 1 month, 3 months, and 1 year. As shown in figure 10.20, the results reflect a progressive increase in the number of subjects who reported an excellent result (36%, 50%, 62%) and a progressive decrease in the number of subjects who reported a poor result (16%, 12%, 8%) over the 1-year period of the study. The results suggest that the reduced level of shock resulting from using the viscoelastic heel inserts enabled the body to help repair the damaged intervertebral joints. The extent of repair may depend on, among other factors, the age of the subject at the start of treatment. This is reflected in the number of young (14-25 years), middle-aged (26-45 years), and older-aged (46-75 years) subjects reporting excellent and good results (figure 10.21). The trend of decreased pain was similar in the three age groups, but the amount of improvement decreased with age.

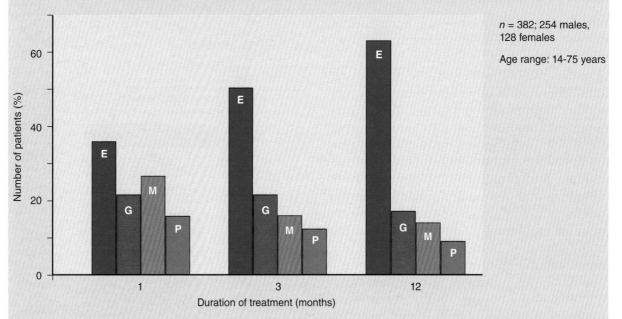

FIGURE 10.20 Effect of continuous use of viscoelastic shoe inserts on the severity of chronic low back pain. E = excellent result, no pain and no restriction on habitual physical activity; G = good result, only slight pain periodically and no restriction on habitual physical activity; M = moderate result, significant decrease in pain and little or no restriction on habitual physical; P = poor result, either no decrease in pain or decrease not maintained.

Adapted from *Archives of Physical Medicine and Rehabilitation*, Vol. 66, J. Wosk and A.S. Voloshin, "Low back pain: Conservative treatment with artificial shock absorbers," pp. 146-147, Copyright 1985, with permission from Elsevier.

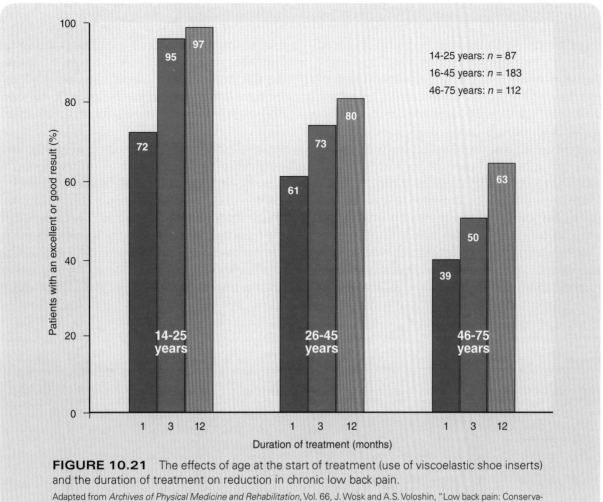

FIGURE 10.21 The effects of age at the start of treatment (use of viscoelastic shoe inserts) and the duration of treatment on reduction in chronic low back pain.

Adapted from *Archives of Physical Medicine and Rehabilitation*, Vol. 66, J. Wosk and A.S. Voloshin, "Low back pain: Conservative treatment with artificial shock absorbers," pp. 146-147, Copyright 1985, with permission from Elsevier.

KEY POINT

Heel strike during walking is the most common source of impact loading on the human body; repeated exposure to high levels of shock resulting from heel strike can be a major cause of joint degeneration. Viscoelastic shoe inserts can compensate to a certain extent for the decreased shock-absorbing capacity of degenerated joints.

Energy Absorption During Active Loading

The amount of strain energy absorbed by the various components of a system during deformation depends on the toughness and volume of the components. The tougher the component and the greater its volume, the more strain energy it is likely to absorb. Of the various musculoskeletal tissues and related tissues such as the heel pad, muscle has the least capacity for absorbing strain energy, tendon has the highest capacity (approximately 65 times that of muscle), and the capacity of bone (approximately 55 times that of muscle) is between that of muscle and tendon (Evans 1971). Whereas the energy-absorbing capacity of muscle is very low compared with tendon, the musculotendinous unit has the highest energy-absorbing capacity of all musculoskeletal and associated tissues (Evans 1971). During the midstance phase in running, the system of support surface, shoe (including sock and insole), and musculoskeletal system (including associated tissues such as heel pad and skin) deforms in response to the impulse of the (active) ground reaction force. The leg is forcibly flexed and the arch of the foot is forcibly flattened such that strain energy is stored in the bones (bending of long bones and compression of short bones), joints (compression of epiphyses), musculotendinous units (eccentric actions of leg extensor and foot arch support musculotendinous units), and ligaments supporting the arch of the foot. Figure 10.22 shows a model of the movement of the lower leg and foot from foot flat (figure 10.22, *a* and *c)* to peak force (figure 10.22, *b* and *d)* during the active phase, which illustrates the strain on the calf muscles and arch support mechanisms.

When an athlete is running at a middle-distance pace (7-8 m/s), his gravitational potential energy and kinetic energy are estimated to decrease by about 100 J during the absorption phase and then increase by about 100 J during the propulsion phase (Ker et al 1987). The same researchers also estimate that during the absorption phase, approximately 35 J, 17 J, and 5 J are stored in the calf muscle–tendon units, arch support mechanisms, and shoe, respectively. The heel pad is estimated to store approximately 3 J (Cavanagh et al 1984), and a specially designed indoor running track can store in the region of 7 J (McMahon and Green 1978). These findings are summarized in the first column of table 10.2, and the resilience of the materials (reported in the same sources) is shown in the second column of the table. The third column shows the theoretical potential for the release of strain energy in the form of

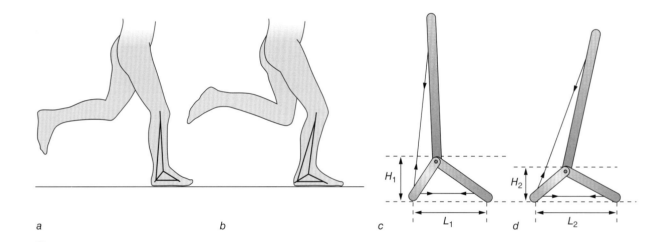

FIGURE 10.22 Storage of strain energy in the calf muscles and arch support mechanisms during midstance in running.

Table 10.2 Energy Storage and Resilience in System Components During the Ground Contact Phase in Middle-Distance Running

Material	Energy absorbed (J)	Resilience (%)	Energy returned (J)
1. Calf muscles	35	93[a]	32.5
2. Arch support mechanism	17	78	13.2
3. Heel pad	3	20	0.6
4. Shoe	5	60	3.0
5. Track	7	90	6.3
6. Other[b]	33	64[c]	21.1
Total	100		76.7

[a]Resilience of Achilles tendon; [b]energy stored in other parts of the body; [c]average of 1, 2, and 3.

Data for energy absorbed and resilience reported in Cavanagh et al. 1984, Ker et al. 1987, and McMahon and Greene 1978.

useful mechanical work during the propulsion phase. Whereas the estimate for the release of useful strain energy is likely higher than that which occurs in the real situation, the values indicate the considerable energy-saving potential of the system.

Clearly, the energy-absorbing capacity and resilience of musculotendinous units are likely to be major determinants of the quality of a person's performance in games and sports. However, musculotendinous units can only function efficiently when the muscles are not fatigued. A fatigued muscle is limited in its ability to apply tension to its tendons, which limits the ability of the tendons to absorb strain energy. Consequently, muscle fatigue reduces the energy-absorbing capacity of a musculotendinous unit; the greater the degree of muscle fatigue, the greater the reduction in energy-absorbing capacity. In any movement involving the absorption of gravitational potential energy and kinetic energy, such as in the absorption phase of running or when landing from a jump, the extent to which a particular musculotendinous unit has to lengthen to make its contribution to energy absorption is determined by the force the muscle can produce. The greater the force of the muscle, the shorter the distance it must lengthen to absorb a given amount of energy. Fatigue reduces the amount of contractile force a muscle can produce; a fatigued muscle has to lengthen farther than a nonfatigued muscle to absorb a given amount of energy.

It has been shown that a muscle yields at the same degree of strain irrespective of its level of fatigue. Consequently, a fatigued muscle is more likely to be stretched beyond its yield point than a nonfatigued muscle (Mair et al 1996). Furthermore, when the level of effort remains fairly constant, as when one is running at a constant speed, the gradual reduction in energy-absorbing capacity of musculotendinous units due to fatigue inevitably results in increased strain on associated structures—especially bones and joints—because the amount of energy absorbed in each absorption phase of the activity is likely to remain fairly constant. Consequently, the greater the degree of muscular fatigue, the greater the risk of injury to all components of the musculoskeletal system (Grimston and Zernicke 1993). Case study 7 examines the effect of lower-extremity fatigue on shock attenuation and lower-limb joint mechanics, with applications to athletes.

> ### KEY POINT
>
> The musculotendinous unit has the highest energy-absorbing capacity of all the musculoskeletal tissues, but the capacity to absorb energy decreases with increasing fatigue. The gradual reduction in energy-absorbing capacity of musculotendinous units due to fatigue increases the strain on associated structures, especially bones and joints.

CASE STUDY 7 FATIGUE AND SHOCK ABSORPTION

Coventry E, O'Connor KM, Hart BA, Earl JE, Ebersole KT. 2006. The effect of lower extremity fatigue on shock attenuation during single-leg landing. *Clinical Biomechanics* 21:1090-1097.

When an athlete lands from a height, as in sports such as volleyball, basketball, netball, and gymnastics, the shock experienced by the musculoskeletal system as a result of the impact of the landing must be attenuated to reduce the risk of injury. Shock is attenuated passively by noncontractile tissues including bone, cartilage, and synovial fluid. Shock is attenuated actively through eccentric actions of musculotendinous units and coordinated joint movements. Active mechanisms, in particular, eccentric actions of musculotendinous units, are thought to be the major contributors to shock attenuation. Consequently, it is thought that muscle fatigue is likely to decrease the capacity of the body to attenuate shock and therefore increase the risk of injury. Previous studies of the

(continued)

effect of muscle fatigue on shock attenuation have produced equivocal results, largely because of the complex interaction of the passive and active shock attenuation mechanisms. The purpose of the present study was to determine the effect of lower-extremity fatigue on shock attenuation and lower-limb joint mechanics (joint movement and energy absorption in joints). The subjects were eight male subjects (age 25.8 ± 2.4 years; mass 81.6 ± 6.8 kg; height 184 ± 0.07 cm).

After adequate practice, each subject performed a fatigue landing protocol (FLP) to exhaustion on the dominant leg. The FLP consisted of repeated trials of a continuous sequence of movements: landing from a drop height (80% of maximum two-leg countermovement jump), a maximum effort one-leg countermovement jump, 5 one-leg squats to 90° of knee flexion. As subjects landed from the drop height in each sequence of the FLP, the investigators assessed shock attenuation by means of accelerometers attached to the anteromedial aspect of the tibia and to the forehead. Lower-extremity joint movements were assessed by means of an electromagnetic tracking system, and the energy absorbed at each joint was assessed by combining ground reaction force data (landing on a force platform) with joint movement data.

The FLP induced exhaustion, but there was no significant change in shock attenuation. With increasing fatigue, hip and knee flexion increased and ankle plantar flexion decreased at touch-down. The energy absorbed by the hip extensors increased and the energy absorbed by the ankle plantar flexors decreased. The results suggest that the body is able to adapt to one-leg fatigue by altering limb kinematics to use the larger proximal muscles to absorb energy, which prevents or minimizes any increase in shock.

Application

The results of the study are consistent with earlier investigations of the effect of muscle fatigue on shock attenuation and lower-limb kinematics, especially in middle-distance and long-distance runners (Yoshikawa et al 1994, Fyhrie et al 1998, Milgrom et al 2000). In particular, the results indicate that as leg fatigue increases, there is a progressive increase in the energy absorbed by the stronger muscle groups (in general, the hip extensors are stronger than the knee extensors, which are stronger than the ankle plantar flexors) to prevent a decrease in shock attenuation. A decrease in shock attenuation has been associated with an increase in the incidence of stress fractures in runners. Consequently, physical conditioning for runners should ensure adequate strength training for all three leg extensor muscle groups (hip extensors, knee extensors, ankle plantar flexors), and runners should avoid excessive mileage that is likely to fatigue the hip extensors, because this is likely to decrease shock attenuation and increase the risk of stress fractures in the bones of the lower limbs.

Summary

The musculoskeletal system responds to changes in loading in two ways: actively, in the form of muscular control of joint movements, and passively, by passive deformation. The combined effect of these responses is to prevent or reduce potentially harmful loads. The effectiveness of the active response is limited by muscle latency such that high rates of loading result in a completely passive response in the initial phase of loading. Furthermore, high passive loads tend to produce shock and repeated exposure to shock can cause joint degeneration. The musculoskeletal system normally adapts its structure to more readily withstand the time-averaged loads exerted on it; this is the subject of the next chapter.

Review Questions

1. Describe the stress–strain characteristics of bone and ligament.
2. Differentiate between toughness, fragility, and brittleness.
3. Differentiate between hysteresis, resilience, and damping.
4. Describe the effects of strain rate dependence on the mechanical characteristics of viscoelastic materials.
5. Differentiate between passive and active loading.
6. Differentiate between impact and shock.
7. Describe the effects of heel strike and forefoot strike on the level of shock produced and mechanism of energy absorption.
8. Describe the effects of muscle fatigue on the stress experienced by other musculoskeletal components.

Structural Adaptation of the Musculoskeletal System

In response to the loads imposed on the musculoskeletal system as a result of normal everyday physical activity (ranging from activities of daily living to manual labor and training for sport), the musculoskeletal components (muscles, tendons, ligaments, cartilage, bone) experience strain. A certain level of strain is necessary to ensure normal growth and development, but excessive strain results in abnormal growth and development and injury. Under normal circumstances the musculoskeletal components continuously adapt their size, shape, and structure to the time-averaged strain. This chapter describes the effects of changes in time-averaged loading on the size, shape, and structure of the musculoskeletal components.

11

OBJECTIVES

After reading this chapter, you should be able to do the following:

1. Describe the relationship between genetic and environmental influences on growth and development.
2. Differentiate between response to loading and adaptation to loading.
3. Differentiate between anatomical alignment and functional alignment in normal joints.
4. Describe the effects of age on the modeling capacity of bone.
5. Describe structural adaptation in regular fibrous tissues, at tendon and ligament insertions, and in muscle.

Adaptation

Two types of factors influence growth and development of the human body: genetic and environmental factors. The genes determine the pattern of growth and development in a set of "genetic instructions" called the **genotype.** However, the body is constantly subjected to a variety of environmental influences including nutritional state, changes in body temperature, and the physical stress imposed by movement of the body. These environmental influences affect the genotype by modifying the timing, rate, extent, and type of growth and development that occur, such that the body becomes adapted to function effectively in relation to the environmental conditions (Malina et al 2004, Huang et al 2004, Wackerhage and Rennie 2006, Seynnes et al 2008). **Adaptation** to environmental conditions is continuous throughout life.

The combination of size, shape, and structure of the body that results from the effect of environmental conditions on the genotype is called the **phenotype.** The phenotype reflects both the general effects of the genotype and the specific effects of the environmental influences. We can actually see and measure the phenotype. Identical twins have identical genotypes and tend to look similar to each other throughout life. However, they obviously have different phenotypes in that they are not subjected to exactly the same environmental influences. This is usually more apparent during adulthood, when differences in lifestyle, especially in terms of nutrition and physical exercise, can result in marked differences in phenotype. Figure 11.1 illustrates the concept of adaptation.

Key Terms

genotype The pattern of growth and development determined by the genes.

adaptation The changes in structure and function that occur as a result of a change in environmental conditions.

phenotype The appearance and form (size, shape, and structure) of the body that result from the interaction of the genotype and environmental influences.

Biopositive and Bionegative Effects of Loading

The effects of loading on the musculoskeletal system tend to be either bionegative or biopositive (Nigg et al 1981). **Bionega-** tive effects result from insufficient loading and excessive loading (figure 11.2). Prior to maturity, insufficient loading can result in

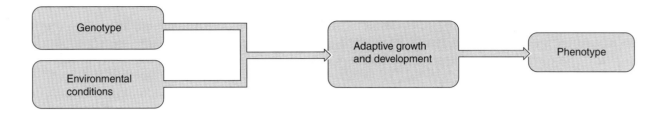

FIGURE 11.1 Model of adaptation.

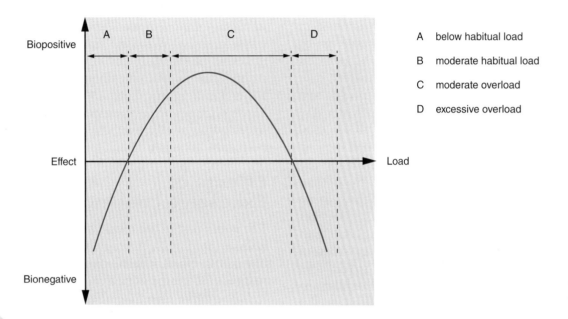

A below habitual load

B moderate habitual load

C moderate overload

D excessive overload

FIGURE 11.2 The relationship between load and effect on the musculoskeletal system.

Adapted from B.M. Nigg, J. Denoth, and P.A. Neukomm, 1981, Quantifying the load on the human body: Problems and some possible solutions, In *Biomechanics, Vol. VIIB,* edited by A. Morecki et al. (Baltimore: University Park Press), 95. By permission from B.M. Nigg.

abnormal growth and development of the musculoskeletal system; in adults, insufficient loading can decrease the functional capacity of musculoskeletal components. Like insufficient loading, excessive loading can result in abnormal growth and development of the musculoskeletal system prior to maturity. In addition, excessive loading can injure musculoskeletal components at any age. **Biopositive effects**—including normal growth and development of the musculoskeletal system prior to maturity and increased functional capacity of musculoskeletal components following physical training—result from moderate loading.

Effects of Insufficient Loading

A certain amount of loading is essential to promote normal growth and development of the musculoskeletal system (Zernicke and Loitz 1992, Lieber 1992). This is clearly demonstrated in children with diseases of the nervous system, such as polio, which prevent normal muscular activity and normal patterns of loading. The muscles of such children are usually poorly developed and the bones are usually abnormal in shape and much smaller than in a child who does not have polio (Jurimae and

Jurimae 2001, Malina et al 2004). To grow and develop normally, the musculoskeletal system needs the mechanical stimulation provided by a normal pattern of motor skill development—the postures, movements, and levels of physical activity associated with an active childhood.

After maturity, the maintenance of the musculoskeletal system still depends on the mechanical stimulation provided by regular exercise. Lack of use results in decreased functional capacity and in **atrophy**—decrease in mass—of musculoskeletal components. With regard to muscle, loss of strength is rapid in totally inactive muscles. For example, studies of bed rest and unilateral limb suspension have shown that a strength loss is evident after only 1 day of inactivity and that strength loss proceeds at the rate of 1% to 1.5% per day (Muller 1970, de Boer et al 2007, Tesch et al 2008). Immobilization of a limb in a plaster cast results in a loss of strength of about 22% in the first 7 days (Muller 1970, Booth 1987, Urso et al 2006). Similarly, lack of weight bearing due to bed rest (Donaldson et al 1970; Baecker et al 2003), immobilization in a plaster cast (Anderson and Nilsson 1979), denervation of muscle (Brighton et al 1985), and space flight (Whedon 1984, Adams et al 2003) substantially decrease bone mass and, therefore, the strength and toughness of bone.

Effects of Moderate Loading

Moderate loading generally has a biopositive effect on growth, development, and maintenance of the musculoskeletal system. The moderate range can be subdivided into a **moder-**

KEY POINT

To grow and develop normally, the musculoskeletal system needs the mechanical stimulation provided by a normal pattern of motor skill development. After maturity, the maintenance of normal musculoskeletal function depends on the mechanical stimulation provided by regular physical activity.

ate habitual range—loading associated with normal daily activity (see figure 11.2)—and a **moderate overload range**—loading associated with physical training. The moderate habitual range promotes normal growth and development of the musculoskeletal system prior to maturity and maintains a healthy level of musculoskeletal function consistent with normal daily activity after maturity. Loading within the moderate overload range has a biopositive effect on the musculoskeletal system resulting in **hypertrophy**—increase in mass—of musculoskeletal components and, consequently, increased functional capacity in terms of strength and toughness of musculotendinous units, ligaments, cartilage, and bone (Zernicke and Loitz 1992, Lieber 2002). These changes occur in response to moderate overloading at all ages, but they are far more noticeable in adults than in children.

Effects of Excessive Loading

The upper part of the moderate overload range is associated with a progressive decrease in biopositive effect, and the **excessive overload range** is associated with a progressive increase in bionegative effect (see figure 11.2). Excessive overload in the form of continuous or highly repetitive nonimpact loading, such as can occur in swimming, rowing, and cycling, can result in severe angular deformities in joints or reduced bone lengths prior to maturity and degenerative joint disease in adults (Frost 1990, Forwood 2001, Duyar 2008). Excessive overload in the form of impact loading, such as can occur following heel strike or foot strike in running, or hitting the ball in baseball, tennis and volleyball, can result in **injury**—structural damage to one or more components of the musculoskeletal system, resulting in decreased functional capacity (Petersen and Renström 2000).

The basic relationship between load and effect on the musculoskeletal system, shown in figure 11.2, is similar for everyone. However, the actual amount (magnitude and frequency) of loading corresponding to insufficient loading, moderate habitual loading, moderate overloading, and excessive overloading for a particular person depends on her stage of maturity and level of physical conditioning. In other words, a particular amount of loading that corresponds to moderate overload for one person may correspond to insufficient loading or excessive loading for someone else.

Key Terms

bionegative effects   Changes in the structure and function of the musculoskeletal system resulting in abnormal growth and development and decreased functional capacity.

biopositive effects   Changes in the structure and function of the musculoskeletal system resulting in normal growth and development and increased functional capacity.

atrophy   Decrease in mass of musculoskeletal components, especially in reference to muscles, tendons, and ligaments.

moderate habitual range   The range of loading on the musculoskeletal system associated with normal daily activity.

moderate overload range   The range of loading on the musculoskeletal system that is associated with physical training and results in biopositive effects.

hypertrophy   Increase in mass of musculoskeletal components, especially in reference to muscles, tendons, and ligaments.

excessive overload range   The range of loading on the musculoskeletal system that is associated with physical training and results in bionegative effects.

injury   Structural damage to one or more components of the musculoskeletal system resulting in decreased functional capacity.

KEY POINT

Bionegative effects result from insufficient and excessive loading. Biopositive effects result from moderate loading.

Physical Training and Injury

An athlete's performance largely depends on her physical fitness and motor ability. The need for a certain level of physical fitness—cardiorespiratory endurance, local muscular endurance, strength, and flexibility—reflects the physiological demands of the sport. Similarly, the need for a certain level of motor ability—ability to perform specific skills—reflects the technical demands of the sport. A high level of motor ability (skill) requires a high level of neuromuscular coordination in the form of a combination of speed, agility, balance, and power.

The physiological and technical demands of different sports vary considerably. For example, performance in long-distance running events requires cardiorespiratory endurance, whereas performance in springboard diving requires more technical ability. A training program—exercises designed to improve an athlete's performance—should reflect the balance between the physiological and technical demands of the sport. However, whether a particular training exercise is designed to improve physical fitness or technique, the effect on the musculoskeletal system is basically the same—the musculoskeletal system experiences overload (figure 11.3). The effect of training depends on the level of overload.

A fairly high level of physical effort is necessary to generate a large enough physiological stimulus to improve cardiorespiratory endurance in an already well-trained endurance athlete. The intensity of such effort almost certainly subjects the musculoskeletal system to loads at the upper end of the moderate overload range and probably to loads in the lower end of the excessive overload range. This level of loading inevitably results in minor damage (microtrauma) to the musculoskeletal tissues, especially muscle and connective tissue. Given adequate rest, these tissues not only heal but also adapt their structures over time to more readily withstand loads imposed on them during training. However, when rest periods are inadequate, the rate at which microtrauma occurs outpaces the processes

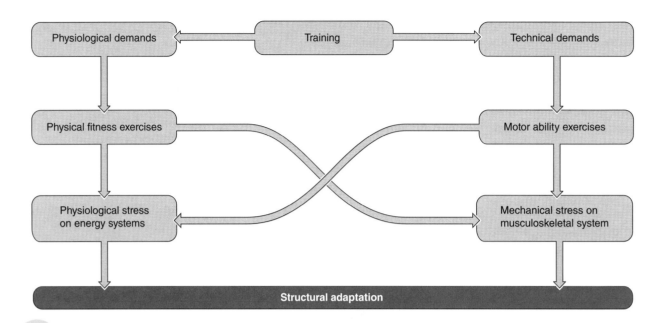

FIGURE 11.3 The training process.

of repair and structural adaptation such that microtrauma gradually accumulates and eventually results in a **chronic injury**—an injury that develops over a period of time (cumulative microtrauma) and is characterized by a gradual increase in pain and functional impairment (Watkins and Peabody 1996).

The main cause of chronic injuries appears to be **overtraining,** a training and competition schedule involving training sessions that are too long or rest periods between training sessions that are too short (Jones et al 1994). Proper progression in the amount of exercise and, therefore, the level of loading on the musculoskeletal system, is essential to avoid injury. The amount of exercise should increase gradually so the musculoskeletal and cardiorespiratory systems have sufficient time to adapt to the gradual increase in the intensity of the training exercises. Although it is possible to significantly increase cardiorespiratory fitness relatively quickly, similar changes in the musculoskeletal system take much longer, especially if the person has been sedentary for a number of years (Micheli 1982, Hootman et al 2001).

Jumper's knee is a common chronic injury in sports such as volleyball and basketball that involve high-frequency jumping and landing. The condition is characterized by pain at one or more of three sites: the insertion of the quadriceps tendon to the upper pole of the patella and the insertion of the patellar ligament to the lower pole of the patella or tibial tuberosity (Ferretti et al 1990). Ferretti and colleagues (1990) describe three stages of jumper's knee. Stage 1 is characterized by pain after a practice or after a game. Stage 2 is characterized by pain at the beginning of activity that disappears after warm-up and reappears after completion of activity. Stage 3 is characterized by fairly persistent pain that usually is severe enough to prevent participation in games and sports. Continuing to train or play at this stage can result in complete failure of the patellar ligament (Ferretti et al 1990). Complete rupture of the ligament is usually referred to as an **acute injury**—an injury that occurs suddenly (sudden macrotrauma) and is severe enough in terms of pain or functional disablement to prevent further participation, at least temporarily (Watkins and Peabody 1996). Acute injuries range from minor (e.g., a minor muscle tear) to severe (e.g., a complete ligament rupture or complete bone fracture). It has been shown that many acute injuries, especially severe muscle, tendon, and ligament injuries, are the sudden end result of progressive degeneration of the structure concerned (Ferretti et al 1990). Not surprisingly, Ferretti and colleagues (1990) suggested that continuing to train or play when suffering from a minor injury significantly increases the risk of a severe acute injury.

Key Terms

chronic injury An injury that develops over a period of time (cumulative microtrauma)

and is characterized by a gradual increase in pain and functional impairment.

overtraining A training and competition schedule involving training sessions that are too long or rest periods between training sessions that are too short, resulting in decreased functional capacity.

acute injury An injury that occurs suddenly (sudden macrotrauma) and is severe enough in terms of pain or functional disablement to prevent further participation, at least temporarily.

Response and Adaptation of Musculoskeletal Components to Loading

The response of a musculoskeletal component to loading refers to the immediate changes in stress and strain experienced by the component. Provided the load (or change in load) is not prolonged, the stress and strain experienced by the component in **response to loading** are unlikely to result in any structural change in the external form (size and shape) or internal architecture of the component, even if the load is within the moderate overload range. For example, it is unlikely that the changes in stress and strain experienced by the elbow extensor musculotendinous units during the performance of a single press-up will structurally change the external form or internal architecture of the elbow extensor musculotendinous units. However, if the press-up exercise is performed more frequently, for example, as part of a fitness training program, and the load on the elbow extensor musculotendinous units is within the moderate overload range during each repetition, the elbow extensor musculotendinous units may experience structural changes in their external form or internal architecture in a way that enables them to more readily withstand the time-averaged increase in load. The structural change in the external form or internal architecture of musculoskeletal components that occurs as a result of changes in the time-averaged loads exerted on them is referred to as **structural adaptation** (Carter et al 1991).

Key Terms

response to loading The immediate changes in stress and strain experienced by musculoskeletal components following a change in loading.

structural adaptation The structural change in the external form or internal architecture of musculoskeletal components that occurs as a result of changes in the time-averaged loads exerted on them.

Optimal Strain Environment

The period of growth and development prior to maturity is associated with a gradual increase in the range and complexity of movement patterns and an almost continuous increase in body weight. Consequently, the total load on the musculoskeletal system increases during this period. To a certain extent this increase in loading is balanced by an increase in the size of the musculoskeletal system. However, the period of growth and development is also associated with considerable changes in the relative size and shape of the various body segments (figure 11.4). For example, up to about 2 years of age the length of the lower limbs is about the same length as the upper limbs. Thereafter, the lower limbs increase in length and weight more rapidly than the upper limbs. The change in the complexity of movement patterns, change in relative size of body segments, and gradual increase in body weight combine to produce an almost continuous change in the pattern of loading on the musculoskeletal system.

The musculoskeletal system adapts to these changes and continues to adapt to such changes throughout life. It is well established that all components of the musculoskeletal system adapt (or try to adapt) their external form and internal architecture to the time-averaged loads exerted on them to maintain an optimal strain environment (Taber 1995, Frost 2003, Doschak and Zernicke 2005). The **optimal strain environment** (OSE) of each component appears to be maintained by a negative-feedback system that is similar in operation to a thermostat; a change in time-averaged load producing a level of strain outside the strain limits of the OSE results in a change in the external form or internal architecture of the component so that the OSE is restored. For example, in bone, an increase in the time-averaged load above the upper limit of the OSE range will result in an increase in the cross-sectional area of the bone shaft or a realignment of the trabeculae in one or both epiphyses such that OSE is restored. Similarly,

a decrease in the time-averaged load below the lower limit of the OSE range will result in a decrease in the cross-sectional area of the shaft of the bone such that OSE is restored. Prolonged absence of load or minimal load, for example, because of paralysis or prolonged immobilization following injury, will result in fairly rapid adaptation in the form of reduced mass. However, the loss of mass eventually ceases at a level referred to as the **genetic baseline**—the genetically predetermined mass that results from growth and development in the absence of loading (Rubin 1984; Frost 2003).

Key Terms

optimal strain environment   The genetically predetermined strain range within which each musculoskeletal component normally functions; structural adaptation maintains the optimal strain environment of each component.

genetic baseline   The genetically predetermined rate of growth and development of musculoskeletal components in the absence of loading.

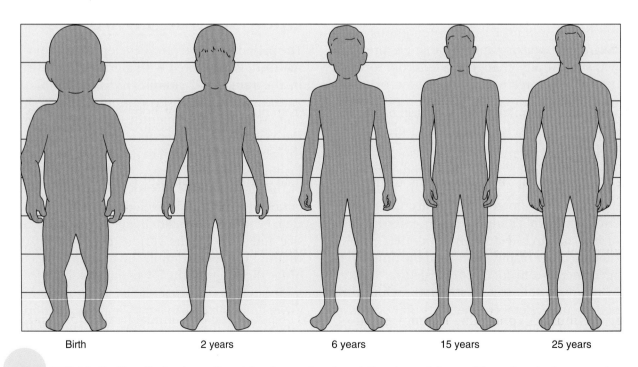

| Birth | 2 years | 6 years | 15 years | 25 years |

FIGURE 11.4   The effects of growth and development on the relative size and shape of the various body segments.

Structural Adaptation in Bone

The last 40 years have produced much of the present knowledge concerning the adaptation of musculoskeletal components to changes in time-averaged load (Frost 1988a, 1988b, 1990, 2003). However, the fundamental concepts concerning the adaptation of bone were established more than 100 years ago (Gross and Bain 1993). In 1892, Julius Wolff (1836-1902) summarized the contemporary views of bone adaptation to changes in time-averaged load in what came to be known as **Wolff's law** (Wolff 1988). Wolff's law, shown to be more or less correct, hypothesized that

- bone adapts its external form and internal architecture to the load exerted on it,
- adaptation in bone provides optimal strength with minimal bone mass, and
- adaptation in bone occurs in an ordered and predictable manner.

Key Term

Wolff's law   Bone adapts its external form and internal architecture to the time-averaged load exerted on it in an ordered and predictable manner to provide optimal strength with minimal bone mass.

Stereotypical Loading and Optimal Bone Mass

As described in chapter 3, the adaptation of bone to environmental influences, in particular to time-averaged load, is referred to as modeling. In normal growth and development, modeling has been estimated to account for 20% to 50% of the dimensions of mature bones (Frost 1988b, Schoenau and Frost 2003). Some of the load experienced by bone is due to the weight of body segments. However, this source of loading is small relative to the loads exerted by muscles. Consequently, it is reasonable to assume that modeling in a particular bone reflects the time-averaged loads exerted by the muscles controlling the movement of the bone, and thus modeling strengthens the bone to withstand the habitual load exerted by the muscles. Because bones, especially long

bones, are stronger in axial compression than any other form of loading (Frost 2003), the habitual form of loading exerted by muscles on long bones is mainly axial compression—compression along the long axis of the bones. It is generally thought that the muscles controlling the movement of a particular long bone act synergistically stereotypically—they exert an axial compression load even though the magnitude of the load can vary (Rubin 1984, Frost 2003). Stereotypical loading produces a restricted strain environment in which the type of strain experienced by a bone is similar whenever it is loaded, even though the amount of strain can vary.

A bone requires less bone mass in a restricted-strain environment than it would in a broad-strain environment. It is generally agreed that all bones function in restricted-strain environments and that modeling optimizes a bone's strength—provides maximum strength with minimal bone mass—for its particular restricted-strain environment (Gross and Bain 1993, Frost 2003).

KEY POINT

In normal growth and development, modeling accounts for an estimated 20% to 50% of the dimensions of mature bones.

Bone Modeling Throughout Life

From birth to maturity, bone has the capacity to model external form and internal architecture. However, the capacity to model external form gradually decreases and virtually ceases at maturity. The capacity to model internal architecture also decreases with age but is retained to some extent throughout life. Bone generally adapts to changes in time-averaged loads in the same way as other musculoskeletal components—by increasing or decreasing bone mass to maintain an optimal strain environment. In bone, the optimal strain environment is characterized by minimal flexure (or bending) strain and an even distribution of stress (usually compression stress) across articular areas. Minimal flexure strain is maintained by modeling in accordance with the phenomenon of flexure–drift. An even distribution of stress across articular areas is maintained by modeling in accordance with the phenomenon of chondral modeling (see later in this chapter) (Frost 1979).

Flexure–Drift Phenomenon

At birth, long bone shafts and the interarticular regions of many other bones such as the vertebrae are straight. However, prior to skeletal maturity the bone shafts adopt a narrow-waisted shape, such that the diameter of the shaft is smaller at the middle than at the ends. The change from a straight to a narrow-waisted shape reflects the shaft's modeling to minimize flexure strain. This is illustrated in figure 11.5 with respect to the development of a vertebra.

Figure 11.5a shows a frontal cross section through the body of a vertebra of a young child. In an unloaded state, as in figure 11.5a, the sides of the body of the vertebra are basically straight. Figure 11.5b shows the vertebra subjected to a vertical compression load. The compression load reduces the vertical height of the vertebra (from h_1 to h_2), and the increased pressure on the cancellous bone and surrounding marrow bends the side walls of the vertebra outward. The greater the compression load, the greater the pressure exerted by the cancellous bone and marrow on the side walls. As the side walls bend in response to the pressure exerted by the cancellous bone and marrow, the compression load adds to the bending load exerted on the side walls. The type of load exerted on the side walls by the compression load (which occurs after the side walls have started to bend) is referred to as a cantilever (indirect) bending load. The type of bending load exerted on the side walls by the

pressure of the cancellous bone and marrow is referred to as a static (direct) bending load (figure 11.5c).

When a bone (or part of a bone) is subjected to flexure strain above the upper limit of the optimal strain environment, **flexure–drift** occurs—bone is absorbed (removed) from the convex-tending surface of the bone and new bone is deposited on the concave-tending surface of the bone. The flexure–drift phenomenon is sometimes referred to as the flexure–drift law (Frost 1979). In the bodies of vertebrae, the flexure–drift phenomenon results in bone absorption from the periosteal surfaces of the side walls and deposition of new bone on the endosteal surfaces such that each vertebra adopts a narrow-waisted shape (figure 11.5, d and e). When this shape is subjected to a vertical compression load, the side walls experience a cantilever bending load that bends them farther inward and a static bending load exerted by the cancellous bone and marrow that bends them outward. Thus, the cantilever and static bending loads oppose each other, which minimizes the amount of flexure strain on the side walls (figure 11.5, f and g). Consequently, whereas the thickness of the compact bone in the side walls of the vertebrae reflects the time-averaged compression load exerted on them, the shape of the walls reflects the time-averaged bending load exerted on them.

Trabeculae in cancellous bone adapt to axial loads (loads in line with the trabeculae) and bending loads in the same way as compact bone. The thickness of trabeculae reflects the magnitude of the time-averaged compression or tension loads experienced by the trabeculae, and the orientation of the trabeculae reflects the normal line of action of the compression or tension loads. A change in time-averaged loading on trabeculae can change the thickness or orientation of the trabeculae, as shown in figure 11.6.

Key Term

flexure–drift Modeling that maintains minimal flexure strain in bones.

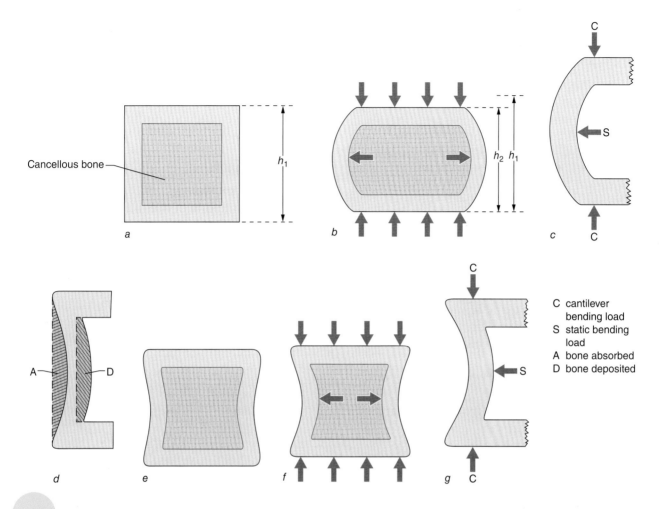

FIGURE 11.5 The flexure–drift phenomenon in relation to the development of the body of a vertebra.

C cantilever
  bending load
S static bending
  load
A bone absorbed
D bone deposited

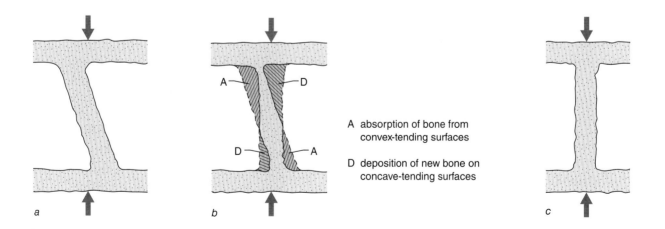

A absorption of bone from
  convex-tending surfaces

D deposition of new bone on
  concave-tending surfaces

FIGURE 11.6 Adaptation of trabeculae to a change in time-averaged loading. The normal loading on trabeculae is axial loading. *(a)* Change in load on the bone resulting in nonaxial load on the trabecula. *(b)* Adaptation of the trabecula to restore axial loading. *(c)* Axial load on the trabecula restored.

In accordance with the flexure–drift phenomenon, the shafts of long bones and the interarticular regions of many other bones such as the vertebrae develop narrow-waisted shapes, minimizing the flexure strain on the bones. Trabeculae in cancellous bone also model in accordance with the flexure–drift phenomenon.

Chondral Modeling Phenomenon

All bones that develop from hyaline cartilage via endochondral ossification (chapter 3) experience **chondral modeling**—the effect of loading on the rate and amount of new bone formed by hyaline cartilage. Chondral modeling applies to the following regions of bones:

- Articular cartilage: growth of epiphyses (figure 11.7*a*)

- Epiphyseal plates: growth of metaphyses (figure 11.7*a*)

- The layers of hyaline cartilage at the insertion of tendons and ligaments: growth of apophyses, nonapophyseal insertions of tendons, insertions of ligaments (figure 11.7*a*)

- Apophyseal plates: growth of shaft adjacent to apophyseal plate (figure 11.7*a)*

- The layers of hyaline cartilage at the bone–cartilage interfaces in symphysis joints: growth of bony end plates (figure 11.7*b*)

- The hyaline cartilage surrounding sesamoid bones: growth of sesamoid bones

The effect of change of load on rate of growth in these regions is shown in the **chondral growth–force relationship** (Frost 1979) (figure 11.8). The relationship has the following main features:

- Genetic baseline rate of growth: A certain amount of growth occurs in response to zero load.

- Ascending tension limb: Increasing tension (within the moderate range) tends to increase rate of growth.

- Ascending compression limb: Increasing compression (within the moderate range) tends to increase rate of growth.

- Descending compression limb: Excessive compression tends to decrease the rate of growth and can result in the cessation of growth.

Key Terms

chondral modeling   Modeling of endochondral regions of the skeleton.

chondral growth–force relationship   The relationship between load (type and magnitude) and rate of growth in endochondral regions of the skeleton.

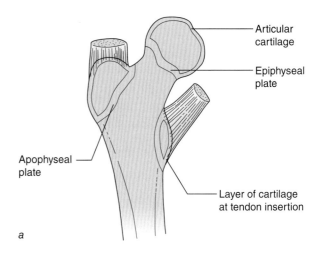

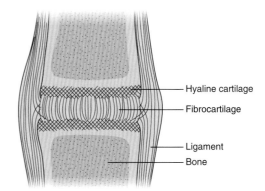

FIGURE 11.7   Regions of bone at which chondral modeling occurs. *(a)* Coronal section through the proximal end of a femur. *(b)* Coronal section through a typical intervertebral joint; only the hyaline cartilage region experiences chondral modeling.

Growth of Epiphyses and Metaphyses in Bones Forming Synovial Joints

In a long bone, the size and shape of the epiphyses and metaphyses and the orientation of the epiphyses of a bone to its shaft are determined by chondral modeling in articular cartilage and epiphyseal plates. Epiphyseal plates normally load to the left of the peak of the compression component of the chondral growth–force relationship, whereas articular cartilage normally loads around the peak (see figure 11.8). Consequently, the rate of growth of epiphyses is greater than that of metaphyses when congruence between the articular surfaces is maximum.

However, incongruence tends to have a more marked effect on rate of growth of the epiphyses than on the associated metaphyses. Incongruence decreases the rate of growth of the epiphyses below that of the associated metaphyses such that the orientation of the epiphyses to the shaft is largely determined by metaphyseal growth. Metaphyseal growth not only determines the orientation of the epiphyses of a bone to its shaft but also affects the alignment of bones at joints.

Changes in the alignment of the long bones of the legs are particularly noticeable from birth to about 6.5 years of age (Salenius and Vankka 1975) (figure 11.9). At birth, most children have genu varum—bowleg—with an angle between the femur and tibia of about 15° (figure 11.9a). Between birth and about 2 years of age, the genu varum gradually decreases to zero; the femur and tibia become directly in line (figure 11.9b). During the third year, genu valgum—knock-knee—develops. This reaches a maximum of about 12° of valgus between the femur and tibia at about 3 years of age (figure 11.9c). During the next 3.5 years the genu valgum gradually decreases, reaching about 5° of valgus at about 6.5 years of age (figure 11.9d). The changes in the degree of genu varum and genu valgum from birth to approximately 6.5 years of age reflect the considerable changes in body weight, relative size of body segments, and complexity of movement patterns occurring during this period (Wearing et al 2006).

When a synovial joint is maximally congruent, the loading on articular cartilage and epiphyseal plates tends to be evenly distributed. Incongruence results in an unequal distribution of load across articular cartilage and epiphyseal plates. If prolonged, such unequal loading results in modeling to restore maximal congruence. However, the actual changes that occur depend on the extent of the changes in the patterns of loading on the articular cartilage and epiphyseal plates. If the changes in loading remain within the normal range, as indicated by the chondral growth–force relationship, then maximal congruence is usually restored, perhaps with some degree of abnormal alignment of the bones. However, changes in loading that are outside the normal range are likely to aggravate the condition, resulting in progressively worsening misalignment.

Modeling of Metaphyses

A functionally normal joint is a congruent joint that transmits loads across the articulating surfaces in a normal manner. An anatomical misalignment at the

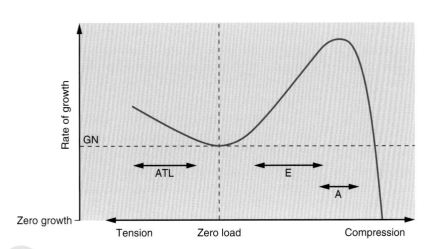

FIGURE 11.8 The chondral growth–force relationship: GN = genetic baseline rate of growth; ATL = normal range of loading on apophyseal plates, tendon insertions, and ligament insertions; E = normal range of loading on epiphyseal plates; A = normal range of loading on articular cartilage.

Springer/*Calcified Tissue International*, Vol. 28, 1979, p. 184, "A chondral modeling theory," H.M. Frost, figure 2, Copyright Springer; with kind permission from Springer Science and Business Media.

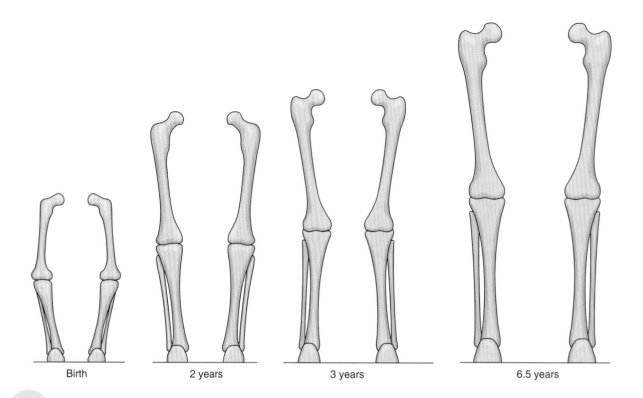

| Birth | 2 years | 3 years | 6.5 years |

FIGURE 11.9 Changes in the alignment between the femur and the tibia during the period from birth to 6.5 years.

knee, or any other joint, will be functionally normal if the misalignment stabilizes (does not get progressively worse). In these cases, the anatomical misalignments represent normal modeling in response to abnormal patterns of loading imposed on the skeleton by the combined effects of muscle forces and body weight. The skeletal adaptations ensure normal transmission of loads across the joints. The anatomical misalignments must be normal functionally or the joints will become painful. Figure 11.10 illustrates the effect of a hip abductor–adductor muscle imbalance on modeling of the metaphyses of the distal femoral epiphysis and the proximal tibial epiphysis. A hip abductor–adductor muscle imbalance can result, for example, from a decrease in physical activity in association with an increase in age or body weight (Johnson et al 2004) or repetitive actions in sport combined with inadequate abductor–adductor strength training (Tyler et al 2001, Heinert et al 2008).

Figure 11.10*a* illustrates normal balance between the abductor and adductor muscles at the hip such that the resultant horizontal force at the knee is zero, that is, there is no tendency to pull the femoral epiphysis medially (which would increase genu valgum) or laterally (which would decrease genu valgum). This situation is associated with normal alignment between the femur and tibia and an even distribution of load across the epiphyseal plates (figure 11.10*b*). Figure 11.10*c* shows the loading (even distribution) on the epiphyseal plates in relation to the chondral growth–force curve. Figure 11.10*d* shows the same knee with an abductor–adductor imbalance such that there is a net, medially directed, horizontal force at the knee tending to increase the degree of genu valgum. Figure 11.10*e* shows the unequal pattern of loading on the epiphyseal plates associated with the muscle imbalance. Figure 11.10*f* shows the range of loading on the epiphyseal plates in relation to the chondral growth–force curve.

Because the unequal range of loading is within the normal range, the rate of growth of the lateral aspects of the metaphyses is increased and the rate of growth of the medial aspects of the metaphyses is decreased such

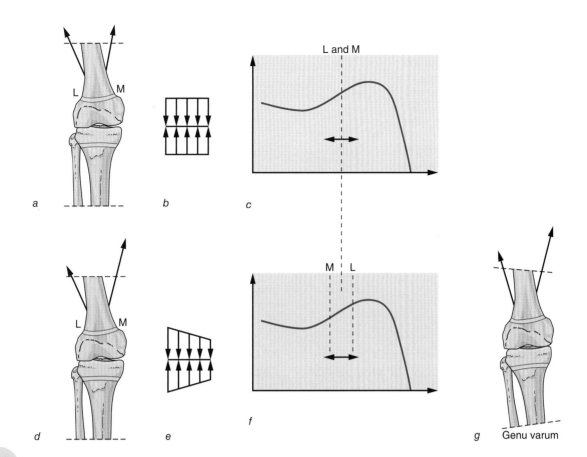

FIGURE 11.10 Chondral growth–force relationship: effect of a hip abductor–adductor muscle imbalance on modeling of the metaphyses of the distal femoral epiphysis and the proximal tibial epiphysis. L = lateral; M =medial. *(a)* Normal balance between the abductor and adductor muscles at the hip such that the resultant horizontal force at the knee is zero. *(b)* Even distribution of load across the epiphyseal plates. *(c)* Loading (even distribution) on the epiphyseal plates in relation to the chondral growth–force curve. *(d)* The same knee with an abductor–adductor imbalance such that there is a net, medially directed horizontal force at the knee tending to increase the degree of genu valgum. *(e)* Uneven distribution of load on the epiphyseal plates associated with the muscle imbalance. *(f)* The range of uneven load (within the normal range) on the epiphyseal plates in relation to the chondral growth–force curve. *(g)* Restoration of functional alignment at the expense of an abnormal anatomical alignment.

that normal congruence is restored (with net zero horizontal force at the knee) at the expense of an abnormal alignment between the femur and tibia, that is, much reduced genu valgum or even slight genu varum relative to most people (figure 11.10*g*).

Whether or not a particular joint is anatomically misaligned during childhood, the only time it can become painful (excluding injuries and pathological conditions not due to loading) is during adulthood, when the bones are no longer capable of modeling in response to abnormal loading. In most adults, abnormal patterns of loading are the result of an increasingly sedentary lifestyle in which body weight gradually increases and muscle strength

gradually decreases. Figure 11.11 shows the likely effect of a hip abductor–adductor muscle imbalance prior to and after skeletal maturity on the orientation of the femur to the tibia in the coronal plane. Figure 11.11*a* shows normal muscle balance (between the abductors and adductors on the femur), normal anatomical alignment (normal angle between the femur and tibia), and normal functional alignment (even distribution of load across the epiphyseal plates of the distal femoral epiphysis and the proximal tibial epiphysis). Figure 11.11*b* shows the same knee following the development of a hip abductor–adductor muscle imbalance that results in an uneven distribution of load across the epiphyseal plates. In response to

FIGURE 11.11 Effect of a hip abductor–adductor muscle imbalance prior to and after skeletal maturity on the orientation of the femur to the tibia in the coronal plane. *(a)* Normal muscle balance, normal anatomical alignment, and normal functional alignment. *(b)* Abnormal muscle balance resulting in abnormal functional alignment. *(c)* Restoration of normal functional alignment at the expense of the development of slight genu varum by modeling of the metaphyses. *(d)* Change in muscle balance after maturity, resulting in worsening anatomical and functional alignments.

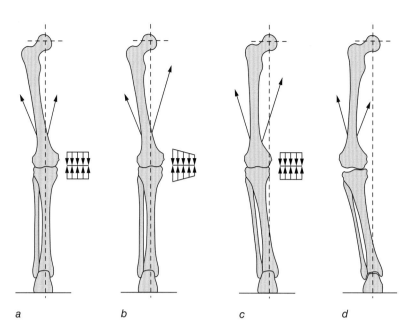

the uneven distribution of load across the epiphyseal plates, modeling of the metaphyses occurs so that functional alignment is restored at the expense of the development of slight genu varum (figure 11.11*c*). Following skeletal maturity, a further change in the abductor–adductor muscle balance is likely to result in anatomical and functional misalignments that cannot be compensated by modeling (figure 11.11*d*).

Some of the world's best athletes have marked genu varum or genu valgum, but they train and perform without injury (Micheli 1982). Similarly, some of the world's best ballet dancers have severely pronated feet, but they practice and perform without injury (Micheli 1982). In these cases, it must be assumed that the anatomical abnormalities are functionally normal and arose during childhood. Anatomical misalignments that develop after maturity are the result of a combination of muscle weakness, ligament laxity, and progressive bone collapse (Frost 1979).

> **KEY POINT**
>
> Anatomically misaligned but functionally normal joints represent normal modeling in response to abnormal patterns of loading imposed on the skeleton by the combined effects of muscle forces and body weight.

Modeling of Articular Surfaces Because articular surfaces in synovial joints are not physically attached to each other, even minor incongruences result in large changes in the load experienced by different parts of the articular surfaces. This is especially the case in joints with pulley-shaped articular surfaces such as the ankle joint (figure 11.12). Under normal circumstances, the subtalar joint contributes to inversion and eversion of the foot (figure 11.12, *a* and *b*). However, if movement at the joint is absent or limited, inversion and eversion of the foot twist the talus in the tibiofibular mortise, resulting in excessive loading on those parts of the articular surfaces of the ankle joint that remain in contact (figure 11.12*c*). The excessive loading on the impinging areas reduces or halts growth in these areas, while growth of the unloaded areas proceeds at the genetic baseline rate (see figure 11.8). Consequently, the shapes of the articular surfaces of the ankle joint adapt to the abnormal loading conditions by forming a rounded surface in the frontal plane rather than a trochlear-shaped surface, and the ankle joint as a whole resembles a ball-and-socket joint rather than a hinge joint (figure 11.12*d*) (Frost 1979).

Modeling of Vertebral Bodies The load experienced by the articular regions of vertebral bodies in intervertebral joints in

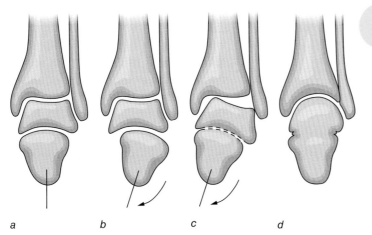

a   b   c   d

FIGURE 11.12 Effect of modeling of the articular surfaces of the ankle joint. *(a)* Normal orientation of the tibia, fibula, talus, and calcaneus in the coronal plane. *(b)* Normal movement of the calcaneus in inversion of the foot. *(c)* Effect of restricted movement of the subtalar joint during inversion of the foot; the talus is twisted in the tibiofibular mortise resulting in excessive loading on those parts of the articular surfaces of the ankle joint that remain in contact. *(d)* Modeling of the articular surfaces of the ankle joint in response to the abnormal loading.

Springer/*Calcified Tissue International*, Vol. 28, 1979, p. 187, "A chondral modeling theory," H.M. Frost, fig. 4, Copyright Springer, with kind permission from Springer Science and Business Media.

upright postures tends to be evenly distributed compression, as shown in figure 11.13, *a* and *b*. Bending the vertebral column in any direction produces a pattern of loading on the discs and the hyaline cartilage interfaces between the end plates of the vertebrae and the fibrocartilage of the discs, ranging from relatively high compression on the concave-tending side of the column to relatively high tension on the convex-tending side of the column (figure 11.13, *c* and *d*). This loading pattern in response to bending the column is normal. However, if part of the column is subjected to higher-than-normal, unidirectional, bending stress during the period of growth and development, the vertebral bodies in the affected region can become wedge shaped to restore a more normal pattern of loading (figure 11.13, *e* and *f*). Wedging of vertebrae occurs in the thoracic and lumbar regions in the median and frontal planes (see chapter 5).

Effects of Age on Modeling Capacity

The modeling capacity of bone decreases with age. Whereas the ability to adapt internal architecture to changes in patterns of loading is somewhat retained throughout life, the ability to adapt size and shape decreases more rapidly and is negligible in the adult (Frost 1979, Lanyon 1981). Consequently, during childhood a change in the normal pattern of loading on bones usually results in adaptive bone growth to restore normal loading. However, in an adult, adaptive bone growth cannot accommodate a change in the normal pattern of loading that results, for example, from an

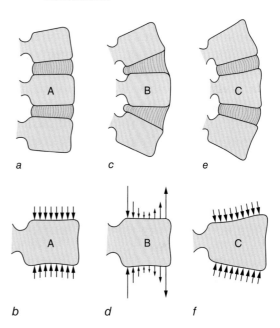

a   c   e

b   d   f

FIGURE 11.13 Effect of modeling of vertebrae. *(a, b)* Evenly distributed compression on vertebrae in normal upright posture. *(c, d)* Bending of the vertebral column, resulting in high compression on the concave-tending side of the column and high tension on the convex-tending side of the column. *(e, f)* Wedging of vertebrae caused by modeling in response to prolonged unidirectional bending stress.

increase in body weight or a change in the muscle strength balance about a joint. Unless corrected by specific forms of treatment such as physiotherapy or orthotics, the abnormal pattern of loading increases the likelihood of degenerative joint disease. An orthosis is a musculoskeletal support—an external appliance such as leg calipers, a foot arch support, or heel raise designed to improve musculoskeletal function by helping to maintain normal transmission of loads in joints (Redford 1987). Orthotics is the theory and practice of development, manufacture, and the use of orthoses (Redford 1987).

The following examples illustrate the rate at which the modeling capacity of bone decreases with age and the possible effects of a lack of modeling capacity in the adult.

• *Tibial torsion.* Tibial torsion is an abnormality of the tibia in which the lower end of the tibia is excessively rotated internally or externally with respect to the upper end about the long axis of the shaft. Internal and external tibial torsion are characterized by abnormal toeing in and toeing out, respectively (figure 11.14). The normal degree of rotation of the lower end of the tibia with respect to the upper end is 15° to 20° of external rotation (the foot is turned out 15-20° with respect to the knee; figure 11.14*b*). Tibial torsion can be corrected easily in an infant by applying a constant torsion load to the tibia by means of special splints (Frost 1979). The torsion load is applied in the direction opposite to that of the abnormality. The bone adapts to the torsion stress by modeling to relieve the torsion stress. This process gradually reduces the tibial torsion, provided the torsion load and stress are maintained.

However, the effect of this type of treatment decreases fairly rapidly with age. For example, 30° to 40° of correction of an internal tibial torsion can be produced in 10 weeks of treatment when applied to a 5-month-old child. The same amount and duration of treatment have a negligible effect in a 5-year-old child (Frost 1979).

• *Bone fracture.* Following an injury to a bone involving two or more fractures, the chances of the broken pieces knitting together in exactly the same orientation to each other as existed prior to the injury are remote. Consequently, after healing, the line of action of the muscles crossing the joints formed by the two ends of the bone is altered. This can result in abnormal loading on the articulating surfaces

KEY POINT

Anatomically misaligned but functionally normal joints that develop in childhood can become painful during adulthood when the bones are no longer capable of modeling in response to abnormal loading. In most adults, abnormal patterns of loading result from an increasingly sedentary lifestyle in which body weight gradually increases and muscle strength gradually decreases.

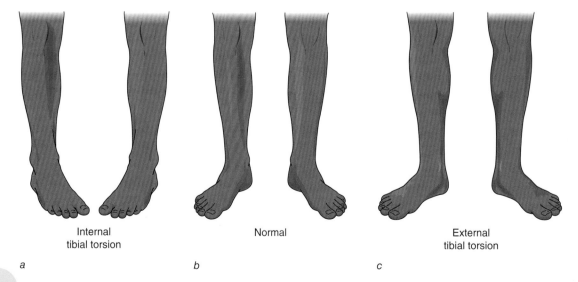

Internal tibial torsion
Normal
External tibial torsion

a   *b*   *c*

FIGURE 11.14 Tibial torsion. *(a)* Internal tibial torsion: excessive internal rotation of the distal end of the tibia with respect to the proximal end. *(b)* Normal degree of rotation of the distal end of the tibia with respect to the proximal end: 15° to 20° of external rotation. *(c)* External tibial torsion: excessive external rotation of the distal end of the tibia with respect to the proximal end.

of the joints. In a child, the articulating surfaces usually restore normal loading by modeling. In an adult, the abnormal loading increases the likelihood of degenerative joint disease, especially in weight-bearing joints (Radin 1986).

Structural Adaptation in Regular Fibrous Tissues

Regular fibrous tissues, and particularly ligaments and tendons, are designed to provide strength in tension. The effect of changes in time-averaged tension load on regular fibrous tissues depends on the change in magnitude and type of load. When subjected to continuous tension at a level exceeding the optimal strain environment, regular fibrous tissues experience creep (chapter 10). When subjected to intermittent tension at a level exceeding the optimal strain environment, regular fibrous tissues hypertrophy to restore the optimal strain environment. For example, increasing the strength of a muscle increases the time-averaged tension load exerted on the associated tendons, which experience an increase in strain. The tendons adapt to the increased strain by increasing their cross-sectional areas to a level that restores the optimal strain environment. The capacity of regular fibrous tissues to adapt to increases in time-averaged tension load is called the **stretch–hypertrophy rule** (Frost 1979).

Just as an increase in time-averaged intermittent tension on regular fibrous tissue results in hypertrophy, a decrease in time-averaged tension results in atrophy to restore the optimal strain environment. Atrophy in regular fibrous tissue involves a decrease in cross-sectional area and shortening (Akeson et al 1977, Frost 1990). Immobilization, which can occur because of paralysis or injury, results in considerable atrophy in ligaments. Noyes (1977) studied the effects of immobilization and subsequent rehabilitation on the strength of the anterior cruciate ligament (ACL) in rhesus monkeys. The results of the study showed that immobilization in a whole-body cast for 8 weeks produced an average decrease of 39% in the strength of the ACL, and after 5 months and 12 months of rehabilitation the average strength of the ACL was still only 79%

and 91%, respectively, of the preimmobilization level. In humans, rehabilitation of ligaments following immobilization and injury is a lengthy process (Frost 1979, 1990, Sharma and Maffulli 2006).

Under normal circumstances, the load experienced by ligaments is markedly reduced by ligament–muscle stretch reflex mechanisms; stretching a ligament tends to result in reflex contraction of associated muscles, which relieves the load on the ligaments (Cohen and Cohen 1956, Kondradsen et al 1993, Roberts et al 2007). For example, hyperextension of the knee results in reflex contraction of the hamstrings that is triggered, at least in part, by stretch reflex loops between the knee ligaments (cruciate and collateral) and the hamstrings. Injury to ligaments can impair the stretch reflex mechanism by delaying the elicitation of muscular contraction. This can occur directly in the form of damage to nerves or indirectly in the form of laxity in the ligaments such that greater movement is necessary in the joint to stretch the ligaments sufficiently to elicit the stretch reflex. The greater the delay in the elicitation of the stretch reflex, the greater the likelihood of ligaments being excessively overloaded or associated joints being dislocated. Injury to ligaments can also impair their blood supply, which can reduce the rate of hypertrophy.

It is perhaps not surprising that many ligament reconstructions based on grafts from other parts of the body fail because it takes time for development of an adequate blood supply, which is essential for hypertrophy, and stretch reflex loops, which protect the ligaments from overload (Frost 1973).

> **KEY POINT**
>
> Under normal circumstances, the load experienced by ligaments is markedly reduced by ligament–muscle stretch reflex mechanisms. Injury to ligaments can impair the functioning of the stretch reflex mechanisms by delaying the elicitation of muscular contraction, which increases the likelihood of excessively overloading ligaments or dislocating associated joints.

Structural Adaptation at Ligament and Tendon Insertions

All ligaments and tendons insert onto bone via a hyaline cartilage interface, and thus the bone adapts to the tensile load exerted via the tendons and ligaments in accordance with the chondral modeling phenomenon (figure 11.15). Just before entering the layer of hyaline cartilage, the ligament or tendon separates into bundles of collagen fibers. Each bundle of fibers passes through and is embedded within all three layers of the region of insertion: hyaline cartilage, calcified cartilage, and subchondral bone. In the layer of subchondral bone, the collagen fiber bundles become encrusted with mineral salts. At this stage each bundle is referred to as a **Sharpey fiber** (Frost 1973, Hamrick 1999).

The ultimate shear stress between a Sharpey fiber and the surrounding subchondral bone determines the fiber's resistance to being pulled out of the insertion. The deeper the penetration of the fiber, the greater the area of fiber presented to the surrounding bone and therefore the lower the shear stress on the fiber. The lower the shear stress on the fiber, the greater the amount of tension required to pull out the fiber. It is therefore not surprising that each ligament and tendon insertion consists of a very large number of Sharpey fibers; the larger the number of fibers, the lower the shear stress on each fiber. Furthermore, the fibers are distributed over a relatively large area (much larger than that generally indicated on bony specimens and models) and merge with adjacent fibrous tissues. This is referred to as **fiber fan-out** (Frost 1973, Myers 2001). The larger the area of insertion, the lower the tensile stress on the region of insertion.

The combination of fiber fan-out and Sharpey fibers produces strong attachments that, in the case of apophyses, are stronger than the apophyseal plate. Avulsion fractures occur along the apophyseal plate—the tendon stays intact with the apophysis and the apophysis is displaced along with the tendon. This can reasonably be compared to pulling up a plant that has an extensive root system; usually a lot of soil is still attached to the roots. In this sense a tendon or ligament can be described as having a very extensive root system. The insertions of ligaments and tendons adapt to changes in time-averaged tension loads by varying the degree of fiber fan-out and the number of Sharpey fibers. Increases in loading increase the area of fiber fan-out and the number of fibers, and decreases in loading reduce the area of fan-out and number of fibers.

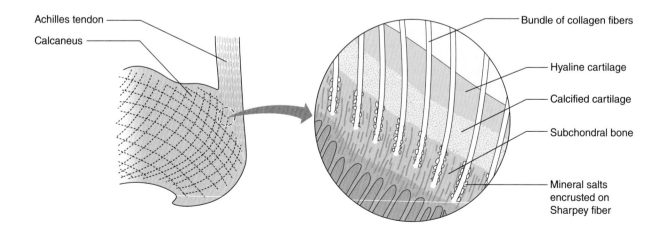

FIGURE 11.15 Sharpey fibers.

From H.M. Frost, *Orthopedic biomechanics*, Vol. 5, 1973, Courtesy of Charles C. Thomas Publisher, Ltd, Springfield, IL.

stretch–hypertrophy rule   The capacity of regular fibrous tissues to hypertrophy when subjected to a time-averaged increase in intermittent tension that exceeds the optimal strain environment.

Sharpey fiber   A terminal bundle of collagen fibers in a ligament or tendon insertion encrusted with mineral salts.

fiber fan-out   The area of distribution of Sharpey fibers in a ligament or tendon insertion.

Growth, Development, and Aging of Tendons

Tendons have a growth–development phase (the period between birth and skeletal maturity) and an aging phase (the period between skeletal maturity and death) (Frost 1990, Smith et al 2002). At birth, the matrix of all of the tendons is similar (irregular collagenous connective tissue) and unadapted to load bearing. During the growth–development phase, the matrix of each tendon adapts to the time-averaged load exerted on it. In addition to being influenced by the type (magnitude and frequency) of load, matrix adaptation is significantly influenced by growth factors (noncollagenous proteins that are only present during the growth–development phase). The type and concentration of growth factors present in the matrix during the growth–development phase depend to a large extent on the type of load experienced by the tendon.

Various combinations of load and growth factors appear to produce two types of tendons: energy-absorbing tendons and joint-positioning tendons. Energy-absorbing tendons are highly resilient and are found in musculotendinous units whose main function is to act like springs by absorbing energy during braking actions and then recycling or dissipating the stored energy following the braking phase. This type of tendon is found in the hip, knee, and ankle extensor musculotendinous units, which are associated with apophyseal tendinous attachments. Joint-positioning tendons, stiffer and less resilient than energy-absorbing tendons, are found in musculotendinous units whose primary function is to position body segments. These units include the ankle dorsiflexors and the wrist and finger flexors and extensors.

Joint-positioning tendons rarely experience strain-induced tendinopathy (damage to tendons due to excessive strain). In contrast, the incidence of strain-induced tendinopathy in energy-absorbing tendons is relatively high and increases with age and intensity of exercise.

Whereas energy-absorbing tendons generally adapt positively to changes in time-averaged loading that exceed the optimal strain environment during the growth–development phase, the capacity of energy-absorbing tendons to adapt to changes in time-averaged loading that exceed the optimal strain environment after skeletal maturity appears to be minimal. Furthermore, excessive exercise during the aging phase appears to accelerate the aging process, resulting in tendon matrix degeneration and increased risk of strain-induced tendinopathy. This may seem counterintuitive, because muscle and bone adapt positively to changes in time-averaged loading that exceed the optimal strain environment (within the moderate overload range) throughout life. However, the response of bone is also very age dependent, with the highest activity during puberty and the lowest activity in old age.

It seems that after an optimized tendon is formed at skeletal maturity (as a result of its history of time-averaged loading and growth factors), the capacity for matrix adaptation is minimal. Consequently, appropriate exercise that results in a progressive increase in time-averaged loading within the moderate overload range is necessary during the growth–development phase to optimize tendons, and suboptimization of tendons during the growth–development phase cannot be overcome in the aging phase.

Structural Adaptation in Muscle

As in other musculoskeletal components, a time-averaged increase in intermittent load that exceeds the optimal strain environment of a muscle increases the strength of the muscle, and a time-averaged decrease in load that falls below the optimal strain environment decreases the strength of the muscle. Similarly, a time-averaged increase in constant load that exceeds the optimal strain environment of a muscle increases the length of the muscle, and a time-averaged decrease in load that falls below the optimal strain environment decreases the length of the muscle.

Strength Changes

Unlike the noncontractile components of the musculoskeletal system, connective tissue and bone, which passively transmit loads applied to them, a muscle only functions (i.e., produces tension) when it is stimulated to contract. Furthermore, the amount of tension that a muscle produces depends on the degree of activation: the greater the activation, the greater the tension. Consequently, strength gains are the result of neuromuscular adaptations, that is, changes in structure and function of both the neural and muscle components of the neuromuscular system (Häkkinen 1994).

Change in structure is mainly reflected in hypertrophy of musculotendinous units, that is, an increase in cross-sectional area and stiffness of tendons and an increase in the cross-sectional area of muscle fibers. The latter is due to an increase in the number of actin and myosin filaments, which increases the potential number of cross bridges and, therefore, the capacity for tension generation (Reeves et al 2006, Narici and Maganaris 2006). Change in function is mainly reflected in increased capacity for synchronous recruitment of motor units and increased rate of discharge of motor units (Moritani 1993, Enoka 1997, Aagaard et al 2002).

The number of motor units recruited and, by inference, the level of stimulation of a muscle correspond to the net sum of all the action potentials of all the motor units in a muscle at a particular point in time (Kamen and Caldwell 1996). It is not possible to measure the sum of all the action potentials directly, but the sum can estimated from the **electromyogram (EMG),** a recording of the electrical activity in the muscle by means of electrodes located on the skin (surface electrodes) overlying the target muscles or electrodes inserted into the target muscles (intramuscular electrodes).

In previously untrained people who start a prolonged strength-training program, the first few weeks of training result in a significant increase in strength without any significant hypertrophy. This initial increase in strength, reflected in an increase in the integrated EMG (IEMG: area under the EMG), is attributed to neural adaptation (Moritani 1993). After this initial phase, further increases in strength are associated with a progressive increase in the contribution of muscular hypertrophy.

The pattern of strength gain in response to strength training in women is similar to that in men, an initial phase dominated by neural adaptation followed by a longer period in which hypertrophy progressively dominates. However, in women the initial neural adaptation phase is shorter than in men, and ultimate strength gains in women tend to be lower than in men attributable largely to hormonal differences, especially lower levels of testosterone in women than in men.

In women and men, decreases in strength subsequent to a period of detraining (lower time-averaged loading) follow a similar two-phase pattern: a rapid and significant decrease in strength (associated with decreased IEMG)

> **KEY POINT**
>
> Changes in strength result from changes in muscle mass and motor unit activation. Initial strength increases are primarily caused by neural adaptation, followed by a longer period where strength increases are caused by hypertrophy. Initial strength decreases are primarily caused by neural adaptation, followed by a longer period where strength decreases are caused by atrophy.

without any significant atrophy (decrease in muscle mass) followed by more gradual decrease in strength that is associated with significant atrophy. As with strength gain, the initial period of strength loss is attributed to neural adaptation (Moritani 1993).

Strength in normal, healthy, untrained women and men normally reaches its peak between the ages of 20 and 30 years, remains fairly constant or declines slightly between the ages of 30 and 50 years, and then progressively declines, with an increase in age associated with a marked increase in the rate of decline. For most people, strength declines 30% to 40% between the ages of 30 and 70 years (Häkkinen 1994). Loss of strength decreases the ability of people to undertake activities of daily living and, in particular, decreases speed of movement and speed of reaction to postural perturbations, which significantly increases the risk of falling (Grabiner and Enoka 1995). Loss of postural stability and an inability to walk without assistance are the main reasons that elderly people are permanently admitted to nursing homes (Schultz et al 1992).

As shown in figure 11.16, the age-related decline in strength tends to be associated with three interrelated changes: change in hormone balance, progressive degeneration in the structure and function of the neuromuscular system, and a decrease in physical activity.

Whereas it is reasonable to assume that some of the deterioration in the structure and function of the neuromuscular system that occurs with age is a natural consequence of aging, the extent of such change is not yet clear. It is clear that in normal, healthy people, the neuromuscular system retains the capacity to adapt positively to strength training and motor ability training throughout life (Reeves et al 2006, Narici and Maganaris 2006, Baker et al 2007, Capodaglio et al 2007). In the absence of intervention, age-related decline in the structure and function of the neuromuscular system is associated with a decrease in muscle mass, referred to as **senile sarcopenia** (Narici and Maganaris 2006); a decrease in the ability to control the amount of force produced, especially in manipulative tasks; and a decrease in coordination (figure 11.16).

Key Terms

electromyogram (EMG) A recording of the electrical activity in the muscle by means of electrodes located on the skin (surface electrodes) overlying the target muscles or electrodes inserted into the target muscles (intramuscular electrodes).

senile sarcopenia The loss of muscle mass associated with aging.

Muscle Extensibility Changes

Muscle fiber length determines a muscle's extensibility, and the number of sarcomeres in each myofibril determines the length of each muscle fiber. Changes in the length of muscle fibers and in the extensibility of muscles occur as a result of changes in time-averaged range of movement of the muscles (Goldspink 1992, Herzog 2000). Muscles adapt to changes in time-averaged range of movement by increasing or decreasing the number of sarcomeres in each myofibril to maintain optimal function of muscle fibers in terms of length–tension characteristics. For example, immobilizing the ankles of cats in full dorsiflexion—with the soleus muscles in a stretched position—has been found to increase the number of sarcomeres in the fibers of the soleus muscle by about 20% in 4 weeks. The same researchers found that immobilizing the ankles of cats in full plantar flexion—the soleus muscles in a shortened position—decreased the number of sarcomeres in the fibers of the soleus muscle by about 40% in 4 weeks (Williams and Goldspink 1973, 1978).

To maximize performance (by increasing the potential for strength of muscle–tendon units) and reduce the risk of muscle and tendon injuries (by increased muscle extensibility), flexibility training should be undertaken to ensure that muscles are long enough to

KEY POINT

Muscles adapt to changes in time-averaged range of movement by increasing or decreasing the length of muscle fibers to maintain optimal function in terms of length-tension characteristics.

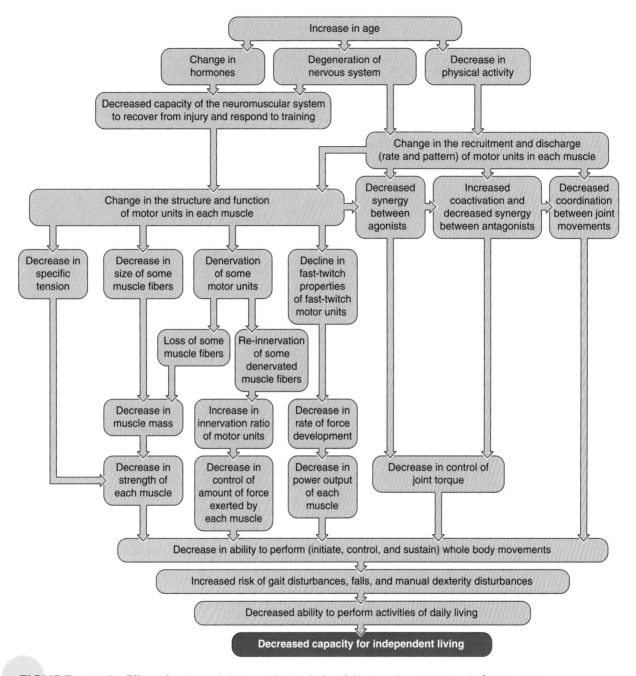

FIGURE 11.16 Effect of aging and decrease in physical activity on voluntary control of movement.

facilitate full ranges of whole body movement. Lengthening a muscle through appropriate flexibility training will increase the length of its myofibrils, which will increase its potential for strength development and reduce the risk of overstretching the muscle.

Summary

Changes in time-averaged loading affect the musculoskeletal system. A certain level of loading is necessary to ensure normal growth and development; however, excessive loading, like insufficient loading, is associated with abnormal growth and development. Each musculoskeletal component has an optimal strain environment, and changes in time-averaged loading result in structural adaptation to restore the optimal strain environment. Decreases in time-averaged loading result in atrophy and decreased functional capacity, and moderate increases in time-averaged loading result in hypertrophy and increased functional capacity. Injury occurs as a result of excessive loading of an acute or chronic nature, and to prevent or minimize the risk of injury in a particular activity it is necessary to understand the factors that contribute to loading; this is the subject of the final chapter.

Review Questions

1. Differentiate between biopositive and bionegative effects of loading.
2. Differentiate between moderate habitual load and moderate overload.
3. Differentiate between acute and chronic injury.
4. Differentiate between response to loading and adaptation to loading.
5. Describe the chondral growth–force relationship.
6. Differentiate between cantilever and static bending loads.
7. Differentiate between anatomical and functional alignment in normal joints.
8. Describe the effects of age on the modeling capacity of bone.
9. Describe the structure of ligament and tendon insertions.
10. Describe the effects of changes in loading and range of movement on muscle.

Etiology of Musculoskeletal Disorders and Injuries

Musculoskeletal disorders and injuries are the most common cause of chronic disability throughout the world and affect most people at some time in their lives. The underlying cause of many musculoskeletal disorders is excessive overload of a chronic or acute nature. In any particular body posture or movement, the load on the musculoskeletal system depends on the influence of a number of risk factors and their interaction. Without a clear understanding of the relevant risk factors it is difficult to identify the cause of a disorder, and treatment may be confined to dealing with signs and symptoms. This chapter describes the three main types of risk factors that influence the level of loading on the musculoskeletal system.

12

Kinetic Chain

All postures and movements involve the transmission of forces across a number of joints. For example, when a person is sitting, the weight of the upright trunk is transmitted to the chair seat by the joints of the vertebral column and hips. Similarly, bending over a bath to turn on the taps involves the transmission of forces across all of the joints of the body, from feet to head and fingers. Because most actions, like turning on bath taps, can be accomplished with a range of slightly different movement patterns, the forces transmitted across some or all of the joints will vary depending on the pattern adopted; that is, the forces transmitted across the joints in the segmental chain are interdependent. The segmental chain used to accomplish a particular action is referred to as a **kinetic chain** to indicate that forces are transmitted across all of the joints in the chain (Steindler 1955, McMullen and Uhl 2000; Kibler 2005). Because the forces transmitted across the joints in a kinetic chain are interdependent, it follows that abnormal movement in one joint may cause overloading in one or more of the other joints in the chain. Consequently, in any particular movement or sequence of movements, the movement of the various joints must be properly coordinated so none of the separate joints are excessively overloaded.

Key Term

kinetic chain The segmental chain used to accomplish a particular action.

Compensatory Movements

All joints have one to three degrees of freedom (chapter 4). Consequently, for most whole-body movements, an infinite number of possible movement patterns can accomplish a given task. It follows that when movement in one or more joints in a particular kinetic chain is restricted, perhaps because of muscle weakness or anatomical abnormalities, the other joints in the chain can usually compensate by altering their own ranges of movement so that the task can still be accomplished (Voight et al 2006, Shum et al 2007). For example, stiffness in one knee is often compensated in walking by tilting the pelvis upward on the affected side to allow the leg to swing forward without the foot catching on the floor. However, for any given task there is one most efficient movement pattern (or range of movement patterns within fairly strict limits), that is, a movement pattern that minimizes the amount of muscle force required and therefore minimizes the load on the musculoskeletal system as a whole (Kugler and Turvey 1987, Holt 1998, Nigg 2001). Even a slight amount of **compensatory movement** in a joint alters the normal pattern of loading in the joint (Riegger-Krugh and Keysor 1996). Furthermore, the greater the degree of compensatory movement in a joint, the greater the loading on the joint and its supporting structures. If compensatory movements are excessive and prolonged (continuous or repetitive), the joints concerned and their supporting structures may be excessively overloaded, resulting in a musculoskeletal disorder.

Key Term

compensatory movement The change occurring in the movement of one or more joints in a kinetic chain in response to restricted movement in other joints so that a given task can still be carried out.

Risk Factors for Musculoskeletal Disorders

A **musculoskeletal disorder** is any temporary or permanent abnormality of the musculoskeletal system resulting in pain or discomfort. Disorders of the musculoskeletal system affect most people at some stage in their lives (Woods 2005). Some disorders occur dramatically and their cause is obvious, for example, a broken leg caused by the impact of a motor vehicle accident or a hard tackle in football. In these contexts *injury* may be the more appropriate term. Many other disorders of the musculoskeletal system, such as chronic low back pain and osteoarthritis of the hip joint, develop over time, and their causes are usually less obvious (Connelly at al 2006).

Without a thorough understanding of the underlying cause of a musculoskeletal disorder, clinicians may confine treatment to dealing with the signs and symptoms of the disorder. However, if the cause of the disorder is not removed, the disorder can worsen and result in more severe pain or discomfort. The underlying cause of many musculoskeletal disorders is excessive overload of an acute or chronic nature. The actual level of overload depends on the influence of associated risk factors. There are three main categories of risk factors: movement factors, intrinsic (or internal) factors, and extrinsic (or external) factors (Nigg and Segesser 1988, Bahr and Holme 2003, Meeuwisse et al 2007).

Movement risk factors concern the characteristics of the movements comprising the activity, in particular the speed of movement and the degree of physical contact between the person and her surroundings. The latter includes, for example, impacts with other participants, playing surfaces, walls, and implements such as sticks, bats, and balls. In general, the greater the movement speed and the greater the degree of physical contact, the higher the risk of injury (Backx et al 1991, Meeuwisse et al 2007). Consequently, some activities are inherently more dangerous than others. **Intrinsic risk factors** are personal, physical, and psychological characteristics that distinguish people from each other, and **extrinsic risk factors** concern environmental conditions and the manner in which activities are administered (figure 12.1). In addition to being categorized as movement, intrinsic, and extrinsic factors, risk factors can be divided into modifiable factors, such as physical fitness and amount of training, and nonmodifiable factors, such as age and gender. Injury prevention strategy should focus on the modifiable factors in a manner that is appropriate to the nonmodifiable factors.

Whereas the event that results in a disorder or injury (such as a tackle in football) or the point in time when a disorder or injury was first apparent (such as during a game or training session with no obvious inciting event) can be pinpointed, the underlying cause of the disorder or injury will be the result of a complex interaction of movement factors, intrinsic factors, and extrinsic factors (Bahr and Holme 2003, Meeuwisse et al 2007).

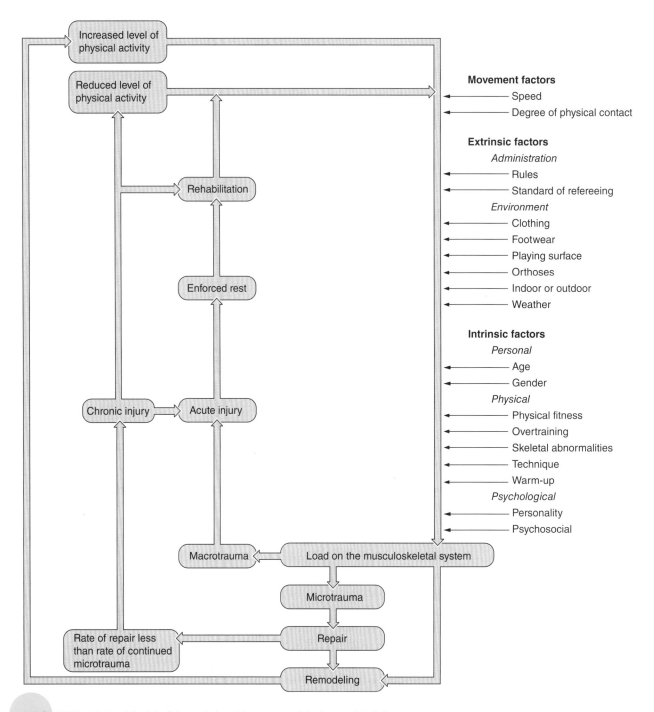

FIGURE 12.1 Model of the relationships among injuries and risk factors.

Key Terms

musculoskeletal disorder Any temporary or permanent abnormality of the musculoskeletal system resulting in pain or discomfort.

movement risk factors The characteristics of a movement that influence the level of loading on the musculoskeletal system.

intrinsic risk factors The personal, physical, and psychological characteristics of a person that influence the level of loading on the musculoskeletal system.

extrinsic risk factors The environmental conditions and administrative procedures that influence the level of loading on the musculoskeletal system.

Intrinsic Risk Factors

Intrinsic risk factors can be categorized as personal factors (age and gender), physical factors (physical fitness, overtraining, skeletal abnormalities, technique, and warm-up), and psychological factors (personality and psychosocial) (Jones et al 1993, Willems et al 2005, Mahieu et al 2006).

Age and Gender

Age-related injuries fall into two major categories: those occurring prior to maturity and those occurring after maturity. Compared with skeletally mature people (i.e., adults), people who are skeletally immature (i.e., children and adolescents) have more pliable bone and softer cartilage, and their ligaments are stronger than the associated centers of bone growth (Micheli 1983, Caine et al 2006). Consequently, loading that causes a ligament or tendon tear in an adult can cause an epiphyseal or apophyseal avulsion in a child or adolescent. Unless properly diagnosed and treated, such damage to epiphyseal or apophyseal regions can result in abnormal bone growth (Barmeda et al 2003, LaLonde and Letts 2005).

Knee injuries and knee pain are the most common orthopedic complaints in children and adolescents (Kannus et al 1988, Ingram et al 2008). For example, Osgood-Schlatter's disease (inflammation and pain at the insertion of the patellar ligament to the tibial tuberosity) is a frequent complaint of young athletes, especially volleyball and basketball players who experience a high frequency of high-intensity jumping and landing. Just as rupture of the patellar ligament can follow the final stage of chronic jumper's knee in an adult, avulsion of the tibial tuberosity can follow chronic Osgood-Schlatter's disease in an immature athlete. Unless properly treated, avulsion of the tibial tuberosity can result in patella alta, which in turn can cause disorders of the patellofemoral joint (Kujala et al 1989, Ozer et al 2002).

Epidemiological data suggest that the incidence of knee disorders, in particular anterior cruciate ligament injuries and patellofemoral syndrome, is higher in females than in males, especially in those who participate in sports involving high-frequency jumping and landing, such as volleyball, basketball, and soccer (Arendt and Dick 1995). The cause of these injuries is not clear, but suggested risk factors include knee alignment (Q angle), muscle balance, and muscle strength (Sommer 1988; Arendt 1996; Fagenbaum and Darling 2003).

> **KEY POINT**
>
> Compared with adults, children have more pliable bone and softer cartilage, and their ligaments are stronger than the associated centers of bone growth. Consequently, loading that causes a ligament or tendon tear in an adult can cause an epiphyseal or apophyseal avulsion in a child or adolescent.

Physical Fitness

The muscles control joint movements and thus not only determine the pattern of movement in a given situation but also protect the joints and joint-supporting structures by helping maintain normal load transmission in joints. The level of conditioning of muscles—the endurance (resistance to fatigue), strength (capacity for producing force), and extensibility (capacity to extend) characteristics of the muscles—affects the risk of injury to all components of the musculoskeletal system. Increases in the endurance, strength, and extensibility of muscles will reduce the risk of injury.

Zohar (1973) compared the process of conditioning the body for a particular physical activity to the process of constructing a building or a bridge. An engineer who designs a bridge must ensure that every part of the structure, irrespective of its size and role, has the required strength and compliance to withstand the forces to which it is subjected. Furthermore, the engineer builds in a suitable safety margin for each component to allow for unusually high loads. Zohar suggested that in many physical conditioning programs, certain parts of the musculoskeletal system, especially those not readily targeted by training equipment, are frequently neglected and become weak links in the kinetic chain. The neglected areas include the axial rotator muscles of the shoulders and hips and the muscles responsible for pronation and supination of the feet. Zohar also suggested that those parts of the musculoskeletal system that are conditioned are frequently not conditioned to a level to incorporate an appropriate safety margin.

For example, in a study of the muscular control of the ankle joint in running, Reber and colleagues (1993) found that during running the tibialis anterior muscle had a higher sustained level of activity than the other muscles, making it more susceptible to fatigue, and that there was a significant increase in the activity of the peroneus brevis muscle as the pace of running increased. Inappropriate conditioning of these muscles may account for the relatively high incidence of fatigue-related lower-leg injuries in running.

KEY POINT

The level of conditioning of the muscles—their endurance, strength, and extensibility—considerably affects the risk of injury to all components of the musculoskeletal system. Increases in the endurance, strength, and extensibility of muscles will reduce the risk of injury.

Imbalances in Strength and Extensibility Normal joint movement involves a high degree of coordination between the members of the various antagonistic pairs of muscles that control joint movement. The coordination between the opposing groups in each antagonistic pair depends on a functional balance between the groups in terms of strength and extensibility. A strength imbalance between the groups can occur if one group is trained more than the other (such as in a strength-training program in which the hamstrings are neglected relative to the quadriceps) or experiences significantly greater relative loading than the other during sport performance (such as jumping and landing in volleyball, which load the quadriceps more than the hamstrings) (Grace 1985, Hasegawa et al 2002).

Strength imbalances can also occur between similar groups of muscles on opposite sides of the body, such as the right and left quadriceps;

this happens, for example, in athletes who favor one leg over the other in jumping and landing (Knapik et al 1991, Hasegawa et al 2002). The normal balance of strength between similar groups of muscles on opposite sides of the body may be viewed in absolute terms. However, the normal balance in strength between the opposing muscle groups in each antagonistic pair must be viewed in relative terms. For example, the cross-sectional area and, therefore, potential strength of the plantar flexors of the ankle joint are approximately seven times greater than those of the dorsiflexors (Silver et al 1985, Wickiewicz et al 1983). Even allowing for differences in leverage, there could not be a strength balance between the two muscle groups in absolute terms. The main function of the dorsiflexors is to provide toe clearance during the swing phase in walking, whereas the plantar flexors are responsible for supporting body weight in the upright posture and propelling the body forward and upward in activities such as walking, running, and jumping. Consequently, the plantar flexors and dorsiflexors are different functionally, and the strength ratio of 7:1 in favor of the plantar flexors is probably functionally normal. Therefore, strength imbalance between the plantar and dorsiflexors may be reflected in a strength ratio markedly different from 7:1. Similarly, the cross-sectional area of the quadriceps is approximately two and a half times that of the hamstrings (Wickiewicz et al 1983). Consequently, strength imbalance between the quadriceps and hamstrings may be reflected in a strength ratio markedly different from 2.5:1.

Just as differences in the (time-averaged) relative load on antagonistic muscle groups can result in strength imbalances, highly repetitive and exclusive movement patterns can produce extensibility imbalances between antagonistic muscle groups (Bach et al 1985, Herzog 2000). Strength imbalances combined with extensibility imbalances can result in **functional imbalances** (also referred to as muscle imbalances), which predispose the person to musculoskeletal disorder or injury (Grace 1985, Hasegawa et al 2002). Functional imbalance is manifest in a number of ways, including

- imbalances between left and right sides of the body (Knapik et al 1991);
- imbalances between antagonistic muscle groups (Bach et al 1985, Francis and Bryce 1987); and
- imbalances between muscle groups of different joints that work together to bring about particular movements (Sommer 1988).

Sommer (1988) investigated the effect of fatigue on the movement of the hips, knees, and ankles in jumping and landing in volleyball and basketball players. The results showed that as fatigue increased, the tendency for the knee joints to be forced into abduction during the powerful phases of leg extension (jumping) and leg flexion (landing) increased. Sommer suggested that this was attributable to functional imbalances between the muscles controlling the hip joints (the gluteal muscles) and that the effects of the imbalances became more pronounced as fatigue increased. The likely effects of knee abduction during intense activity of the knee extensor muscles would be

- lateral tracking of the patella,
- strain on the medial ligaments and other medial supporting structures, and
- asymmetric loading on the quadriceps tendon and patellar ligament, especially at their insertions on the patella.

KEY POINT

A strength imbalance between antagonistic muscle groups can occur if one group is trained more than the other or experiences significantly greater relative loading than the other during sport performance. Highly repetitive and exclusive movement patterns can produce extensibility imbalances between antagonistic muscle groups. Strength imbalances combined with extensibility imbalances result in functional imbalances (muscle imbalances) that predispose the person to injury.

Lateral tracking can result in patellar chondromalacia, and asymmetric loading on the quadriceps tendon and patellar tendon can result in jumper's knee—inflammation and pain at the upper and lower poles of the patella (Ferretti et al 1990).

Key Term

functional imbalances Strength imbalances combined with extensibility imbalances that predispose a person to musculoskeletal disorder or injury.

Growth-Related Imbalances in Strength and Extensibility Imbalances in strength and extensibility can occur at any age, but some imbalances seem to be growth related. Longitudinal growth in the bones of the limbs and in the vertebrae is always more advanced than that of the associated soft tissues—musculotendinous units and fibrous supporting structures (Frontera et al 2007). Consequently, during the growth period the soft tissues have to continually adapt their length to the length of the bones. For most of the growth period, soft tissue adaptation can keep pace with the rate of bone growth. However, during a period of rapid bone growth, such as the adolescent growth spurt, the rate of bone growth can outpace that of soft tissue adaptation. In these circumstances, there may be a significant increase in the tightness of the soft tissues, resulting in a loss of joint flexibility, increased compression loading on articular surfaces, and increased tension loading on ligament and tendon insertions. All of these factors can increase the risk of injury to the tissues concerned, especially when the loss of flexibility is asymmetric. For example, adolescents tend to develop hollowback during the adolescent growth spurt. The hollowback seems to be caused by asymmetric tightness in the fibrous structures supporting the lumbar region. The fibrous supporting structures on the posterior aspect of the lumbar region (the lumbodorsal fascia) are thicker and more extensive than those on the anterior aspect. Consequently, in response to increased growth in the vertebrae—an increase in the height of the vertebral bodies—the posterior soft tissues may take longer to adapt (by structural elongation) than the anterior soft tissues, resulting in a temporary hollowback. As long as the hollowback persists, the lumbar region is in a permanent state of extension, which increases the compression loading on the posterior aspects of the vertebrae and discs and changes the form of loading on the anterior aspects of the vertebrae and discs from compression to tension.

Because hollowback restricts forward bending of the lumbar region (lumbar flexion), there is a tendency to develop a compensatory roundback (round shoulders) in the thoracic region. In this condition the vertebral column in the region of the roundback is flexed, increasing the compression loading on the anterior aspects of the vertebrae and discs and changing the form of loading on the posterior aspects from compression to tension. If the abnormal loading in the hollowback and roundback regions is prolonged or excessive, it can result in abnormal growth of the affected vertebrae and discs, resulting in permanent hollowback and roundback (Scheuermann's disease). However, in most adolescents the normal shape of the vertebral column and normal loading on the vertebrae and discs are gradually restored as the posterior soft tissues gradually elongate. Nevertheless, during the period of hollowback and roundback, the vertebrae and discs, as well as the soft tissue–supporting structures, are more likely to be excessively overloaded by activities that would not normally cause excessive overload. For example, sport training involving repetitive flexion, extension, or rotation of the trunk can result in a variety of injuries such as ruptured discs and fractures of the vertebrae (Frontera et al 2007).

Another example of a growth-related soft tissue imbalance is genu valgum (knock-knee), which occurs during the adolescent growth spurt. As the bones of the upper and lower legs grow in length, the soft tissues that support and move the knee joints must adapt their lengths to accommodate the increase in bone length. During the adolescent growth spurt, the rate of bone growth exceeds that of the soft tissues such that they become tighter, especially when the knees are in the extended position. The

tightness is likely to be asymmetric because the fibrous supporting structures on the lateral aspect of the knee (in particular, the iliotibial tract) are thicker and more extensive than those on the medial aspect. Consequently, the lower leg is pulled outward (abduction of the knee joint), resulting in temporary genu valgum. Genu valgum increases the compression load on the articular surfaces of the lateral condyles of the femur and tibia and unloads the medial condyles. In addition, the patella is displaced laterally, increasing the compression load on the lateral articular surfaces of the patella and patellar surface of the femur and unloading the medial articular surfaces.

In most adolescents, a growth-spurt-related genu valgum gradually disappears as the lateral soft tissues elongate. However, if the genu valgum is prolonged or excessive, the abnormal loading is likely to result in abnormal growth of the articular surfaces between the femur and tibia, causing permanent genu valgum or abnormal growth of the articular surfaces between the patella and patellar surfaces of the femur, resulting in recurrent partial or complete lateral dislocation of the patella.

Persistent postural defects including hollowback, roundback, genu valgum, and genu varum are associated with increased risk of injury, especially muscle and ligament injuries of the low back and lower limbs, in football and basketball (Shambaugh et al 1991, Watson 1995, Winslow and Yoder 1995, Frontera et al 2007). It is reasonable to assume that proper conditioning, especially muscle strength and extensibility, may compensate to a certain extent for the increased risk of injury associated with persistent postural defects.

KEY POINT

Imbalances in strength and extensibility can occur at any age, but some imbalances seem to be growth related. During a period of rapid bone growth, such as an adolescent growth spurt, the rate of growth of bone can outpace that of the associated soft tissues. In these circumstances there can be a significant increase in the tightness of the soft tissues, increasing the risk of injury or disorder.

Overtraining

Normal daily activities such as walking, carrying a shopping bag, pushing a child's buggy, or climbing a flight of stairs involve a relatively low level of physical effort; the load imposed on the musculoskeletal system by these activities is well within the moderate habitual loading range (see figure 11.2). In sharp contrast, the intensity of effort required to improve cardiorespiratory fitness can excessively overload the musculoskeletal system, especially in a previously sedentary person. Consequently, the first few weeks of a training program designed to improve the cardiorespiratory fitness of a previously inactive person should be mainly concerned with conditioning the musculoskeletal system. This involves a level of effort less than that required to produce a cardiorespiratory training effect.

A number of studies have examined the effects of fitness training on previously sedentary adults. Even in studies in which the amount of exercise was carefully controlled to minimize the risk of injury, the injury rate of the participants was still 50% to 90%, and all of the injuries occurred in the first 6 weeks of training (Jones et al 1994, Hootman et al 2001). In these studies, injury was defined as a conditioning-related musculoskeletal disorder severe enough to necessitate reducing training intensity, changing exercise mode, or discontinuing training for at least a week. These studies show that it is easy to overtrain (or overuse) the musculoskeletal system when it is in a deconditioned state.

However, overtraining is not confined to the previously inactive person. Reviews of injuries to recreational joggers and runners have shown that 55% to 65% of all injuries are due to overtraining (Rolf 1995, Taunton et al 2003, Hreljac 2004, Wen 2007). Even well-trained athletes can suffer from overtraining, especially if they train too frequently or abruptly increase their amount of training (Fredericson and Anuruddh 2007, van Gent et al 2007). For example, in a study of knee injuries in elite volleyball players, the incidence of injury was found to be directly related to the frequency of play (including training sessions and games); the incidence of injury in those who played once or twice a week, three times a week, four times a week, and more than four times a week was 3.3%, 14.6%, 29.1%, and 41.8%, respectively (Ferretti et al 1984). In ballet and all forms of modern dance, the major sites of injury are the knee, ankle, and low back, and the most frequently reported cause of injury is overtraining due to long practice sessions (Byhring and Bø 2002, Bronnar et al 2003, Gamboa et al 2008).

> **KEY POINT**
>
> Overtraining refers to a training schedule in which the amount of training (combination of intensity, frequency, and duration of training) in a number of consecutive training sessions exceeds the capacity of the person to adequately recover (physically and mentally) between training sessions. The symptoms of overtraining include persistent fatigue, persistent muscle, tendon and joint soreness, decreased performance, and immunodeficiency (decreased resistance to infectious diseases).

Skeletal Abnormalities

Abnormalities in the length or shape of bones can cause excessive overloading of the musculoskeletal system by causing compensatory movement in associated joints as, for example, with leg length inequality, femoral anteversion, and tibial varum.

Leg Length Inequality Ideally, the right and left sides of the body should be equally loaded during standing, walking, and other daily activities. However, most people have some degree of leg length inequality (LLI), which, unless compensated, results in abnormal loading of the hips, pelvis, and low back (Gibson et al 1983, Gurney 2002, Brady et al 2003). In most people, LLI is less than 5 mm. A person can easily compensate for this discrepancy by making minor adjustments in the movement of the feet to effectively shorten a long leg (via pronation of the foot) and lengthen a short leg (via supination of the foot) to minimize the LLI and bring about normal loading (Gibson et al 1983). However, when the degree of LLI is too large to be completely compensated by foot movements, the pelvis invariably tilts downward on the shorter side when standing and walking (figure 12.2). The pelvic tilt considerably reduces the load-bearing articular surface area in the hip joint of the longer leg, which considerably increases the compression stress on the articular surfaces (Brand and Yack 1996, Wretenberg et al 2008). To compensate for the abnormally tilted pelvis, the lower part of the vertebral column bends toward the longer leg so that the trunk can remain fairly upright.

The compensatory lateral bending of the lower part of the vertebral column is called functional scoliosis (Nakamura et al 2004). Functional scoliosis is brought about by asymmetric activity in the muscles on the left and right sides of the low back. The degree of asymmetry in muscle activity depends on the extent of the LLI; an LLI as small as 10 mm leads to a considerable asymmetric increase in the activity of several muscle groups, making it impossible for the person to maintain a relaxed standing position. The increased muscular activity increases energy expenditure and fatigue (Friberg 1983, Blustein and D'Amico 1985, Kakushima et al 2003). The functional scoliosis not only bends the low back toward the longer leg but also rotates the low back about a vertical axis. The combination of lateral bending and axial rotation causes the intervertebral discs to be compressed on the concave side of the column, stretched on the

other side, and twisted about a vertical axis. This is thought to be the pattern of loading most likely to damage the intervertebral discs (Friberg 1983, Kakushima et al 2003). Furthermore, this pattern of loading causes the discs to bulge outward at the rear on the long leg side, increasing the likelihood of impingement of the adjacent spinal nerves. The impingement of spinal nerves can result in pain such as sciatica and, subsequently, muscle spasm and more pain.

The most common symptom associated with LLI is low back pain (Giles and Taylor 1981, Nadler et al 1998). The extent of abnormal loading on low back muscles, intervertebral discs, and hip joint articular surfaces that results from LLI and the subsequent functional scoliosis depends on the extent of LLI. However, the extent of abnormal loading also depends on the type of movement being performed; the more dynamic the activity, the greater the load. For example, a relatively small LLI may not result in excessive loading when walking but may result in excessive loading when running given the greater muscle and joint reaction forces.

The long-term effects of LLI on the musculoskeletal system depend on the age of the person at the time the LLI occurs. If LLI occurs during childhood, the initial functional scoliosis becomes a structural scoliosis; the vertebrae, pelvic bones, the upper ends of both femurs (and their supporting structures) adapt by chondral modeling to create functionally normal joints at the hips, within the pelvis, and between the vertebrae (Frost 1973, Brady et al 2003). Wedge-shaped vertebrae (Scheuermann's disease) are a well-recognized feature of structural scoliosis. Even though a structural scoliosis minimizes the effects of LLI, the loading imposed by dynamic activities is more likely to become excessive in a person with scoliosis than in a person without scoliosis. LLI occurring after maturity or, to be more precise, becoming excessive after maturity is usually the result of fracture of a femur or tibia (Gibson et al 1983, Clark et al 2006). Because the bones are no longer able to model in response to the abnormal loads imposed by the LLI, the long-term effects can be degenerative joint disease,

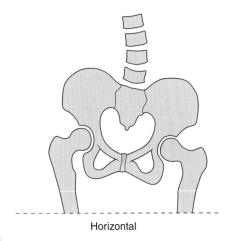

Horizontal

FIGURE 12.2 Effect of leg length inequality on the orientation of the articular surfaces in the hip joints.

especially in the hip joint of the longer leg, and damage to intervertebral discs causing spinal nerve impingement and muscle spasm.

Femoral Anteversion   Turnout—the ability to turn both feet out so that they are in line with each other—is a fundamental posture in ballet and other forms of theatrical dance (Ende and Wickstrom 1982, Hardaker et al 1985) (figure 12.3b). In normal upright posture, most people have about 10° to 20° of out-toeing so that the feet have only to be turned out through another 70° to 80° to achieve a full 90° turnout (figure 12.3, *a* and *b*). Because the knee and ankle joints are not designed to rotate about a vertical axis, turnout

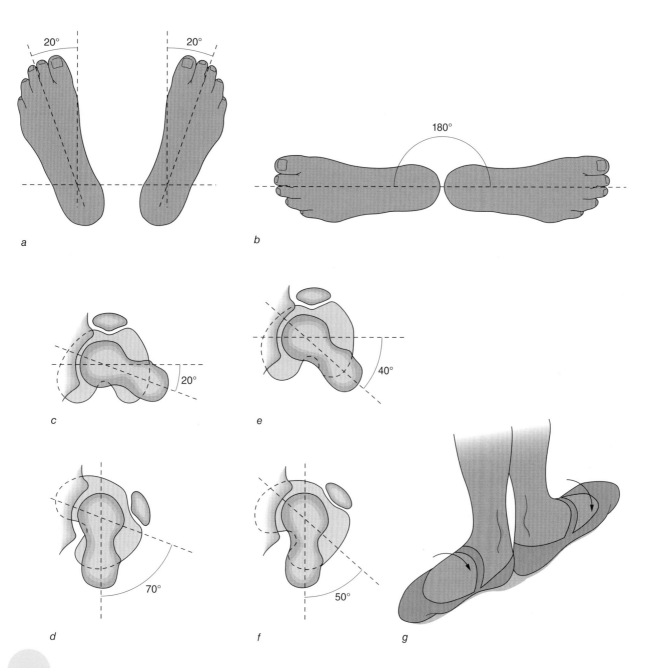

FIGURE 12.3   Effect of femoral anteversion on turnout. *(a)* Natural out-toeing in normal upright standing posture. *(b)* Full turnout. *(c)* Normal upright posture with femoral anteversion of right leg of 20°. *(d)* With femoral anteversion of 20°, outward rotation of the hip is limited to approximately 70°. *(e)* Normal upright posture with femoral anteversion of right leg of 40°. *(f)* With femoral anteversion of 40°, outward rotation of the hip is limited to approximately 50°. *(g)* Rolling in.

of each foot ideally should be achieved entirely by outward rotation of the whole leg about the hip joint. However, the normal range of outward rotation of the hip is 60° to 70° (Hardaker et al 1985). It follows that most dancers can only achieve full turnout by twisting the knee and ankle joints outward (up to 20°) (Hamilton et al 2006). Whereas the knee and ankle joints may be able to tolerate small amounts of twisting, excessive twisting can cause excessive overloading of the articular surfaces and supporting structures of these joints. This situation is likely to arise if outward rotation of the hip is restricted.

Movement in the hip, as in other joints, can be restricted by tight ligaments or bony abnormalities. For example, the range of outward rotation of the hip joint is largely determined by the orientation of the neck of the femur to the shaft of the femur. In most people, the neck of the femur is directed slightly backward from the hip joint such that it makes an angle of about 20° with the frontal plane (figure 12.3c). This angle is called the angle of femoral anteversion: The smaller the anteversion angle, the greater the amount of outward rotation of the hip, and vice versa.

An anteversion angle of about 20° allows 60° to 70° of natural rotation of the hip (figures 12.3, c and d). However, an anteversion angle of 40° limits outward rotation of the hip to 40° to 50° (figure 12.3, e and f). In this case the only way full turnout can be achieved is by forcibly twisting the knees and ankles outward, which can sublux the tibiofemoral and patellofemoral joints and excessively overload parts of the articular surfaces and supporting structures. Twisting the foot (forced eversion) can sublux the ankle and subtalar joints and excessively overload parts of the articular surfaces as well as the arch support structures; this movement of the foot is called rolling in (figure 12.3g).

Tibial Varum Downhill skiing is essentially a series of linked turns used to control the rate of descent and direction of travel (Matheson and Macintyre 1987). The skill of turning depends largely on the skier's ability to control the degree of edging of the skis—the angle of the skis to the surface of the snow—by varying degrees of pronation, dorsiflexion, knee

and hip flexion, and torso counterrotation. The skier's ability to perform these movements effectively and efficiently depends considerably on the degree to which the ski boots accommodate the natural alignment of the lower legs and feet. In normal upright standing posture most people's lower legs are vertical, with the feet flat on the floor and the ankle and subtalar joints in a neutral position (figure 12.4a). Most ski boots accommodate this type of alignment (figure 12.4b). However, there is considerable individual variation in the alignment of the lower legs and feet, especially in the degree of tibial varum—when the lower leg is angled medially from the knee (figure 12.4c). In many ski boots the angle between the cuff and the boot is adjustable. However, if the boot is not adjustable, it may not properly accommodate a marked tibial varum such that in upright standing posture the weight of the skier is transmitted to the floor via the lateral edges of the boots (figure 12.4d). When skiing in these boots the

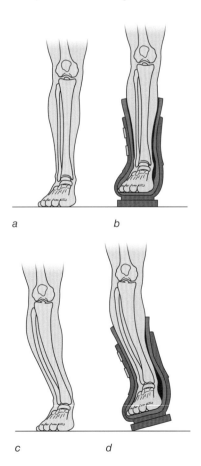

FIGURE 12.4 The inability of a nonadjustable ski boot to accommodate a marked tibial varum.

skier may find it difficult to keep the skis flat on the snow when traversing, which will predispose him to falling as a result of catching the outside edges of the skis. The skier may also find it difficult to press the medial edges of the skis into the snow—essential for turning. To try to keep the skis flat and medially edge the skis, the skier is likely to abduct the knees or adopt a wide-tracked stance. These movements can be strenuous and result in rapid fatigue with a consequent increase in the risk of injury.

Relative to normal lower-leg and foot alignment (figure 12.4*a)*, the foot of a person with a marked tibial varum is in a relatively everted orientation in upright stance (figure 12.4*c)* and, presumably, in weight-bearing movements such as walking and running. Consequently, the muscles that evert the ankle are likely to be relatively shorter and the muscles that invert the ankle are likely to be relatively longer in a person with tibial varum than in a person with a normal lower-leg and foot alignment. The effect of tibial varum on the functional stability of the ankle joint (provided by the everters and inverters of the ankle) is not clear, but there is evidence that women with increased tibial varum are at greater risk of ankle ligament injury (Beynnon et al 2001).

KEY POINT

Skeletal abnormalities can cause excessive overloading of the musculoskeletal system by causing compensatory movement in associated joints.

Technique

Sports and physically active leisure pursuits basically consist of two types of movements:

1. Movements that involve generating or increasing speed of movement of the whole body (in running and jumping activities) and of an arm or leg to propel an implement (in throwing and striking activities)

2. Movements that involve decreasing speed of movement of the whole body (in landing from a jump) and of an implement (in catching a ball)

The movement pattern—the way that the body segments move in relation to each other—of a person when performing a particular movement or sequence of movements is usually referred to as that person's technique. Good technique is a movement pattern that not only is effective in terms of performance (the desired outcome such as height jumped, sprinting speed, accuracy of a tennis serve, or number of points scored in gymnastics) but also minimizes the risk of injury by appropriate distribution of the overall load throughout the kinetic chain. In terms of performance, the quality of technique is far more important in some sports than in others. For example, in gymnastics, diving, and figure skating, the quality of the technique is the performance criterion (Watkins 1987a). In other sports, such as track-and-field athletics and swimming, the performance criterion is the end result of a particular technique, for example, the time to run 100 m (Watkins 1987b). However, in terms of minimizing the risk of injury, good technique is essential, irrespective of how performance is judged. A person is unlikely to achieve full potential in a sport if she can occasionally perform well but is frequently injured because she has a poor technique that excessively overloads certain muscles and joints. Inappropriate use and summation of muscular effort or abnormal joint movements characterize poor technique, result in localized overload, and increase risk of injury.

Inappropriate Use and Summation of Muscular Effort To generate maximum speed in the body as a whole or in a body part, it is necessary to involve as much of the body's musculature as possible; the more muscles and joints used, the greater the force application as a result of increased range of movement. In propulsive actions, the impulse of the thrust is reduced if some muscles (and joints) that could contribute to the action are not used or not used maximally. Furthermore, some joints in the kinetic chain can be excessively overloaded in an effort to compensate for the lack of involvement of the underused joints. For example, in the volleyball spiking action, the objective is to hit the ball downward into the opponent's court. The speed of the ball on leaving the hand depends on the force applied to the ball and the time of force application during ball contact (Coleman et al 1993, Forthomme et al 2005). To apply a large force to the ball, the hand must be moving as quickly as possible on impact. Ideally—with good technique—hand speed is generated largely by the hip and trunk flexor muscles, with little movement in the shoulder (figure 12.5, *a* and *b*). In other words, the large hip and trunk flexor muscles are the sources of power in the spiking action. This minimizes the load on the shoulder and arm muscles and enables the arm to exert control over the hand movement prior to ball contact. Lack of involvement of the hip and trunk muscles is usually compensated by excessive shoulder activity, which with repeated use can cause a variety of disorders of the shoulder joint and its supporting structures (figure 12.5, *c* and *d*).

A tennis serve is similar to the volleyball spiking action in that the power of the movement is largely produced by the hips and trunk (as well as the legs) rather than the shoulder and arm muscles, which are responsible for controlling the racket. Forehand and backhand strokes in tennis are both similar to the spike and tennis serve in the way speed of movement is generated in the racket head (hips and trunk) and the way the racket head is controlled (shoulder and arm muscles). Tennis elbow refers to a range of disorders of the elbow and its supporting structures that occur as a result of lack of involvement of the hips or the trunk. To compensate and generate normal racket head speed, the shoulder and arm muscles have to exert large forces that excessively overload the insertions of muscles and ligaments, especially those around the elbow. Repeated use leads to progressive microtrauma and tennis elbow syndrome (Groppel 1986).

The previous examples from volleyball and tennis demonstrate the importance of the legs and trunk in movements involving the development of large forces. This is not surprising, because the largest and potentially strongest muscles in the body are those of the legs and trunk. Because the speed of movement resulting from a propulsive action depends on the

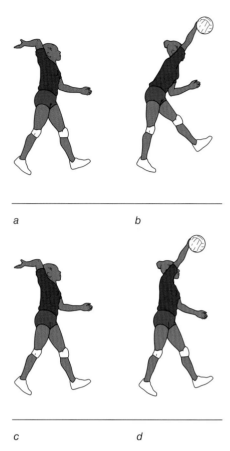

FIGURE 12.5 Spiking actions in volleyball. *(a, b)* Good technique; emphasis is on hip and trunk flexion. *(c, d)* Poor technique; emphasis is on shoulder flexion.

amount of force produced throughout the whole action, it is important to involve as many muscle groups as possible, not just the strongest groups. This is achieved by sequential and overlapping involvement of the various muscle groups starting with the strongest muscles and finishing with the weaker muscles. This process maximizes the summation of force throughout the kinetic chain, resulting in maximum speed of movement (Hopper 1973, Putnam 1993, Gordon and Dapena 2006). This mode of operation can be compared to a multistage rocket in which the first stage (most powerful stage) is responsible for initiating upward movement of the rocket. The thrust produced by the first stage decreases as its fuel starts to run out, so the first stage is eventually replaced by the second stage (next most powerful stage) to maintain upward thrust on the rocket. As the thrust produced by the second stage decreases, the third stage takes over and so on through the remaining stages. In a human kinetic chain, there is considerable overlap between the stages; that is, two or more muscular groups are likely to be contributing to propulsion simultaneously.

Using and summating muscular effort are as important in decreasing the speed of movement of the body or an implement such as a cricket ball as they are in generating or increasing speed of movement. The greater the deceleration of the body (or implement), the greater the load on the body and therefore the greater the risk of injury, especially to bones and joints. For example, some actions that basketball players use, such as landing from vertical jump with the legs too straight, plant-and-cut (sudden deceleration on a straight or near straight leg with intent to change direction in one step), and one-step stop (attempting to stop and reverse direction in one step), are major risk factors for anterior cruciate ligament injury (Henning et al 1994, Boden et al 2000). Controlling joint flexion can reduce deceleration in these movements and consequently reduce the risk of injury.

Abnormal Joint Movements Abnormal joint movements occur when a joint is forced to (a) move about axes abnormal for the joint or (b) move beyond the normal maximum range of movement about a normal axis. As an example of the former, abduction of the knee joint during the leg kick in breaststroke can result in breaststroker's knee. Moving beyond the normal range of movement about a normal axis can be seen in hyperextension of the knee joint resulting from a tackle in football, which can damage the knee ligaments, and in combined adduction and internal rotation of the shoulder joint, which can result in shoulder impingement in a swimmer or thrower.

- *Breaststroker's knee.* The incidence of knee pain in swimmers who specialize in breaststroke is high and has given rise to the term *breaststroker's knee* (Vizsolyi et al 1987, Costill et al 1992, Shapiro 2001). In the early stages of its development, the condition is characterized by pain and tenderness in the supporting structures on the medial side of the knee. Later, the front and lateral side of the knee become involved, resulting in generalized knee pain.

Breaststroker's knee is caused by a particular kind of kicking action called the whip kick (figure 12.6). At the end of the recovery phase, as the heels approach the buttocks, the knees separate (abduction of the hips), the lower legs turn outward, and the ankles dorsiflex to present as large an area as possible to the water at the start of the kicking phase. The whip kick is initiated by inward rotation of the hips, which rotates the lower legs farther outward. This is closely followed by the propulsion phase, which involves simultaneous hip extension, hip adduction, and knee extension. These actions drive the feet backward along semicircular paths. During this action, the lower legs are forcibly abducted and outwardly rotated at the knees, largely by the pressure of water against the lower legs and feet. Abduction

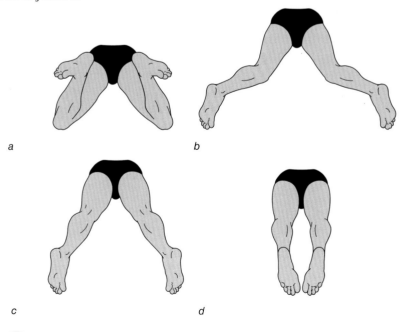

FIGURE 12.6 Whip-kick leg action in breaststroke.

and outward rotation of the lower legs at the knees are abnormal movements and result in the symptoms described here. The load on the knees is greatest at the start of the propulsion phase, when the knees experience a whiplike action, and during the middle phase of the kick, when the lower legs present a large surface area to the water (figure 12.6*b*). By reducing the amount of hip abduction and internal rotation of the hips at the start of the propulsion phase, the feet move directly backward rather than in a semicircular path, which considerably reduces the thrust of the leg kick but also reduces the amount of forced abduction and outward rotation of the knees.

The abnormal pattern of loading on the knees caused by the whip kick is similar to that produced by "screwing the knee" in ballet dancing. This action is caused by inadequate outward rotation of the hips (see earlier section on femoral anteversion and turnout) and occurs when the knees are straightened from a position in which the knees are bent and the feet fully turned out (Ende and Wickstrom 1982). The situation can also arise in the propulsion phase of throwing a javelin; if the foot of the rear leg is turned outward as the hip and knee joints extend, the knee will be forcibly abducted and outwardly rotated (figure 12.7).

- *Shoulder impingement syndrome.* Overhead movements of the arms in swimming, throwing, and racket games are brought about by movements in three joints: the shoulder, sternoclavicular, and acromioclavicular joints (as described in chapter 6). In abducting the arm directly overhead, 180° from the resting position, the shoulder, sternoclavicular, and acromioclavicular joints contribute approximately 120°, 40°, and 20°, respectively (Peat 1986). Consequently, when range of movement in the sternoclavicular and acromioclavicular joints is restricted, the shoulder joint must be hyperabducted to achieve full abduction of the arm. When this occurs, the supporting structures of the shoulder joint come into contact with the coracoacromial arch, resulting in a range of painful disorders referred to as shoulder impingement syndrome (Miniaci and Fowler 1993, Koester et al 2005).

FIGURE 12.7   Rear leg action in throwing the javelin .

Even with normal ranges of movement in the shoulder, sternoclavicular, and acromioclavicular joints, impingement between the shoulder joint structures and the coracoacromial arch is still likely to occur when shoulder abduction is accompanied by internal rotation. This situation can arise during the arm recovery phase in butterfly and freestyle swimming because of a high elbow position (shoulder abduction) and the hand lagging behind the elbow (internal rotation of shoulder) (figure 12.8*a*). The likelihood of impingement can be reduced by leading with the hand (external rotation of shoulder; figure 12.8*b*) and increasing body roll (decreases shoulder abduction; figure 12.8, *c* and *d*) during arm recovery (Penny and Smith 1980, Lewis et al 2005).

KEY POINT

Technique refers to the movement pattern of a person during a particular movement or sequence of movements. Good technique is a movement pattern that not only maximizes performance but also minimizes the risk of injury by appropriately distributing the overall load throughout the kinetic chain. Poor technique is characterized by inappropriate use and summation of muscular effort and abnormal joint movements, both of which result in localized overload and increased risk of injury.

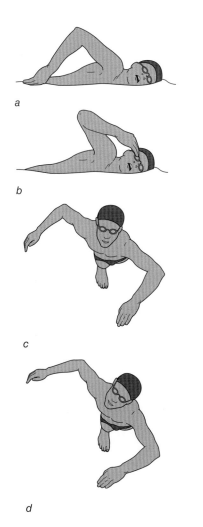

a

b

c

d

FIGURE 12.8 Effect of arm action and body roll during arm recovery on risk of shoulder impingement.

Warm-Up

Warm-up refers to preliminary exercise that prepares the body physically and mentally for the more strenuous activity that is to follow (Watkins 1982, Van Mechelen et al 1993, Soligard et al 2008). The rationale underlying warm-up is that the cardiorespiratory and musculoskeletal systems need a certain amount of time to fully respond to an increase in the level of physical activity. A gradual increase in the intensity of effort enables the cardiorespiratory system to gradually increase the supply of blood and oxygen to the heart and muscles. The increase in muscle temperature accompanying the increased level of muscular activity decreases the viscosity of the muscles, which increases their mechanical efficiency. The joints

(synovial and symphysis) also benefit from a warm-up by becoming more stable. This stability is due to an increase in the thickness of cartilage that results from an increase in the circulation of fluid through the cartilage. For people engaging in physically active leisure, a 10 to 15 min warm-up consisting of light to moderate whole-body exercise, such as jogging and stretching the joints, is recommended (Wilmore et al 2008).

> **KEY POINT**
>
> Warm-up is preliminary exercise that prepares the body both physically and mentally for the more strenuous activity to follow. A warm-up can decrease the risk of injury.

Psychological Factors

Psychological factors can be described as either personality variables or psychosocial variables (Kerr and Fowler 1988, Linton 2000, Williams 2001). Personality variables refer to fairly stable characteristics such as locus of control and level of self-esteem. Psychosocial variables refer to the effect that changes in a person's personal and social life can have on her response in a given situation. In the context of sport there is good evidence that psychological stress affects attentional and physiological responses in the form of, for example, above-normal tension in muscles, defects in fine motor control, distraction, and reduction in peripheral vision, all of which can predispose the person to injury (Petrie 1993, Kelley 1990, Cupal 1998, Galambos et al 2005).

> **KEY POINT**
>
> There is good evidence that psychological stress affects attentional and physiological responses, which can predispose a person to injury.

External Risk Factors

Extrinsic risk factors can be categorized as environmental conditions (in particular, clothing, footwear, playing surfaces, and orthoses) and administration of the sport (rules of competition, standard of refereeing) (Bahr and Holme 2003).

Clothing, Footwear, and Playing Surfaces

Protective clothing designed to reduce impact loading is a feature of many sports. This clothing includes helmets (e.g., American football, hockey, skiing, horse riding, rock climbing, cricket), headguards (amateur boxing, rugby), shoulder pads (American football), elbow pads (volleyball), gloves (boxing, cricket), torso padding (hockey), knee pads (volleyball), and leg pads (hockey, football, soccer, cricket).

Footwear is important in many sports, not just for protection but also to enhance performance. Developments in footwear have been concerned with reducing impact loading in activities such as running and those involving high-frequency jumping and landing such as basketball and volleyball and providing support for the ankle to reduce the risk of ankle sprains. Soft-soled shoes reduce impact force peak and the rate of loading (chapter 10). Ankle sprains, especially inversion sprains, are common in sport and can cause significant, chronic disability (Boruta et al 1990, Gross and Liu 2003, Beynnon et al 2005). Research has been carried out regarding the effect of high-top shoes on the risk of inversion sprains. Whereas high-top shoes appear to increase ankle stability in laboratory tests, there is insufficient evidence from randomized controlled trials of players using high-top shoes to show whether they reduce the incidence of ankle injuries (Pedowitz et al 2008).

Two main mechanical characteristics of playing surfaces influence the risk of injury: surface hardness and frictional properties (Nigg and Segesser 1988). Just as soft-soled shoes reduce impact force peak and rate of loading compared with hard-soled shoes, soft playing surfaces reduce impact force peak and rate of loading compared with hard surfaces. Ferretti and colleagues (1984) and Watkins and Green (1992) found a clear and positive relationship between the hardness of the playing surface and the incidence of injuries in volleyball. Evidence regarding the effects of friction on injury risk is limited, but the frequency of

football and tennis injuries is reported to be higher on high-friction surfaces than on low-friction surfaces (Nigg and Segesser 1988, Pasanen et al 2008, Villwock et al 2009). The risk of injury increases when friction is too low, because players can slip on the playing surface. The risk of injury also increases when friction is too high, because the foot experiences high deceleration at foot strike, resulting in high muscle forces and joint reaction forces at the knee and ankle and increased risk of injury.

KEY POINT

Clothing, footwear, and playing surfaces have a major influence on injury risk. Developments in footwear have been concerned with reducing impact loading and providing support for the ankle to reduce the risk of ankle sprains. Playing surfaces with moderate levels of hardness and friction are associated with reduced risk of injury.

Orthoses

Orthoses are used in sport and nonsport activities to assist the joint-supporting structures in maintaining normal joint transmission of loads by reducing the risk of abnormal movements and restricting the amount of abnormal movement when it occurs. Not surprisingly, most research on sports orthoses has been concerned with knee and ankle orthoses, reflecting the relatively high incidence of chronically debilitating sport-related knee and ankle injuries. There are two basic types of sports orthosis: those that support joints directly and those that support joints indirectly. Orthoses providing direct support range from well-established taping and strapping methods used to support the ankle, knee, and elbow (Anderson et al 1992) to jointed semirigid appliances such as knee braces (Zachazewski and Geissler 1992). Taping, strapping, and braces are intended to assist the muscles in stabilizing vulnerable joints by restricting range of motion in abnormal directions while simultaneously allowing normal joint movements. With regard to taping and strapping (which use nonrigid material), the general finding is that strapping that is comfortable at the start quickly stretches or loosens in some other way (depending on the level of activity) so that its effectiveness is significantly reduced (Anderson et al 1992, Gerard 1998). Knee braces have been found to prevent abnormal abduction and adduction at the knee but have also been found to have adverse effects on normal ranges of movement at the knee (Osternig and Robertson 1993, De Vita et al 1996). Orthoses providing indirect support commonly come in the form of shock-absorbing insoles (Nigg et al 1999) and forefoot or rearfoot valgus or varus wedges, which are used to maintain the correct alignment of the forefoot or rearfoot to the ankle and subtalar joints to maintain normal pronation and supination (McKenzie et al 1985, Matheson and Macintyre 1987, Petrov et al 1988). The effects of these types of orthoses have been found to be small and nonsystematic (in terms of shock absorption and supination and pronation) and appear to be related to athletes' attempts to optimize comfort and energy expenditure (Nigg 2001).

Administration of the Sport

The rules of competition and the extent to which the rules are enforced—the standard of refereeing—clearly affect the risk of injury to participants in any given sport. The rules of competition are designed to ensure fair play and safe competition (Sperryn 1988) and should reflect the overall risk of injury in a particular sport or activity. Consequently, rule changes may be necessary to reduce the risk of injury to an acceptable level. For example, in American football, after it became clear that spearing—the action of a player driving his head (protected by a helmet) into an opponent's body—was associated with a high incidence of severe neck injuries, the practice of spearing was banned in high school and college football in 1976. Following the ban on spearing, the frequency of injuries resulting in permanent cervical quadriplegia decreased from 34 in the 1976 season to 5 in the 1984 season (Torg et al 1985). Case study 8 examines injury data in high school sports and presents possible methods of reducing the incidence of injury.

CASE STUDY 8   INJURIES IN HIGH SCHOOL SPORTS

Rechel JA, Yard EE, Comstock RD. 2008. An epidemiologic comparison of high school sports injuries sustained in practice and competition. *Journal of Athletic Training* 43(2):197-204.

More than 7 million United States high school athletes competed in interscholastic sports during the 2005-2006 school year. Whereas there is evidence to suggest that the incidence (rate) of injuries in interscholastic sports decreased in the decade 1996-2006, more than 1.4 million injuries were sustained by high school athletes during the 2005-2006 school year. To reduce the incidence of injuries, injury prevention programs need to be based on accurate epidemiological data concerning sport-, sex-, and exposure-specific patterns of injuries. Such data were not previously available for boys and girls in the major high school sports. The purpose of the present study was to determine the epidemiology (incidence, location, type, diagnosis, severity) of practice and competition injuries in high school athletes participating in five boys' sports (football, soccer, basketball, wrestling, baseball) and four girls' sports (soccer, volleyball, basketball, softball) in the 2005-2006 school year.

Injury data (number, location, type, diagnosis, severity) and exposure data (number of practice sessions and competitions) were collected from 100 nationally representative high schools via the high school RIO (reporting information online) service. To be included in the study, an injury had to satisfy three criteria: (1) occurred during participation in organized high school practice or competition, (2) required medical attention, and (3) resulted in restriction of the injured person's participation for 1 or more days following the day of the injury. An athlete exposure (AE) was defined as one athlete participating in one practice or one competition. Incidence of injury (reflecting the risk of injury in each participant group) was expressed as the number of injuries per 1,000 AEs.

(continued)

During the year, 4,350 injuries (2,110 in practice, 2,240 in competition) were recorded in 1,730,764 AEs (1,246,499 practice, 484,265 competition), resulting in an overall incidence of injury of 2.51. The incidence of injury in competition (4.63) was greater than in practice (1.69). The incidence of injury in competition was greater than in practice in all nine groups. In competition, the highest incidence was for boys' football (12.09) followed by girls' soccer (5.21), boys' soccer (4.22), boys' wrestling (3.93), girls' basketball (3.60), boys' basketball (2.98), girls' volleyball (1.92), girls' softball (1.78), and boys' baseball (1.77) (figure 12.9). Compared with injuries sustained during practice, higher proportions of competition injuries were head, face, and neck injuries, particularly in boys' soccer and girls' basketball. Competition injuries were more likely to be concussions, especially in boys' soccer and girls' basketball. Higher proportions of competition injuries caused the athletes to miss more than 3 weeks of play, particularly in baseball and volleyball.

Application

The results of the study highlight a number of injury prevention measures:

- Incorporating drills of potentially high-risk situations into practice under controlled conditions to decrease the incidence of injury in competition

- Using appropriate protective equipment in practice and competition

- Requiring coaches and athletic trainers to be adequately qualified and up-to-date with regard to the physical preparation and skill level required to achieve an appropriate balance between performance and safety

- Paying particular attention in physical preparation to reducing the incidence of sprains and strains of the lower extremity, because these were the most common types of injuries across all sports in practice and competition

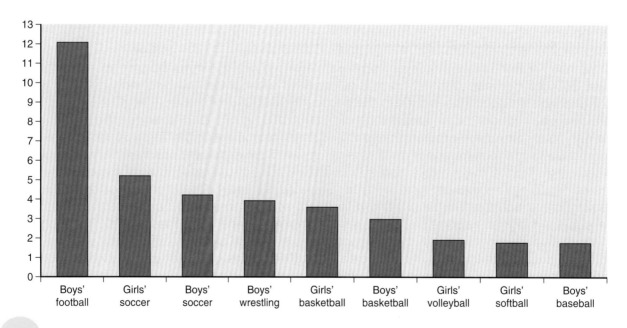

FIGURE 12.9 Incidence of injury in competition in American high school sports during the 2005-2006 school year. Incidence of injury is expressed as the number of injuries per 1,000 athlete exposures.

Summary

Risk factors influence the level of load exerted on the musculoskeletal system. All postures and movements involve the transmission of forces across a number of joints. The segmental chain used to accomplish a particular action is referred to as a kinetic chain, and the forces transmitted across the joints in a kinetic chain are interdependent. When movement in one or more joints in a kinetic chain is restricted, the other joints can usually compensate by altering their own ranges of movement. However, when compensatory movements are excessive and prolonged, the joints concerned and their supporting structures can be excessively overloaded, resulting in a musculoskeletal disorder. There are three main groups of risk factors—movement-related factors (concerning speed of movement and the degree of physical contact between the person and her surroundings), intrinsic factors (personal, physical, and psychological characteristics that distinguish people from each other), and extrinsic factors (environmental conditions and the manner in which activities are administered). Excessive overload is likely to be the result of a complex interaction of movement, intrinsic, and extrinsic risk factors. To identify the cause of excessive overload in a particular situation it is necessary to identify the relevant risk factors and their interaction.

Review Questions

1. Differentiate between causes, signs, and symptoms of injury.
2. Describe the main groups of intrinsic risk factors.
3. Describe the main groups of extrinsic risk factors.
4. Describe how strength and extensibility imbalances can occur.
5. Differentiate between absolute and relative strength balance.
6. Describe what is meant by good and bad technique in sports.
7. Differentiate between personality and psychosocial risk factors.
8. Describe the two main types of orthoses used in sports.

Appendix

Origins, Insertions, and Actions of the Major Muscles of the Human Body

Muscle	Origin	Insertion	Action
Upper limb			
Trapezius	Between the superior and inferior nuchal lines either side of the occipital protuberance Ligamentum nuchae Spines of C7 and T1-T12 and adjacent supraspinous ligament	Lateral one third of superior aspect of clavicle Superior aspect of acromion process and spine of scapula	Elevation and rotation of the scapulae
Deltoid	Lateral one third of anterior aspect of the clavicle Lateral aspect of the acromion process Inferior aspect of the border of the spine of the scapula	Deltoid tuberosity of the humerus	Flexion, abduction, and extension of the shoulder joint
Pectoralis major	Medial one third of anterior aspect of the clavicle Lateral border of the sternum and adjoining costal cartilages	Lateral aspect of the bicipital groove of the humerus	Flexion, adduction, and medial rotation of the shoulder joint
Latissimus dorsi	Posterior aspect of the sacrum Posterior third of the iliac crest Spines of T8-T12 and L1-L5	Medial aspect of the bicipital groove of the humerus	Extension, adduction, and medial rotation of the shoulder joint
Supraspinatus	Supraspinous fossa of the scapula	Superior aspect of the greater tuberosity of the humerus	Abduction of the shoulder joint
Infraspinatus	Infraspinous fossa of the scapula	Posterior aspect of the greater tuberosity of the humerus	Lateral rotation of the shoulder joint
Teres minor	Posterior lateral aspect of the scapula	Posterior aspect of the greater tuberosity of the humerus	Lateral rotation of the shoulder joint
Subscapularis	Subscapular fossa of the scapula	Anterior aspect of the lesser tuberosity of the humerus	Medial rotation of the shoulder joint
Teres major	Inferior lateral posterior aspect of the scapula below the origin of the teres minor	Medial aspect of the bicipital groove of the humerus	Extension, medial rotation, and adduction of the shoulder joint

> continued

> *continued*

Muscle	Origin	Insertion	Action
Upper limb *(continued)*			
Biceps brachii	Superior aspect of the glenoid fossa (long head) and tip of the coracoid process (short head)	Radial tuberosity of the humerus	Flexion of the shoulder and elbow joints
Triceps brachii	Superior one third of the lateral border of the scapula below the glenoid fossa (long head) Medial line on the superior posterior half of the humerus (lateral head) Lower two thirds of the posterior medial aspect of the humerus (medial head)	Olecranon process of the ulna	Extension of the elbow Adduction of the shoulder joint
Brachioradialis	Lateral supracondylar ridge of the humerus	Styloid process of the radius	Flexion of the elbow
Brachialis	Anterior inferior half of the humerus	Coronoid process of the ulna	Flexion of the elbow
Pronator teres	Medial supracondylar ridge of the humerus Medial aspect of the coronoid process	Middle third of lateral aspect of the radius	Flexion of the elbow Pronation of the forearm
Supinator	Lateral epicondyle of the humerus Lateral aspect of the coronoid process of the ulna	Proximal lateral one third of the radius	Supination of the forearm
Flexor carpi radialis	Medial aspect of the humerus above the trochlea	Anterior aspect of the bases of the second and third metacarpals	Flexion of the wrist and elbow
Flexor carpi ulnaris	Medial aspect of the humerus above the trochlea	Anterior aspect of the base of the fifth metacarpal	Flexion of the wrist and elbow
Extensor carpi radialis	Posterior aspect of the lateral epicondyle of the humerus	Posterior aspect of the base of the third metacarpal	Extension of the wrist and elbow
Extensor carpi ulnaris	Posterior aspect of the lateral epicondyle of the humerus	Posterior aspect of the base of the fifth metacarpal	Extension of the wrist and elbow
Flexor digitorum sublimis	Medial epicondyle of the humerus Anterior aspect of the ulna distal to the coronoid process Lateral one third of the radius	Four tendons to the sides of the bases of the middle phalanges of the fingers	Flexion of the wrist and fingers
Extensor digitorum communis	Lateral epicondyle of the humerus	Four tendons to the bases of the middle and distal phalanges of the fingers (dorsal surface)	Extension of the wrist and fingers

Muscle	Origin	Insertion	Action
Lower limb			
Iliopsoas (psoas and iliacus)	Transverse processes and sides of the bodies of T12 and L1-L5 Anterior aspect of the ilium	Lesser trochanter	Flexion of the hip joint
Gluteus medius	Superior half of lateral of ilium	Superior aspect of the greater trochanter	Abduction of the hip joint
Gluteus maximus	Posterior lateral aspect of the iliac crest and sacrum	Gluteal ridge on posterior aspect of the femur	Extension, abduction, and lateral rotation of the hip joint
Tensor fasciae latae	Anterior lateral aspect of the anterior superior iliac spine	Iliotibial tract (inserts onto the lateral aspect of the tibia below the lateral condyle)	Flexion and abduction of the hip joint
Sartorius	Anterior aspect of the ilium between the anterior superior and anterior inferior iliac spines	Medial aspect of the tibia below the medial condyle	Flexion of the hip joint
Gracilis	Pubic symphysis	Anteromedial aspect of the tibia below the medial condyle	Adduction of the hip joint
Adductor longus	Pubic crest	Middle third of the linea aspera	Adduction of the hip
Pectineus	Anterior medial aspect of the pubis	Proximal medial third of the femur	Flexion and adduction of the femur
Rectus femoris	Anterior aspect of the ilium between the anterior superior and anterior inferior iliac spines	Patella tendon, which is continuous with the patellar ligament, which attaches onto the tibial tuberosity	Flexion of the hip and extension of the knee
Vastus lateralis	Lateral aspect of the linea aspera and gluteal ridge on the femur	Patellar tendon	Extension of the knee
Vastus medialis	Medial aspect of the linea aspera and spiral line on the femur	Patellar tendon	Extension of the knee
Vastus intermedius	Superior two thirds of the anterior aspect of the femur	Patellar tendon	Extension of the knee
Biceps femoris	Ischial tuberosity Distal half of the linea aspera Lateral supracondylar ridge of the femur	Posterior lateral aspect of the lateral condyle of the tibia, and the head of the fibula	Extension of the hip and flexion of the knee
Semitendinosus	Ischial tuberosity	Anterior medial aspect of the tibia below the medial condyle	Extension of the hip and flexion of the knee
Semimembranosus	Ischial tuberosity	Posterior aspect of the medial condyle of the tibia	Extension of the hip and flexion of the knee
Flexor digitorum longus (FDL)	Lower two thirds of the posterior aspect of the tibia	Under the medial malleolus. The tendon splits into four branches, one to each of the lateral four toes. The tendons attach onto the bases of distal phalanges of the lateral four toes.	Plantar flexion of the lateral four toes Plantar flexion and inversion of the ankle

> continued

> *continued*

Muscle	Origin	Insertion	Action
Lower limb *(continued)*			
Tibialis posterior (TP)	Posterior aspect of the upper half of the interosseus membrane and adjoining borders of the tibia and fibula	Under the medial malleolus. The tendon splits into six branches that attach onto the plantar aspects of the navicular, the first cuneiform, and the bases of the second through fifth metatarsals.	Plantar flexion and inversion of the ankle
Peroneus longus (PL)	Head and upper two thirds of the lateral aspect of the fibula	Under the lateral malleolus. The tendon splits into two branches that attach onto the plantar aspects of the first cuneiform and the base of the first metatarsal.	Plantar flexion and eversion of the ankle
Peroneus brevis (PB)	Lower two thirds of the lateral aspect of the fibula	Under the lateral malleolus to attach onto the tuberosity of the fifth metatarsal	Plantar flexion and eversion of the ankle
Tibialis anterior (TA)	Upper two thirds of the lateral aspect of the tibia	Anterior to the ankle joint via the medial aspect of the foot to attach onto the plantar aspects of the first cuneiform and the base of the first metatarsal	Dorsiflexion and inversion of the ankle
Extensor hallucis longus (EHL)	Middle two thirds of the anteromedial aspect of the fibula	Anterior to the ankle joint to attach onto the superior aspect of the base of distal phalanx of the hallux	Dorsiflexion of the hallux Dorsiflexion of the ankle
Extensor digitorum longus (EDL)	Anterolateral aspect of the lateral condyle of the tibia and the upper two thirds of the anterior aspect of the fibula	Anterior to the ankle joint. The tendon splits into four branches that attach onto the superior aspects of the bases of the middle and distal phalanges of the lateral four toes.	Dorsiflexion of the lateral four toes Dorsiflexion of the ankle
Gastrocnemius	Medial head: superior aspect of the medial condyle of the femur Lateral head: superior aspect of the lateral condyle of the femur	The two heads attach onto the upper Achilles tendon, which is inserted onto the upper half of the posterior aspect of the calcaneus.	Plantar flexion of the ankle Flexion of the knee
Soleus	Upper third of the fibula and adjoining soleal line on upper third of the tibia	Anterosuperior half of the Achilles tendon, which is inserted onto the upper half of the posterior aspect of the calcaneus	Plantar flexion of the ankle

Muscle	Origin	Insertion	Action
Trunk			
External oblique	Borders of the lower eight ribs at the sides of the chest	Anterior half of the iliac crest	Flexion (together) and lateral flexion (single) of the trunk
Internal oblique	Lateral half of the inguinal ligament (links the pubic tubercle and the anterior superior iliac spine)	Eighth, ninth, and tenth costal cartilages and the linea alba (narrow aponeurosis between the left and right parts of the rectus abdominis)	Flexion (together) and lateral flexion (single) of the trunk
Transversus abdominis	Lateral third of inguinal ligament Inner aspect of the iliac crest Costal cartilages of the lower six ribs Lumbar fascia (merges with origin of latissimus dorsi)	Pubic crest and linea alba	Compression of abdomen
Rectus abdominis	Pubic crest	Fifth, sixth, and seventh costal cartilages and the xiphoid process	Flexion of the trunk
Erector spinae	Posterior aspect of sacrum and crest of the ilium Angles of the lower seven ribs Spines of L1-L5 and T9-T12 Transverse processes of T1-T12.	Transverse processes of all of the vertebrae Between the superior and inferior nuchal; lines on the occipital bone	Extension of the trunk

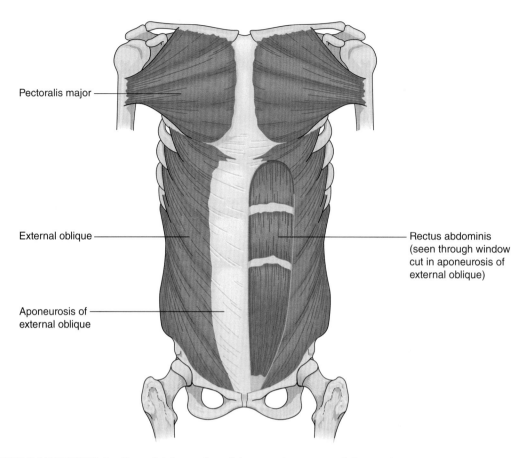

Pectoralis major

External oblique

Aponeurosis of
external oblique

Rectus abdominis
(seen through window
cut in aponeurosis of
external oblique)

APPENDIX FIGURE 1 Superficial muscles of the anterior aspect of the trunk.

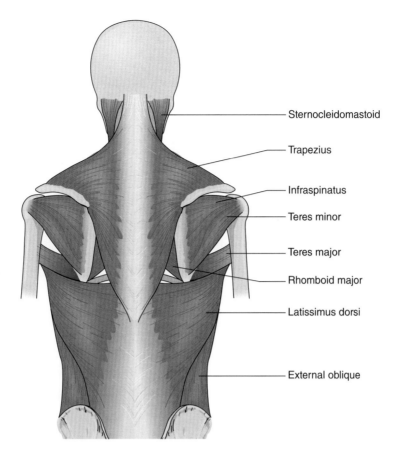

APPENDIX FIGURE 2 Superficial muscles of the posterior aspect of the trunk.

Sternocleidomastoid

Trapezius

Infraspinatus

Teres minor

Teres major

Rhomboid major

Latissimus dorsi

External oblique

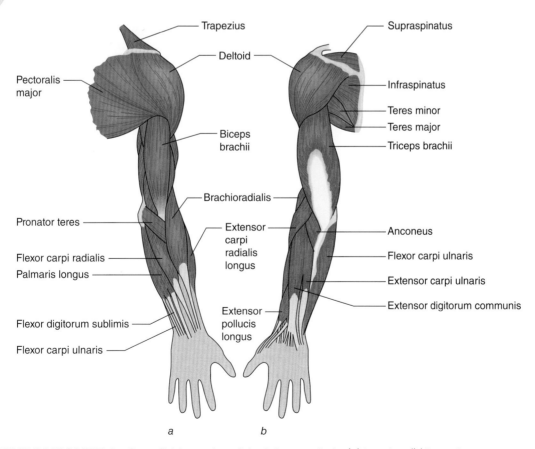

Trapezius

Supraspinatus

Deltoid

Pectoralis major

Infraspinatus

Teres minor

Teres major

Biceps brachii

Triceps brachii

Brachioradialis

Pronator teres

Extensor carpi radialis longus

Anconeus

Flexor carpi radialis

Flexor carpi ulnaris

Palmaris longus

Extensor carpi ulnaris

Extensor digitorum communis

Flexor digitorum sublimis

Extensor pollucis longus

Flexor carpi ulnaris

a   b

APPENDIX FIGURE 3 Superficial muscles of the left upper limb. *(a)* Anterior. *(b)* Posterior.

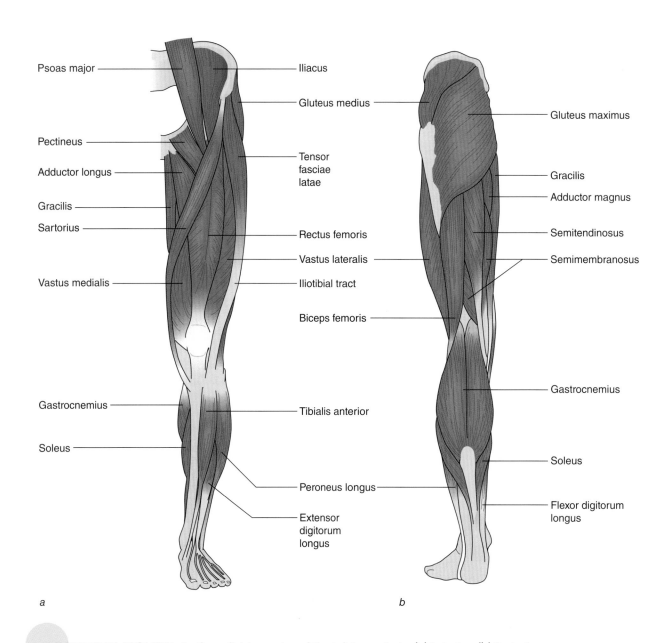

Psoas major — Iliacus

Gluteus medius

Gluteus maximus

Pectineus

Tensor fasciae latae

Gracilis

Adductor longus

Adductor magnus

Gracilis

Semitendinosus

Sartorius

Rectus femoris

Semimembranosus

Vastus lateralis

Vastus medialis

Iliotibial tract

Biceps femoris

Gastrocnemius

Tibialis anterior

Gastrocnemius

Soleus

Soleus

Peroneus longus

Flexor digitorum longus

Extensor digitorum longus

a   b

APPENDIX FIGURE 4   Superficial muscles of the left lower limb. *(a)* Anterior. *(b)* Posterior.

References

Aagaard P, Simonsen EB, Andersen JL, Magnusson P, Dyhre-Puolsen P. 2002. Neural adaptation to resistance training: changes in evoked V-wave and H-reflex responses. *Journal of Applied Physiology* 92:2309–2318.

Adams GR, Caiozzo VJ, Baldwin KM. 2003. Skeletal muscle unweighting: spaceflight and ground-based models. *Journal of Applied Physiology* 95:2185–2201.

Adams M, Bogduk N, Burton K, Dolan P. 2007. *The biomechanics of back pain*. Edinburgh, UK: Churchill Livingstone.

Adams MA, Hutton WC. 1980. The effect of posture on the role of the apophysial joints in resisting intervertebral compressive forces. *Journal of Bone and Joint Surgery* 62B(3):358–362.

Adams MA, Hutton WC.1982. Prolapsed intervertebral disc: a hyperflexion injury. *Spine* 7(3):184–191.

Adams MA, Hutton WC. 1983. The mechanical function of the lumbarapophysial joints. *Spine* 8(3):327–330.

Adams MA, Hutton WC. 1985. The effect of posture on the lumbar spine. *Journal of Bone and Joint Surgery* 67B(4):625–629.

Aglietti P, Insall JL, Cerulli G. 1983. Patellar pain and incongruence. *Clinical Orthopaedics* 176:217-224.

Akeson WH, Amiel D, Mechanic GL, Woo SLY, Harwood FL, Hamer ML. 1977. Collagen cross linking alterations in joint contractures: changes in the reducible cross-links in periarticular connective tissue collagen after nine weeks of immobilization. *Connective Tissue Research* 5:15–19.

Akeson WHC, Frank B, Amiel D, Woo SL. 1985. Ligament biology and biomechanics. In *Symposium on sports medicine: the knee*, ed Finerman G. St. Louis: Mosby, 111–151.

Alberts B, Johnson A, Lewis J, Raff M, Roberts K, Walter P. 2002. *Molecular biology of the cell.* 4th ed. New York: Garland.

Alexander RM. 1968. *Animal mechanics.* London: Sidgwick and Jackson.

Alexander RM. 1975. *Biomechanics.* London: Chapman and Hall.

Alexander RM. 1987. The spring in your step. *New Scientist* 114(1558):42–44.

Alexander RM. 1989. Muscles for the job. *New Scientist* 122(1660):50–53.

Alexander RM. 1992. *The human machine.* London: Natural History Museum Publications.

Alexander RM. 2003. *Principles of animal locomotion.* Princeton, NJ: Princeton University Press.

An KN, Hui FC, Morrey BF, Linscheid RL, Chao EY. 1981. Muscles across the elbow joint: a biomechanical analysis. *Journal of Biomechanics* 14(10):659–669.

An KN, Takahashi K, Harrigan TP, Chao EY. 1984. Determinants of muscle orientations and moment arms. *Journal of Biomechanical Engineering* 106:280–282.

Anderson FC, Pandy MG. 1993. Storage and utilization of elastic strain energy during jumping. *Journal of Biomechanics* 26(12):1413–1427.

Anderson K, Woiyts EM, Lambert PV. 1992. A biomechanical evaluation of taping in reducing knee joint translation and rotation. *American Journal of Sports Medicine* 20(4):416–421.

Anderson SM, Nilsson BE. 1979. Changes in bone mineral content following ligamentous knee injuries. *Medicine and Science in Sports* 11:351–353.

Annear PT, Chakera TMH, Foster DH, Hardcastle PH. 2008. Pars interarticularis stress and disc degeneration in cricket's potent strike force: the fast bowler. *ANZ Journal of Surgery* 62(10):768–773.

Arendt E. 1996. Common musculoskeletal injuries in women. *Physician and Sportsmedicine* 24(7):39–42, 45–48.

Arendt E, Dick R. 1995. Knee injury patterns among men and women in collegiate basketball and soccer. *American Journal of Sports Medicine* 23(6):694–700.

Ascenzi A, Bell GH. 1971. Bone as a mechanical engineering problem. In *The biochemistry and physiology of bone,* ed Boume GH. New York: Academic Press.

Aspden RM. 1987. Intra-abdominal pressure and its role in spinal mechanics. *Clinical Biomechanics* 2:168–174.

Bach DK, Green DS, Jensen GM, Savinar E. 1985. A comparison of muscular tightness in runners and nonrunners and the relation of muscular tightness to low back pain in runners. *Journal of Orthopaedic and Sports Physical Therapy* 6:315–623.

Backx FJG, Berger HJM, Bol E, Erich WBM. 1991. Injuries in high-risk persons and high-risk sports: a longitudinal study of 1818 schoolchildren. *American Journal of Sports Medicine* 19(2):124–130.

Baecker N, Tomic A, Mika C, Gotzmann A, Platen P, Gerzer R, Heer M. 2003. Bone resorption is induced on the second day of bed rest: results of a controlled crossover trial. *Journal of Applied Physiology* 95:977–982.

Bahr R, Holme I. 2003. Risk factors for sports injuries: a methodological approach. *British Journal of Sports Medicine* 37:384–392.

Bailey DA. 1995. The role of mechanical loading in the regulation of skeletal development during growth. In *New horizons in pediatric exercise science,* eds Blimkie CJR, O Bar-Or. Champaign, IL: Human Kinetics, 97–108.

Bailey DA, Martin AD, Houston CS, Howie JL. 1986. Physical activity, nutrition, bone density and osteoporosis. *Australian Journal of Science and Medicine in Sport* 18(3):3–8.

Baker B, Peckham AC, Pupparo F, Sanborn JC.1985. Review of meniscal injury and associated sports. *American Journal of Sports Medicine* 13(1):1–4.

Baker MK, Atlantis E, Fiatarone-Singh MA. 2007. Multi-modal exercise programs for older adults. *Age and Ageing* 36:375–381.

Barmeda A, Gaynor T, Mubarak SJ. 2003. Premature closure following distal tibia physeal fractures. *Journal of Pediatric Orthopaedics* 23:733–739.

Basmajian JV. 1965. Man's posture. *Archives of Physical Medicine and Rehabilitation* 46(1):26-35.

Bauer G. 1960. Epidemiology of fracture in aged persons. *Clinical Orthopaedics* 17:219–225.

Baxter-Jones AG, Kontulainen SA, Faulkner RA, Bailey DA. 2008. A longitudinal study of the relationship of physical activity to bone mineral accrual from adolescence to young adulthood. *Bone* 43(6):1101–1107.

Beard DJ, Dodd CAF, Trundle HR, Simpson AHRW. 1994. Proprioception enhancement for anterior cruciate ligament deficiency. *Journal of Bone and Joint Surgery* 76B:654–659.

Beynnon BD, Renstrom PA, Alosa DM, Baumhauer JF, Vacek PM. 2001. Ankle ligament injury risk factors: a prospective study of college athletes. *Journal of Orthopaedic Research* 19(2):213–220.

Beynnon BD, Vacek PM, Murphy D, Alosa D, Paller D. 2005. First-time inversion ankle ligament trauma: the effects of sex, level of competition, and sport on the incidence of injury. *American Journal of Sports Medicine* 33(10):1485–1491.

Blazina ME, Karlan RK, Jobe FW. 1973. Jumper's knee. *Orthopedic Clinics of North America* 4:665–673.

Blustein SM, D'Amico JC. 1985. Limb length discrepancy: Identification, clinical significance, and management. *Journal of the American Podiatric Medical Association* 75(4):200–206.

Bo Andersen L, Wedderkopp N, Leboeuf-Yde C. 2006. Association between back pain and physical fitness in adolescents. *Spine* 31(15):1740–1744.

Bobbert MF, Schamhardt HC, Nigg BM. 1991. Calculation of vertical ground reaction force estimates during running from positional data. *Journal of Biomechanics* 24:1095–1105.

Boden BP, Dean GS, Faegin JA. 2000. Mechanisms of anterior cruciate ligament injury. *Orthopedics* 23:573–578.

Booth FW. 1987. Physiologic and biochemical effects of immobilisation on muscle. *Clinical Orthopaedics and Related Research* 219:15–20.

Boruta PM, Bishop JO, Braly G, Tullos HS. 1990. Acute lateral ankle ligament injuries: a literature review. *Foot & Ankle* 11:107–113.

Bose K, Kanagasuntheram R, Hosman MBH. 1980. Vastus medialis: an anatomic and physiological study. *Orthopedics* 3:880–883.

Bowden PD, Bowker P. 1995. The alignment of the rearfoot complex axis as a factor in the development of running induced patellofemoral pain. *Journal of British Podiatric Medicine* 50:114–118.

Bowen V, Cassidy JD. 1981. Macroscopic and microscopic anatomy of the sacroiliac joint from embryonic life until eighth decade. *Spine* 6(6):620–627.

Brady RJ, Dean JB, Skinner TM, Gross MT. 2003. Limb length inequality: clinical implications for assessment and intervention. *Journal of Orthopaedic and Sports Physical Therapy* 33(5):221–234.

Brand RA, Yack 1996. Effects of leg length discrepancies on forces at the hip joint. *Clinical Orthopaedics* 333:172–180.

Brighton CT, Katz MJ, Goll SR, Nicholls CE. 1985. Prevention and treatment of sciatic denervation disuse osteoporosis in a rat tibia with capacitatively coupled electrical stimulation. *Bone* 6:87–97.

Brinckmann P. 1985. Pathology of the vertebral column. *Ergonomics* 28(1):77–80.

Brody DM. 1980. Running injuries. *Clinical Symposia* 32(4):1–36.

Bronnar S, Ojofeitimi S, Rose D. 2003. Injuries in a modern dance company: effect of comprehensive management on injury incidence and time loss. *American Journal of Sports Medicine* 32(3):365–373.

Brunet ME, Haddad RJ, Porche EB. 1982. Rotator cuff impingement syndrome in sports. *Physician and Sportsmedicine* 10(12):86–97.

Bullock-Saxton J. 1988. Normal and abnormal postures in the sagittal plane and their relationship to low back pain. *Physiotherapy Practice* 4(2):94–104.

Bunnell WP. 1986. The natural history of idiopathic scoliosis before skeletal maturity. *Spine* 11(8):773–776.

Burstein AH, Wright TM. 1994. *Fundamentals of orthopedic biomechanics*. Baltimore: Williams & Wilkins.

Byhring S, Bø K. 2002. Musculoskeletal injuries in the Norwegian National Ballet: a prospective cohort study. *Scandinavian Journal of Medicine & Science in Sports* 12(6):365–370.

Caine D, DiFiori J, Maffulli N. 2006. Physeal injuries in children's and youth sports: reasons for concern? *British Journal of Sports Medicine* 40:749–760.

Callaghan MJ, Oldham JA. 1996. The role of quadriceps exercise in the treatment of patellofemoral pain syndrome. *Sports Medicine* 21(5):384–391.

Caplan AI. 1984. Cartilage. *Scientific American* 251(4):82–90.

Capodaglio P, Eda MC, Facioli M, Saibene F. 2007. Long-term strength training for community-dwelling people over 75: impact on muscle function, functional ability and life style. *European Journal of Applied Physiology* 100:535–542.

Carter DR, Wong M, Orr TE. 1991. Musculoskeletal ontogeny, phylogeny, and functional adaptation. *Journal of Biomechanics* 24S1:3–16.

Cash JD, Hughston JC. 1988. Treatment of acute patellar dislocation. *American Journal of Sports Medicine* 16(3):244–248.

Casscells W. 1982. Chondromalacia of the patella. *Journal of Pediatric Orthopaedics* 2(5):560–564.

Castle F. 1969. *Five-figure logarithmic and other tables*. London: Macmillan.

Cavanagh PR. 1980. *The running shoe book*. Mountain View, CA: Anderson World.

Cavanagh PR, Lafortune MA. 1980. Ground reaction forces in distance running. *Journal of Biomechanics* 13:397–406.

Cavanagh PR, Valiant GA, Misevich KW. 1984. Biological aspects of modeling shoe/foot interaction during running. In *Sport shoes and playing surfaces*, ed Frederick EC. Champaign, IL: Human Kinetics, 24–46.

Chalmers J, Ho KC. 1970. Geographical variations in senile osteoporosis: the association with osteoporosis. *Journal of Bone and Joint Surgery* 52:667–675.

Chappell JD, Herman DC, Knight BS. 2005. Effect of fatigue on knee kinetics and kinematics in stop-jump tasks. *American Journal of Sports Medicine* 33(7):1022–1029.

Clark CR, Huddleston HD, Schoch EP. 2006. Leg-length discrepancy after total hip arthroplasty. *Journal of the American Academy of Orthopaedic Surgeons* 14(1):38–45.

Clarkson PM, Tremblay I. 1988. Exercise-induced muscle damage, repair, and adaptation in humans. *Journal of Applied Physiology* 65(1):1–6.

Clement DB, Taunton JE, Smart GW, McNicol KL. 1981. A survey of overuse running injuries. *Physician and Sportsmedicine* 9(5):47–58.

Cohen LA, Cohen ML. 1956. Arthrokinetic reflex of the knee. *American Journal of Physiology* 184:433–437.

Colachis SC, Worden RE, Bechtol CO, Strohm BR. 1963. Movement of the sacroiliac joint in the adult male: a preliminary report. *Archives of Physical Medicine and Rehabilitation* 44:490–498.

Coleman, SGS, Benham AS, Northcott SR. 1993. Three-dimensional cinematographical analysis of the volleyball spike. *Journal of Sports Sciences* 11(4):295–302.

Colliton J.1996. Back pain and pregnancy. *Physician and Sportsmedicine* 24(7):89–93.

Combs JA. 1994. Hip and pelvis avulsion fractures in adolescents. *Physician and Sportsmedicine* 22 (7):41–44, 47–49.

Connelly LB, Woolf A, Brooks P. 2006. Cost-effectiveness of interventions for musculoskeletal conditions. In *Disease control priorities in developing countries*, eds Jamieson DT, Breman TG, Measham AR, Alleyne G, Claeson M, Evans DB, Jha P, Mills A, Musgrove P. Oxford, UK: Oxford University Press, 963–980.

Copeland S. 1993. Throwing injuries of the shoulder. *British Journal of Sports Medicine* 27(4):221–227.

Costill DL, Maglischo EW, Richardson AB. 1992. *Swimming*. Oxford, UK: Blackwell Scientific.

Court-Brown C, Caesar B. 2006. Epidemiology of adult fractures: a review. *Injury* 37(8):691–697.

Coventry E, O'Connor KM, Hart BA, Earl JE, Ebersole KT. 2006. The effect of lower extremity fatigue on shock attenuation during single-leg landing. *Clinical Biomechanics* 21:1090-1097.

Cress ME, Buchner DM, Prohaska T, Rimmer J, Brown M, Macera C, DePietro L, Chodzko-Zajko W. 2004. Physical activity programs and behavior counseling in older adult populations. *Medicine and Science in Sports and Exercise* 36(11):1997–2003.

Cummings SR, Black DM, Nevitt MC, Browner W, Cauley J, Enstrud K, Genant HK, Palermo L, Scott J, Vogt TM. 1993. Bone density at various sites for prediction of hip fractures. *Lancet* 341(8837):72–75.

Cummings SR, Melton LJ. 2002. Epidemiology and outcomes of osteoporotic fractures. *Lancet* 359(9319):1761–1767.

Cupal DD. 1998. Psychological interventions in sport injury prevention and rehabilitation. *Journal of Applied Sport Psychology* 10:103–123.

Curtis C, d'Hemecourt P. 2007. Diagnosis and management of back pain in adolescents. *Adolescent Medicine: State of the Art Reviews* 18(1):140–64.

de Boer MD, Maganaris CN, Seynnes OR, Rennie MJ, Narici MV. 2007. Time course of muscular, neural and tendinous adaptations to 23 day unilateral lower-limb suspension in young men. *Journal of Physiology* 583(3):1079–1091.

de Leva P. 1996. Adjustments to Zatsiorsky-Seluyanov's segment inertia parameters. *Journal of Biomechanics* 29(9):1223–1230.

Dempster WT. 1965. Mechanisms of shoulder movement. *Archives of Physical Medicine and Rehabilitation* 46(1):49–70.

De Vita P, Torry M, Glover KL, Speroni D. 1996. A functional knee brace alters joint torque and power patterns during walking and running. *Journal of Biomechanics* 29(5):583–588.

Dickenson RP, Hutton WC, Stott JRR. 1981. The mechanical properties of bone in osteoporosis. *Journal of Bone and Joint Surgery* 63B(2):233–238.

Dickinson JA, Cook SD, Leinhardt TM. 1985. The measurement of shock waves following heel strike while running. *Journal of Biomechanics* 18(6):415–422.

Don Tigny RL. 1985. Function and pathomechanics of the sacroiliac joint: a review. *Physical Therapy* 65(1):35–44.

Donaldson C, Hulley S, Vogel J, Hattner R, Bayers J, McMillan D. 1970. Effect of prolonged bed rest on bone mineral. *Metabolism* 19:1071–1084.

Doschak MR, Zernicke RF. 2005. Structure, function and adaptation of bone-tendon and bone-ligament complexes. *Journal of Musculoskeletal & Neuronal Interactions* 5(1):35–40.

Downing BS, Klein BS, D'Amico JS. 1978. The axis of motion of the rearfoot complex. *Journal of the American Podiatry Association* 68:484–499.

Drummond DS. 1987. Kyphosis in the growing child. *Spine: State of the Art Reviews* 1(2):339–356.

Dunlop RB, Adams MA, Hutton WC. 1984. Disc space narrowing and and the lumbar facet joints. *Journal of Bone and Joint Surgery* 66B(5): 706-710.

During J, Goudfrooij H, Keessen W, Beeker TW, Crowe A. 1985. Towards standards for posture: postural characteristics of the lower back system in normal and pathologic conditions. *Spine* 10(1): 83–87.

Duyar I. 2008. Growth patterns and physical plasticity in adolescent laborers. *Collegium Antropologicum* 32(2):403–412.

Edman KAP. 1992. Contractile performance of skeletal muscle fibres. In *Strength and power in sport*, ed Komi PV. Oxford, UK: Blackwell Scientific.

Elftman H. 1966. Biomechanics of muscle. *Journal of Bone and Joint Surgery* 48A(2):363–377.

Ende LS, Wickstrom J. 1982. Ballet injuries. *Physician and Sportsmedicine* 10(7):101–103, 106–109, 113–115, 118.

Endresen EH. 1995. Pelvic pain and low back pain in pregnant women: an epidemiological study. *Scandinavian Journal of Rheumatology* 24(3):135–141.

Engelhardt M, Reuter J, Freiwald J, Bohme T, Halbsguth A. 1997. Spondylolysis and spondylolisthesis: correlation in sport. *Der Orthopäde* 26(9):755–759.

Englund J. 2005. Chronic compartment syndrome: tips on recognizing and treating. *Family Practice* 54(11):955–960.

Enoka RM. 1994. *Neuromechanical basis of kinesiology*. Champaign, IL: Human Kinetics.

Enoka RM. 1997. Neural adaptations with chronic physical activity. *Journal of Biomechanics* 30(5):447–455.

Evans FG. 1971. Biomechanical implications of anatomy. In *Biomechanics*, ed Cooper JM. Chicago: The Athletic Institute.

Fagan V, Delahunt E. 2008. Patellofemoral pain syndrome: a review on the associated deficits and current treatment options. *British Journal of Sports Medicine* 42:789–795.

Fagenbaum R, Darling WG. 2003. Jump landing strategies in male and female college athletes and implications of such strategies for anterior cruciate ligament injury. *American Journal of Sports Medicine* 31(2):233–240.

Ferretti A, Papandrea P, Conteduca F. 1990. Knee injuries in volleyball. *Sports Medicine* 10(2):132–138.

Ferretti A, Puddu G, Mariani PP, Neri M. 1984. Jumper's knee: an epidemiological study of volleyball players. *Physician and Sportsmedicine* 12(10):97–106.

Ferretti JL, Cointry GR, Capozza RF, Frost HM. 2003. Bone mass, bone strength, muscle-bone interactions, osteopenias and osteoporoses. *Mechanisms of Ageing and Development* 124:269–279.

Fleisig GS, Barrentine SW, Escamilla FR, Andrews JR. 1996. Biomechanics of overhand throwing with implications for injuries. *Sports Medicine* 21(6):421–437.

Fong DT, Hong Y, Chan LK, Yung PS, Chan KM. 2007. A systematic review of ankle injury and ankle sprain in sports. *Sports Medicine* 37:73–94.

Ford DM, Bagnall KM, Clements CA, McFadden KD. 1988. Muscle spindles in the paraspinal musculature of patients with adolescent idiopathic scoliosis. *Spine* 13(5):461–465.

Forthomme B, Croisier J-L, Ciccarone G, Crielaard J-M, Cloes M. 2005. Factors correlated with volleyball spike velocity. *American Journal of Sports Medicine* 33:1513–1519.

Forwood MR. 2001. Mechanical effects on the skeleton: are there clinical implications? *Osteoporosis International* 12:77–83.

Francis RS, Bryce GR. 1987. Relationships between lumbar lordosis, pelvic tilt, and abdominal muscle performance. *Physical Therapy* 67(8):1221–1225.

Fredericson M, Anuruddh M. 2007. Epidemiology and aetiology of marathon running injuries. *Sports Medicine* 37(4–5):437–439.

Freeman M, Wyke B. 1967. Articular reflexes at the ankle joint: an electromyographic study of normal and abnormal influences of ankle joint mechanoreceptors upon reflex activity in the leg muscles. *British Journal of Surgery* 54:990–1001.

Freeman WH, Bracegirdle B. 1967. *An atlas of histology.* London: Heinemann.

Friberg O. 1983. Clinical symptoms and biomechanics of lumbar spine and hip joint in leg length inequality. *Spine* 8(6):643–651.

Frontera WR, Micheli LJ, Herring SA, Silver JK. 2007. *Clinical sports medicine: medical management and rehabilitation.* Oxford, UK: Elsevier Heath Sciences.

Frost HM. 1967. *An introduction to biomechanics.* Springfield, IL: Charles C Thomas.

Frost HM. 1973. *Orthopedic biomechanics.* Vol 5. Springfield, IL: Charles C Thomas.

Frost HM. 1979. A chondral modeling theory. *Calcified Tissue International* 28:181–200.

Frost HM. 1988a. Structural adaptations to mechanical usage: a proposed three-way rule for bone modeling, part I. *Veterinary and Comparative Orthopaedics and Traumatology* 1:7–17.

Frost HM. 1988b. Structural adaptations to mechanical usage: a proposed three-way rule for bone modeling, part II. *Veterinary and Comparative Orthopaedics and Traumatology* 2:80–85.

Frost HM. 1990. Structural adaptations to mechanical usage (SATMU), 4: mechanical influences on intact fibrous tissues. *Anatomical Record* 226:433–439.

Frost HM. 1997. Defining osteopenias and osteoporoses: another view (with insights from a new paradigm). *Bone* 20(5):285–391.

Frost HM. 2003. Bone's mechanostat: a 2003 update. *Anatomical Record* 275A:1081–1101.

Fyhrie DP, Milgrom C, Hoshaw SJ, Simkin A, Dar S, Drumb D, Burr DB. 1998. Effect of fatiguing exercise on longitudinal bone strain as related to stress fracture in humans. *Annals of Biomedical Engineering* 26(4):660–665.

Galambos SA, Terry PC, Moyle GM, Locke SA. 2005. Psychological predictors of injury among elite athletes. *British Journal of Sports Medicine* 39:351–354.

Galea AM, Albers JM. 1994. Patellofemoral pain: beyond empirical diagnosis. *Physician and Sportsmedicine* 22(4):48, 53, 54, 56, 58.

Gamble JG. 1988. *The musculoskeletal system: physiological basics.* New York: Raven Press.

Gamboa JM, Roberts LA., Maring J, Fergus A. 2008. Injury patterns in elite preprofessional ballet dancers and the utility of screening programs to identify risk characteristics. *Journal of Orthopaedic and Sports Physical Therapy* 38(3):126–136.

Gandevia SC, McClosky DI, Burke D. 1992. Kinesthetic signals and muscle contraction. *Trends in Neuroscience* 15:62–65.

Garn S, Newton R. 1988. Kinesthetic awareness in subjects with multiple ankle sprains. *Physical Therapy* 68:1667–1671.

Garrett WE, Safran MR, Seaber AV, Glisson RR, Ribbeck BM. 1987. Biomechanical comparison of stimulated and nonstimulated skeletal muscle pulled to failure. *American Journal of Sports Medicine* 15(5):448–454.

Gerard DF. 1998. External knee support in Rugby Union: effectiveness of bracing and taping. *Sports Medicine* 25(5):313–317.

Gibson PH, Papaigannou T, Kenwright J. 1983. The influence on the spine of leg-length discrepancy after femoral fracture. *Journal of Bone and Joint Surgery* 65B:584–587.

Giles LGF, Taylor JR. 1981. Low-back pain associated with leg length inequality. *Spine* 6(5):510–521.

Giza E, Fuller C, Junge A, Dvorak J. 2003. Mechanisms of foot and ankle injuries in soccer. *American Journal of Sports Medicine* 31:550–554.

Goldsmith W. 1960. *Impact: the theory and physical behaviour of colliding solids.* London: Edward Arnold.

Goldspink G. 1992. Cellular and molecular aspects of adaptation in skeletal muscle. In *Strength and power in sport,* ed PV Komi. Oxford, UK:Blackwell Scientific.

Gollehon DL, Torzilli PA, Warren RF. 1987. The role of the posterolateral and cruciate ligaments in the stability of the human knee: a biomechanical study. *Journal of Bone and Joint Surgery* 69A:233–242.

Gordon AM, Huxley AF, Julian FJ. 1966. The variation in isometric tension with sarcomere length in vertebrate muscle fibres. *Journal of Physiology* 184:170–192.

Gordon BJ, Dapena J. 2006. Contributions of joint rotations to racquet speed in the tennis serve. *Journal of Sports Sciences* 24(1):31–49.

Gozna ER, Harrington IJ. 1982. *Biomechanics of musculoskeletal injury.* Baltimore: Williams & Wilkins.

Grabiner MD, Enoka RM. 1995. Changes in movement capabilities with aging. *Exercise and Sport Sciences Reviews* 23:65–104.

Grace TG. 1985. Muscle imbalance and extremity injury: a perplexing relationship. *Sports Medicine* 2:77–82.

Grana WA, Holder S, Schelberg-Karnes E. 1987. How I manage acute anterior shoulder dislocations. *Physician and Sportsmedicine* 15(4):88–93.

Greene DA, Naughton GA. 2006. Adaptive skeletal responses to mechanical loading during adolescence. *Sports Medicine* 36(9):723–732.

Greenspan A, Pugh JW, Norman A, Norman RS. 1978. Scoliotic index: a comparative evaluation of methods for the measurement of scoliosis. *Bulletin of the Hospital for Joint Diseases* 39(2):117–125.

Gregor RJ. 1993. Skeletal muscle mechanics and movement. In *Current issues in biomechanics*, ed Grabiner MD. Champaign, IL: Human Kinetics, 171–211.

Gregory PL, Batt ME, Kerslake RW. 2009. Comparing spondylolysis in cricketers and soccer players. *British Journal of Sports Medicine* 38:737–742.

Grieve GP. 1976. The sacroiliac joint. *Physiotherapy* 62:384–400.

Grigg P. 1994. Peripheral neural mechanisms in proprioception. *Journal of Sport Rehabilitation* 3:2–17.

Grimston SK, Zernicke RF. 1993. Exercise-related stress responses in bone. *Journal of Applied Biomechanics* 9:2–14.

Groppel JL. 1986. The biomechanics of tennis: an overview. *International Journal of Sport Biomechanics* 2:141–55.

Gross ML, Flynn M, Sonzogni JJ.1994.Overworked shoulders: managing injury of the proximal humeral physis. *Physician and Sportsmedicine* 22(3):81, 82, 85, 86.

Gross MT, Liu HY. 2003. The role of ankle bracing for prevention of ankle sprain injuries. *Journal of Orthopaedic & Sports Physical Therapy* 33(10):572–577.

Gross TS, Bain ST. 1993. Skeletal adaptation to functional stimuli. In *Current issues in biomechanics,* ed Grabiner MD. Champaign, IL: Human Kinetics, 151–169.

Gurney B. 2002. Leg length discrepancy. *Gait & Posture* 15(2):195–206.

Ha KY, Lee JS, Kim KW. 2008. Degeneration of sacroiliac joint after instrumented lumbar or lumbosacral fusion: a prospective cohort study over five-year follow-up. *Spine* 33(11):1192–1198.

Haderspeck K. Schultz A. 1981. Progression of scoliosis: an analysis of muscle actions and bodyweight influences. *Spine* 6(5):447–455.

Hagen KB, Thune O. 1998. Work incapacity from low back pain in the general population. *Spine* 23(19):2091–2095.

Hagino H, Yamamoto K, Nakamura T, Nose T, Ohshiro H. 1991. Epidemiology of osteoporotic limb fractures in tottori prefecture, Japan. *Journal of Bone and Mineral Metabolism* 9(Suppl 1):25–28.

Hagins M, Lamberg E. 2006. Natural breath control during lifting tasks: effect of load. *European Journal of Applied Physiology* 956:453-458.

Hainline B. 1994. Low-back pain in pregnancy. *Advances in Neurology* 64:65–76.

Häkkinen K. 1994. Neuromuscular adaptation during strength training, aging, detraining, and immobilization. *Critical Reviews in Physical and Rehabilitation Medicine* 6(3):161–198.

Hall MG, Ferrell WR, Baxendale RH, Hamblen DL. 1994. Knee joint proprioception: Threshold detection levels in healthy young subjects. *Neuro-Orthopedics* 15:81-90.

Hall MG, Ferrell WR, Sturrock RD, Hamblen DL, Baxendale RH. 1995. The effect of the hypermobility syndrome on knee joint proprioception. *British Journal of Rheumatology* 34:121–125.

Hamilton D, Aronsen P, Løken JH, Berg IM, Skotheim R, Hopper D, Clarke A, Briffa NK. 2006. Dance training intensity at 11–14 years is associated with femoral torsion in classical ballet dancers. *British Journal of Sports Medicine* 40:299–303.

Hamrick MW. 1999. A chondral modeling theory revisited. *Journal of Theoretical Biology* 201(3):201–208.

Hanson PG, Angevine M, Juhl JH. 1978. Osteitis pubis in sports activities. *Physician and Sportsmedicine* 6(10):111–114.

Hardaker WT, Margello S, Goldner JL. 1985. Foot and ankle injuries in theatrical dancers. *Foot & Ankle* 6(2):59–69.

Hasegawa H, Dziados J, Newton RU, Fry AC. Kraemer WJ, Häkkinen K. 2002. Periodized training programmes for athletes. In: *Strength training for sports,* eds Kraemer WK, Häkkinen K. Oxford, UK: Blackwell Scientific, 69–134.

Hawkins RJ, Mohtadi N. 1994. Rotator cuff problems in athletes. In *Orthopaedic sports medicine: principles and practice,* eds DeLee JC, Drez D. Philadelphia: Saunders, 623–656.

Heinert B, Kernozek TW, Greany J, Fater DCW. 2008. Hip abductor weakness and lower extremity kinematics during running. *Journal of Sport Rehabilitation* 17(3):243–256.

Hennig, EM, Staats A, Rosenbaum D. 1994. Plantar pressure distribution paterns of young children in comparison to adults. *Foot & Ankle* 15(10):35–40.

Henning CE, Griffis ND, Vequist SW, Yearout KM, Decker KA. 1994. Sport-specific knee injuries. In *Clinical practice of sports injury prevention and care,* ed Renström PA. Oxford, UK: Blackwell Scientific.

Henwood TR, Riek S, Taaffe DR. 2008. Strength versus muscle power-specific resistance training in community-dwelling older adults. *Journal of Gerontology* 63A(1):83-91.

Herbert, R. 1988. The passive mechanical properties of muscles and their adaptations to altered patterns of use. *Australian Journal of Physiotherapy* 34(3):141–149.

Herrington L, Nester C. 2004. Q-angle undervalued? The relationship between Q-angle and mediolateral position of the patella. *Clinical Biomechanics* 19(10):200–204.

Herzog W. 2000. Muscle properties and coordination during voluntary movement. *Journal of Sports Sciences* 18: 141–152.

Holt KG. 1998. Constraints on the emergence of preferred locomotory patterns. In *Timing of behavior: neural, psychological, and computational perspectives,* eds Rosenbaum DA, Collyer CE. Cambridge, MA: MIT Press, 261–291.

Hontas MJ, Haddad RJ, Schlesinger LC. 1986. Conditions of the talus in the runner. *American Journal of Sports Medicine* 14(6):486–490.

Hootman JM, Macera CA, Ainsworth BE, Martin M, Addy CL, Blair SN. 2001. Association among physical activity level, cardiorespiratory fitness, and risk of musculoskeletal injury. *American Journal of Epidemiology* 154(3):251–258.

Hopper BJ. 1973. *The mechanics of human movement.* London: Crosby Lockwood Staples.

Hopper MA, Robinson P. 2008. Ankle impingement syndromes. *Radiologic Clinics of North America* 46(6):957–961.

Hoshina H. 1980. Spondylolysis in athletes. *Physician and Sportsmedicine* 8(9):75–79.

Hoshino A, Wallace AW. 1987. Impact absorbing properties of the human knee. *Journal of Bone and Joint Surgery* 69B(5):807–811.

Hourigan SR, Nitz JC, Brauer S, O'Neill S, Wong J, Richardson CA. 2008. Positive effects of exercise on falls and fracture risk in osteopenic women. *Osteoporosis International* 19:1077-1086.

Hreljac A. 2004. Impact and overuse injuries in runners. *Medicine and Science in Sports and Exercise* 36:845–849.

Huang H, Kamm RD, Lee RT. 2004. Cell mechanics and mechanotransduction: pathways, probes, and physiology. *American Journal of Physiology: Cell Physiology* 287:C1–C11.

Huberti HH, Hayes WC. 1984. Patellofemoral contact pressures: the influence of Q angle and tendofemoral contact. *Journal of Bone and Joint Surgery* 66A:715–724.

Hughes GH, Watkins J. 2006. A risk-factor model for anterior cruciate ligament injury. *Sports Medicine* 36(5):411–428.

Huijing PA. 1992. Mechanical muscle models. In *Strength and power in sport,* ed Komi PV. Oxford, UK: Blackwell Scientific.

Hutton WC, Cyron BM. 1978. Spondylolysis. *Acta Orthopedica Scandinavica* 49:604-609.

Huxley AF. 2000. Cross bridge action: present views, prospects and unknowns. In *Skeletal muscle mechanics: from mechanisms to function,* ed Herzog W. Chichester, UK: Wiley, 7–32.

Huxley HE, Hanson J. 1954. Changes in the cross striations of muscle during contraction and stretch and their structural interpretation. *Nature* 173:973–077.

Indelicato PA. 1995. Isolated medial collateral ligament injuries in the knee. *Journal of the American Academy of Orthopaedic Surgeons* 3(1):9–14.

Ingram JG, Fields SK, Yard EE, Comstock RD. 2008. Epidemiology of knee injuries among boys and girls in US high school athletics. *American Journal of Sports Medicine* 36(6):1116–1122.

Inman VT. 1976. *Joints of the ankle.* Baltimore: Williams & Wilkins.

Insall JN, Salvati E. 1971. Patellar position in the normal knee joint. *Radiology* 101:101–109.

Ishikawa M, Komi PV, Grey MJ, Lepola V, Bruggemann G-P. 2005. Muscle-tendon interaction and elastic energy usage in human walking. *Journal of Applied Physiology* 99:603–608.

Jackson DW, Wiltse LL, Cirincione RJ. 1976. Spondylolysis in the female gymnast. *Clinical Orthopaedics* 117:68–73.

Janssen I, Heymsfield SB, Wang Z, Ross R. 2000 Skeletal muscle mass and distribution in 468 men and women aged 18-88 yr. *Journal of Applied Physiology* 89:81–88.

Johnson GR. 1988. The effectiveness of shock-absorbing insoles during normal walking. *Prosthetics and Orthotics International* 12(1):91–95.

Johnson ME, Mille ML, Martinez KM, Crombie G, Rogers MW. 2004. Age-related changes in hip abductor and adductor joint torques. *Archives of Physical Medicine and Rehabilitation* 85(4):593–597.

Johnson R. 2003. Osteitis pubis. *Current Sports Medicine Reports* 2(2):98–102.

Jones BH, Bovee MW, Harris JMcA, Cowan DN. 1993. Intrinsic risk factors for exercise-related injuries among male and female army trainees. *American Journal of Sports Medicine* 21(5):705–710.

Jones BH, Cowan DN, Knapik J. 1994. Exercise, training, and injuries. *Sports Medicine* 18(3):202–214.

Jurimae T, Jurimae J. 2001. *Growth, physical activity, and motor development in prepubertal children.* London: Informa HealthCare.

Kaaria S, Kaila-Kangas L, Kirjonen J, Riihimaki H, Luukkonen R, Leino-Arjas P. 2005. Low back pain, work absenteeism, chronic back disorders, and clinical findings in the low back as predictors of hospitalization due to low back disorders: a 28-year follow-up of industrial employees. *Spine* 30(100):211–218.

Kado DM, Browner WS, Palermo L, Nevitt MC, Genant HK, Cummings SR. 1999. Vertebral fractures and mortality in older women: a prospective study *Archives of Internal Medicine* 159:1215–1220.

Kakushima M, Miyamoto K, Shimizu K. 2003. The effect of leg length discrepancy on spinal motion during gait: three-dimensional analysis in healthy volunteers. *Spine* 28(21):2472–2476.

Kalichman L, Kim DH, Li L, Guermazi A, Berkin V, Hunter DJ. 2009. Spondylolysis and spondylolisthesis: prevalence and association with low back pain in the adult community-based population. *Spine* 34(2):199–205.

Kamen G, Caldwell GE. 1996. Physiology and interpretation of the electromyogram. *Journal of Clinical Neurophysiology* 13(5):366–384.

Kandel ER, Schwartz JH, Jessell TM, eds. 2000. *Principles of neural science.* 4th ed. New York: McGraw-Hill.

Kannus P, Nittymaki S, Jarvinen M. 1988. Athletic overuse injuries in children: a 30-month prospective follow-up study at an outpatient sports clinic. *Clinical Pediatrics* 27(7):333–337.

Kapandji IA. 1970. *The physiology of the joints.* Vol 2. Edinburgh, UK: Churchill Livingstone.

Kapandji IA. 1974. *The physiology of the joints.* Vol 3. Edinburgh, UK: Churchill Livingstone.

Kaplan FS. 1983. Osteoporosis. Clinical Symposia. *Ciba-Geigy* 35(4): 32-42.

Kawakami Y, Fukunaga T. 2006. New insights into human skeletal muscle function. *Exercise and Sport Sciences Reviews* 34(1):16–21.

Keim HA. 1982. *The adolescent spine.* New York: Springer-Verlag.

Kelley MJ. 1990. Psychological risk factors and sports injuries. *Journal of Sports Medicine and Physical Fitness* 30(2):202–221.

Kendall FP, McCreary EK, Provance PG, Rodgers MM, Romani WA. 2005. *Muscles testing and function with posture and pain.* Baltimore: Lippincott Williams & Wilkins.

Kenyon C. 1988. The nematode *Caenorhabditis elegans. Science* 240:1448–1452.

Ker RF, Bennett MB, Bibby SR, Kester RC, Alexander RM. 1987. The spring in the arch of the human foot. *Nature* 325(7000):147–149.

Kerr BA, Beauchamp L, Fisher V, Neil R. 1983. Foot-strike patterns in distance running. In *Biomechanical aspects of sport shoes and playing surfaces,* eds Nigg BM, Kerr BA. Calgary, Ontario, Canada: University Printing.

Kerr G, Fowler B. 1988. The relationship between psychological factors and sports injuries. *Sports Medicine* 6:127–134.

Kibler WB. 2005. The kinetic chain. In *Injury prevention and rehabilitation for active older adults,* ed Speer KP. Champaign, IL: Human Kinetics, 49–58.

Kitaoka HB, Luo ZP, An K-N. 1997. Three-dimensional analysis of normal ankle and foot mobility. *American Journal of Sports Medicine* 25(2):238–242.

Kleinrensink GJ, Stoeckart R, Meulstee J, Kaulesar Sukul DMKS, Vleeming A, Snijders CJ, Van Noort A. 1994. Lowered motor conduction velocity of the peroneal nerve after inversion trauma. *Medicine and Science in Sports and Exercise* 26(7):877–883.

Knapik JJ, Bauman CL, Jones BH, Harris JM, Vaughan L. 1991. Preseason strength and flexibility imbalances associated with athletic injuries in female collegiate athletes. *American Journal of Sports Medicine* 19(1):76–81.

Koester M, George MS, Kuhn JE. 2005. Shoulder impingement syndrome. *American Journal of Medicine* 118(5):452–455.

Komi PV. 2003. Stretch-shortening cycle. In *Strength and power in sport,* ed Komi PV. Oxford, UK: Blackwell, 184–202.

Komi PV, Bosco C. 1978. Utilization of stored elastic energy in leg extensor muscles by men and women. *Medicine and Science in Sports* 10(4):261–265.

Konradsen L, Ravn JB, Sørensen AI. 1993. Proprioception at the ankle: the effect of anaesthetic blockade of ligament receptors. *Journal of Bone and Joint Surgery* 75B(3):433–436.

Krueger-Franke M, Siebert CH, Pfoerringer W. 1992. Sports-related epiphyseal injuries of the lower extremity: an epidemiological study. *Journal of Sports Medicine and Physical Fitness* 32(1):106–111.

Kugler PN, Turvey MT. 1987. *Natural law, and the self assembly of rhythmic movement*. Hillside, NJ: Erlbaum.

Kujala UM, Aalto T, Ostermann K, Dahlstrom S. 1989. The effect of volleyball playing on the knee extensor mechanism. *American Journal of Sports Medicine* 17(6):766–769.

Kujala UM, Friberg O, Aalto T, Kvist M, Osterman K. 1987. Lower limb asymmetry and patellofemoral joint incongruence in the etiology of knee exertion injuries in athletes. *International Journal of Sports Medicine* 8(3):214-220.

Kujala UM, Kvist M, Osterman K, Friberg O, Aalto T. 1986. Factors predisposing army conscripts to knee exertion injuries in a physical training program. *Clinical Orthopaedics and Related Research* 210:203–211.

LaBrier K, O'Neill DB. 1993. Patellofemoral stress syndrome: current concepts. *Sports Medicine* 16(6):449–459.

LaLonde KA, Letts M. 2005. Traumatic growth arrest of the distal tibia: a clinical and radiographic review. *Canadian Journal of Surgery* 48:143–147.

Lancourt JE, Cristini JA. 1975. Patella alta and patella infera. *Journal of Bone and Joint Surgery* 57A:1112–1115.

Lanyon LE. 1981. Adaptive mechanics: the skeleton's response to mechanical stress. In *Mechanical factors and the skeleton*, ed Stokes IA. London: John Libbey.

Larson RL.1973. Physical activity and the growth and development of bone and joint structures. In *Physical activity: human growth and development*, ed Rarick GL. New York: Academic Press.

Larson RL, McMahan RO.1966. The epiphyses and the childhood athlete. *Journal of the American Medical Association* 196:607–612.

Lassiter T, Malone T, Garrick J. 1989. Injury to the lateral ligaments of the ankle. *Orthopedic Clinics of North America* 20:629–640.

Lawrence JP, Greene HS, Grauer JN. 2006. Back pain in athletes. *Journal of the American Academy of Orthopaedic Surgeons* 14(13):726–735.

Lenehan KL, Fryer G, McLaughlin P. 2003. The effect of muscle energy technique on gross trunk range of motion. *Journal of Osteopathic Medicine* 6(1):13–18.

Lentz SS. 1995. Osteitis pubis: a review. *Obstetrical and Gynecological Survey* 50(4):310–315.

Letts M, Smallman T, Afanasiev R, Gouw G. 1986. Fracture of the pars interarticularis in adolescent athletes: a clinical-biomechanical analysis. *Journal of Pediatric Orthopaedics* 6 (1):40–46.

Lewis JS, Wright C, Green A. 2005. Subacromial impingement syndrome: the effect of changing posture on shoulder range of movement. *Journal of Orthopaedic and Sports Physical Therapy* 35(2):72–87.

Lieber RL. 1992. *Skeletal muscle structure and function*. Baltimore: Williams & Wilkins.

Lieber RL. 2002. *Skeletal muscle structure, function & plasticity: the physiological basis of rehabilitation*. Philadelphia: Lippincott Williams & Wilkins.

Lieber RL, Bodine-Fowler SC. 1993. Skeletal muscle mechanics: implications for rehabilitation. *Physical Therapy* 73(12):844–856.

Lieber RL, Fazeli BM, Botte MJ. 1990. Architecture of selected wrist flexor and extensor muscles. *Journal of Hand Surgery* 15(2):244–250.

Lindh M. 1989. Biomechanics of the lumbar spine. In *Basic biomechanics of the musculoskeletal system*, eds Nordin M, Frankel VH. Philadelphia: Lea & Febiger.

Linton SJ. 2000. A review of psychological risk factors in back and neck pain. *Spine* 25(9):1148–1156.

Lloyd-Smith R, Clement DB, McKenzie DC, Taunton JE. 1985. A survey of overuse and traumatic hip and pelvic injuries in athletes. *Physician and Sportsmedicine* 13(10):131–141.

Lowe J, Libson E, Ziv, I, Nyska M, Floman Y, Bloom RA, Robin GC. 1986. Spondylolysis in the upper lumbar spine. *Journal of Bone and Joint Surgery* 69B(4):582–586.

Lowe TG, Line BG. 2007. Evidence based medicine: analysis of Scheuermann kyphosis. *Spine* 32(19 Suppl):S115–S119.

Luethi SM, Stacoff A. 1987. The influence of the shoe on foot mechanics in running. In *Current research in biomechanics*, eds Van Gheluwe B, Atha J. Basel, Switzerland: Karger.

Lundberg A, Svensson OK. 1988. The axes of rotation of the talocalcaneal and talonavicular joints. In *Patterns of motion of the ankle/foot complex*, ed Lundberg A. Gotab, Stockholm: Carolina Institute.

Luoto S, Heliovaara M, Hurri H, Alaranta H. 1995. Static back endurance and the risk of low back pain. *Clinical Biomechanics* 10:325–330.

MacIntosh BR, Gardiner PF, Mc Comas AJ. 2006. *Skeletal muscle: form and function*. Champaign, IL: Human Kinetics.

MacKinnon JL. 1988. Osteoporosis. *Physical Therapy* 10:1533–1539.

Mahieu NN, Witvrouw E, Stevens V, Van Tiggelen D, Roget P. 2006. Intrinsic risk factors for the development of Achilles tendon overuse injury: a prospective study. *American Journal of Sports Medicine* 34(2):226–235.

Mair SD, Seaber AV, Glisson RR, Garrett WE. 1996. The role of fatigue in susceptibility to acute muscle strain injury. *American Journal of Sports Medicine* 24(2):137–143.

Malina RM, Bouchard C, Bar-Or O. 2004. *Growth, maturation, and physical activity.* Champaign, IL: Human Kinetics.

Martin JA, Buckwalter JA. 2002. Aging, articular cartilage chondrocyte senescence and osteoarthritis. *Biogerontology* 3(5):257–264.

Martinez SF, Steingard MA, Steingard PM. 1993. Thigh compartment syndrome: a limb threatening emergency. *Physician and Sportsmedicine* 21(3):94, 96, 99, 100, 103, 104.

Masso PD, Meeropol E, Lennon E. 2002. Juvenile onset scoliosis followed up to adulthood: orthopedic and functional outcomes. *Journal of Pediatric Orthopaedics* 22:279–284.

Matheson GO, Macintyre JG. 1987. Lower leg varum alignment in skiing: relationship to foot pain and suboptimal performance. *Physician and Sportsmedicine* 15(9):162–169, 172, 176.

Matthews PB. 1988. Proprioceptors and their contribution to somatosensory mapping: Complex messages require complex processing. *Canadian Journal of Physiology and Pharmacology* 66:430-438.

McArdle WD, Katch FI, Katch VL 1996. *Exercise physiology: energy, nutrition, and human performance.* Philadelphia: Lea & Febiger.

McClinton S, Bennell KL, Crossley KM. 2007. Influence of step height on quadriceps onset timing activation during stair ascent in individuals with patellofemoral pain syndrome. *Journal of Orthopaedic and Sports Physical Therapy* 37:239–244.

McConnell J. 1986. The management of chondromalacia patellae: a long term solution. *Australian Journal of Physiotherapy* 32:215–222.

McGill SM, Norman RW. 1987. Reassessment of the role of intra-abdominal pressure in spinal compression. *Ergonomics* 30(11):1565–1588.

McHugh MP, Tyler PF, Tetro DT, Mullaney MJ, Nicholas SJ. 2006. Risk factors for noncontact ankle sprains in high school athletes: the role of hip strength and balance ability. *American Journal of Sports Medicine* 34:464–470.

McKay GD, Goldie PA, Payne WR, Oakes BW. 2001. Ankle injuries in basketball: injury rate and risk factors. *British Journal of Sports Medicine* 35:103–108.

McKenzie DC, Clement DB, Taunton JE. 1985. Running shoes, orthotics, and injuries. *Sports Medicine* 2:334–347.

McMahon TA, Greene PR. 1978. Fast running tracks. *Scientific American* 239(6):148–163.

McMullen J, Uhl TL. 2000. A kinetic chain approach for shoulder rehabilitation. *Journal of Athletic Training* 35(3):329–337.

Meeuwisse WH, Tyreman H, Hagel B, Emery C. 2007. A dynamic model of etiology in sport injury: the recursive nature of risk and causation. *Clinical Journal of Sport Medicine* 17(3):215–219.

Melnyk M, Faist M, Gothner M, Claes L, Friemert B. 2007. Changes in stretch reflex excitability are related to "giving way" symptoms in patients with anterior cruciate ligament rupture. *Journal of Neurophysiology* 97:474–480.

Melton LJ. 1997. Epidemiology of spinal osteoporosis. *Spine* 22(Suppl):2S–11S.

Micheli LJ. 1982. Lower extremity injuries: overuse injuries in the recreational adult. In *The exercising adult,* ed Cantu RC. Lexington, MA: Collamore.

Micheli LJ. 1983. Overuse injuries in children's sports: the growth factor. *Orthopedic Clinics of North America* 14(2):337–360.

Micheli LJ, Stanitski CL. 1981. Lateral patellar retinacular release. *American Journal of Sports Medicine* 9(5):330–336.

Milgrom C, Finestone A, Levi Y, Simkin A, Ekenman I, Mendelson S, Millgram M, Nyska M, Benjuya N, Burr DB. 2000. Do high impact exercises produce higher tibial strains than running? *British Journal of Sports Medicine* 34(3):195–199.

Miller DI. 1980. Body segment contributions to sport skill performance: two contrasting approaches. *Research Quarterly for Exercise and Sport* 51:219–233.

Miller JAA, Schultz AB, Anderson GBJ. 1987. Load-displacement behaviour of sacroiliac joints. *Journal of Orthopaedic Research* 5:92–101.

Miniaci A, Fowler PJ. 1993. Impingement in the athlete. *Clinics in Sports Medicine* 12(1):91–110.

Moritani T. 1993. Neuromuscular adaptations during the acquisition of muscle strength, power and motor tasks. *Journal of Biomechanics* 26(Suppl 1):95–107.

Morris JM, Lucas DB, Bresler B. 1961. Role of the trunk in the stability of the spine. *Journal of Bone and Joint Surgery* 43A(3):327–351.

Motley G, Nyland J, Jacobs J, Caborn DNM. 1998. The pars interarticularis stress reaction, spondylolysis, and spondylolisthesis progression. *Journal of Athletic Training* 33(4):351–358.

Muller EA. 1970. Influence of training and of inactivity on muscle strength. *Archives of Physical Medicine and Rehabilitation* 51:449–462.

Muschik M, Hahnel H, Robinson PN, Perka C, Muschik C. 1996. Competitive sports and the progression of spondylolisthesis. *Journal of Pediatric Orthopaedics* 16(3):391–412.

Myers TW. 2001. *Anatomy trains: myofascial meridians for manual and movement therapists.* Edinburgh, UK: Churchill Livingstone.

Mynark RG, Koceja DM. 2001. Effects of age on the spinal stretch reflex. *Journal of Applied Biomechanics* 17:188–203.

Nachemson A. 1992. Newest knowledge of low back pain: a critical look. *Clinical Orthopaedics* 279:8–20.

Nadler SF, Wu KD, Galski T, Feinberg JH. 1998. Low back pain in college athletes: a prospective study correlating lower extremity overuse or acquired ligamentous laxity with low back pain. *Spine* 23(7):828–833.

Nakamura M, Higo M, Matsunaga S, Komiya S. 2004. Scoliosis in patients with leg length discrepancy. *Journal of Japanese Paediatric Orthopaedic Association* 13(1):11-14.

Narici MV, Maganaris CN. 2006. Adaptability of elderly human muscles and tendons to increased loading. *Journal of Anatomy* 208:433–443.

National Aeronautics and Space Administration (NASA). 2008. Table of tan (angle). Retrieved February 17, 2009, from http://www.grc.nasa.gov/WWW/K-12/airplane/tabltan.html.

Negrini S, Antoninni GI, Carabalona R, Minozzi S. 2003. Physical exercises as a treatment for adolescent idiopathic scoliosis: a systematic review. *Pediatric Rehabilitation* 6:227–235.

Nelson BW, O'Reilly E, Miller M, Hogan M, Wegner JA, Kelly C. 1995. The clinical effects of intensive, specific exercise on chronic low back pain: a controlled study of 895 consecutive patients with 1-year follow up. *Orthopedics* 18(10):971–981.

Nester CJ. 1997. Rearfoot complex: a review of its interdependent components, axis orientation and functional model. *Foot* 7: 86–96.

Nicholas JA, Marino M. 1987. The relationship of injuries of the leg, foot, and ankle to proximal thigh strength in athletes. *Foot & Ankle* 7(4):218–228.

Nigg BM. 1985. Biomechanics, load analysis, and sports injuries in the lower extremities. *Sports Medicine* 2:367–379.

Nigg BM. 2001. Role of impact forces and foot pronation: a new paradigm. *Clinical Journal of Sport Medicine* 11:2–9.

Nigg BM, Bahlsen HA. 1988. The influence of heel flare and midsole construction on pronation, supination, and impact forces for heel-toe running. *International Journal of Sport Biomechanics* 4:205–219.

Nigg BM, Cole GK, Bruggemann G-P. 1995. Impact forces during heel-toe running. *Journal of Applied Biomechanics* 11:407–432.

Nigg BM, Denoth J, Kerr B, Luethi S, Smith D, Stacoff A. 1984. Load sport shoes and playing surfaces. In *Sport shoes and playing surfaces,* ed Frederick EC. Champaign, IL: Human Kinetics.

Nigg BM, Denoth J, Neukomm PA. 1981. Quantifying the load on the human body: problems and some possible solutions. In *Biomechanics.* Vol. VIIB, eds Morecki A, Fidelus K, Kedzior K, Wit I. Baltimore: University Park Press.

Nigg BM, Morlock M. 1987. The influence of lateral heel flare of running shoes on pronation and impact forces. *Medicine and Science in Sports and Exercise* 19(3):294–302.

Nigg BM, Nurse MA, Stefanyshyn DJ. 1999. Shoe inserts and orthotics for sport and physical activities. *Medicine and Science in Sports and Exercise* 31(7):S421–S428.

Nigg BM, Segesser B. 1988. The influence of playing surfaces on the load on the locomotor system and on football and tennis injuries. *Sports Medicine* 5:375–385.

Nijweide PJ, Burger EH, Klein-Nulend J. 2002. The osteocyte. In *Principles of bone biology,* eds Bilezikian JP, Raisz LG, Rodan GA. San Diego: Academic Press.

Nordin M, Frankel VH. 2001. *Basic biomechanics of the musculoskeletal system.* 3rd ed. Philadelphia: Lippincott Williams & Wilkins.

Norkin CC, Levangie PK. 1992. *Joint structure and function: a comprehensive analysis.* Philadelphia: Davis.

Noth J. 1992. Motor units. In *Strength and power in sport,* ed Komi PV. Oxford, UK: Blackwell Scientific.

Noyes FR. 1977. Functional properties of knee ligaments and alterations induced by immobilization. *Clinical Orthopaedics* 123:210–242.

Nyman T, Grooten WJA, Wiktorin C, Liwing J, Norrman L. 2007. Sickness absence and concurrent low back and neck-shoulder pain: results from the MUSIC-Norrtälje study. *European Spine Journal* 16(5):631–638.

Old JL, Calvert M. 2004. Vertebral compression fractures in the elderly. *American Family Physician* 69(1):111–116.

Osternig LR, Robertson RN. 1993. Effects of prophylactic knee bracing on lower extremity joint position and muscle activation during running. *American Journal of Sports Medicine* 21(5):733–737.

Ozer H, Turanli S, Baltaci G. 2002. Avulsion of the tibial tuberosity with a lateral plateau rim fracture: case report. *Knee Surgery, Sports Traumatology, Arthroscopy* 10:310–312.

Pappas AM.1983. Epiphyseal injuries in sports. *Physician and Sportsmedicine* 11(6):140–148.

Parfitt AM. 2002. The life history of osteocytes: relationship to bone age, bone remodeling, and bone fragility. *Journal of Musculoskeletal & Neuronal Interactions* 2(6):499–500.

Pasanen K, Parkkari J, Rossi L, Kannus P. 2008. Artificial playing surface increases the injury risk in pivoting indoor sports: a prospective one-season follow-up study in Finnish female floorball. *British Journal of Sports Medicine* 42:194-197.

Peat M. 1986. Functional anatomy of the shoulder. *Physical Therapy* 66:1855–1865.

Pedowitz DI, Reddy S, Parekh SG, Huffman GR, Sennett BJ. 2008. Prophylactic bracing decreases ankle injuries in collegiate female volleyball players. *American Journal of Sports Medicine* 36(2):324–327.

Penny JN, Smith C. 1980. The prevention and treatment of swimmer's shoulder. *Canadian Journal of Applied Sport Sciences* 5(3):195–202.

Percy EC, Strother RT. 1985. Patellagia. *Physician and Sportsmedicine* 13(7):43–59.

Perdriolle R, Vidal J. 1985. Thoracic idiopathic scoliosis curve evolution and progression. *Spine* 10(9):785–791.

Perlman M, Leveille D, DeLeonibus J, Hartman R, Klein J, Handelman R, Schultz E, Wertheimer S. 1987. Inversion lateral ankle trauma: differential diagnosis, review of the literature, and prospective study. *Journal of Foot Surgery* 26:95–135.

Peters JW, Trevino SG, Renström PA. 1991. Chronic lateral ankle instability. *Foot & Ankle* 12:182–191.

Petersen L, Renström P. 2000. *Sports injuries; their prevention and treatment.* London: Martin Dunitz.

Petrie TR. 1993. Coping skills, competitive trait anxiety, and playing status: moderating effects on the life stress-injury relationship. *Journal of Sport and Exercise Psychology* 15:261–274.

Petrov O, Roth DW, Weis WW, Rader AJ. 1988. Nonprescription custom insoles for ski boots. *Journal of the American Podiatric Medical Association* 78(8):422–428.

Plum P, Ofeldt T. 1985. *Low back pain: new treatment.* Skaerup, Denmark: Brage.

Pollock ML, Gaesser G, Butcher JD, Despres J-P, Dishman RK, Franklin BA, Garber CE. 1998. ACSM position stand: the recommended quantity and quality of exercise for developing and maintaining cardiorespiratory and muscular fitness, and flexibility in healthy adults. *Medicine and Science in Sports and Exercise* 30(6):975–991.

Putnam CA. 1993. Sequential motions of body segments in striking and throwing skills: descriptions and explanations. *Journal of Biomechanics* 26(Suppl 1):125–135.

Radin EL. 1984. Biomechanical considerations. In *Osteoarthritis: diagnosis and management,* eds Moskowitz RW, Howell DS, Goldberg VM, Mankin HJ. Philadelphia: Saunders.

Radin EL. 1986. Osteoarthrosis: What is known about prevention. *Clinical Orthopedics and Related Research* 222:6-65.

Radin EL, Parker HG, Pugh JW, Steinberg RS, Paul IL, Rose RM. 1973. Response of joints to impact loading, III: relationship between trabecular microfractures and cartilage degeneration. *Journal of Biomechanics* 6:51–57.

Radin EL, Paul IL. 1971. Response of joints to impact loading I: in vitro wear. *Arthritis and Rheumatism* 14:356–362.

Ranawat VS, Dowell JK, Heywood-Waddington MB. 2003. Stress fractures of the lumbar pars interarticularis in athletes: a review based on long-term results of 18 professional cricketers. *Injury* 34(12):915–919.

Reber L, Perry J, Pink M. 1993. Muscular control of the ankle in running. *American Journal of Sports Medicine* 21(6):805–810.

Rechel JA, Yard EE, Comstock RD. 2008. An epidemiologic comparison of high school sports injuries sustained in practice and competition. *Journal of Athletic Training* 43(2):197-204.

Recht MP, Burk DL, Dalinka MK. 1987. Radiology of wrist and hand injuries in athletes. *Clinics in Sports Medicine* 6(4):811–828.

Redford JB. 1987. Orthotics: general principles. *Physical Medicine and Rehabilitation State of the Art Reviews* 1(1):1–10.

Reeves ND, Narici MV, Maganaris CN. 2006. Myotendinous plasticity to ageing and resistance exercise in humans. *Experimental Physiology* 91(3):483–498.

Reid DC. 1987. Preventing injuries to the young ballet dancer. *Physiotherapy Canada* 39(4):231–235.

Reider B. 1996. Medial collateral ligament injuries in athletes. *Sports Medicine* 21(2):147–156.

Renström P, Ljungqvist A, Arendt E, Beynnon B, Fukubayashi T, Garrett W, Georgoulis T, Hewett TE, Johnson R, Krosshaug T, Mandelbaum B, Micheli L, Myklebust G, Roos E, Roos H, Schamasch P, Shultz S, Werner S, Wojtys E, Engrbretsen L. 2008. Non-contact ACL injuries in female athletes: an International

Olympic Committee current concepts statement. *British Journal of Sports Medicine* 42:394–412.

Richter DE, Nash CL, Moskowitz RW, Goldberg VM, Rosner IA. 1985. Idiopathic adolescent scoliosis: a prototype of degenerative joint disease. *Clinical Orthopaedics and Related Research* 193:221–229.

Riegger-Krugh C, Keysor JL. 1996. Skeletal malalignments of the lower quarter: correlated and compensatory motions and postures. *Journal of Orthopaedic and Sports Physical Therapy* 23(2):164–170.

Risch SV, Norvell NK, Pollock ML, Rische E, Langer H, Fulton M. 1993. Lumbar strengthening in chronic low back pain patients: physiologic and psychologic benefits. *Spine* 18:232–238.

Roberts D, Ageberg E, Andersen G, Fridén T. 2007. Clinical measurements of proprioception, muscle strength and laxity in relation to function in the ACL-injured knee. *Knee Surgery, Sports Traumatology, Arthroscopy* 15:9–16.

Roberts TDM. 1995. *Understanding balance: the mechanics of posture and locomotion.* London: Chapman and Hall.

Roels J, Martins M, Mulier JC, Burssens A. 1978. Patellar tendinitis (jumper's knee). *American Journal of Sports Medicine* 6:362–68.

Rolf C. 1995. Overuse injuries of the lower extremity in runners. *Scandinavian Journal of Medicine & Science in Sports* 5(4):181–190.

Rossi F, Dragoni S. 1994. Lumbar spondylolysis and sports: radiological findings and statistical considerations. *La Radiologia Medica* 87(4):397–400.

Rossi-Durand C. 2006. Proprioception and myoclonus. *Neurophysiologie Clinique* 36:299–308.

Roughley PJ. 2004. Biology of intervertebral disc aging and degeneration: involvement of the extracellular matrix. *Spine* 29(23):2691–2699.

Rubin CT. 1984. Skeletal strain and the functional significance of bone architecture. *Calcified Tissue International* 36:S11–S18.

Ryan C, Fricker PA, Hannaford PGA. 1987. Muscle compartment pressure syndrome of the upper limb and shoulder: two case studies. *Australian Journal of Science and Medicine in Sport* 19(3):24–25.

Rydevik B, Brown MD, Lundborg G. 1984. Pathoanatomy and pathophysiology of nerve root compression. *Spine* 9(1):7–15.

Salenius P, Vankka E. 1975. Development of the tibiofemoral angle of children at different ages. *Journal of Bone and Joint Surgery* 57A:259–261.

Sargeant AJ. 1994. Human power output and muscle fatigue. *International Journal of Sports Medicine* 15(3):116–121.

Schoenau E, Frost HM. 2003. The "muscle–bone unit" in children and adolescents. *Calcified Tissue International* 70: 405–407.

Schultz AB, Alexander NB, Ashton-Miller JA. 1992. Biomechanical analyses of rising from a chair. *Journal of Biomechanics* 25:1383–1991.

Schultz A, Haderspeck K, Takashima S. 1981. Correction of scoliosis by muscle stimulation: biomechanical analyses. *Spine* 6(5):468–476.

Seddon JH. 1972. *Surgical disorders of peripheral nerves.* Edinburgh, UK: Churchill Livingstone.

Segan DJ, Sladek EC, Gomez J, McCoy HJ, Cairns DA. 1988. Weight lifting as a cause of bilateral upper extremity compartment syndrome. *Physician and Sportsmedicine* 16(10):73–75.

Seynnes OR, Maganaris CN, de Boer MD, di Prampero PE, Narici MV. 2008. Early structural adaptations to unloading in the human calf muscles. *Acta Physiologica* 193(3):265–274.

Shambaugh JP, Klein A, Herbert JH. 1991. Structural measures as predictors of injury in basketball players. *Medicine and Science in Sports and Exercise* 23(3):522–527.

Shapiro C. 2001. Swimming. In *Sports injury: prevention and rehabilitation,* eds Shamus E, Shamus J. Chicago: McGraw-Hill Professional, 103–154.

Sharma P, Maffulli N. 2006. Biology of tendon injury: healing, modeling and remodeling. *Journal of Musculoskeletal & Neuronal Interactions* 6(2):181–190.

Shum GLK, Crosbie J, Lee RYW. 2007. Movement coordination of the lumbar spine and hip during a picking up activity in low back pain subjects. *European Spine Journal* 16(6):749–758.

Siegler S, Block J, Schneck CD. 1988. The mechanical characteristics of the collateral ligaments of the human ankle joint. *Foot & Ankle* 8:234–242.

Siffert RS. 1987. Lower limb-length discrepancy. *Journal of Bone and Joint Surgery* 67A:1100–1106.

Silliman JF, Hawkins RJ. 1991. Current concepts and recent advances in the athlete's shoulder. *Clinics in Sports Medicine* 10(4):693–705.

Silver DM, Campbell P. 1985. Arthroscopic assessment and treatment of dancers knee injuries. *Physician and Sportsmedicine* 13(11):75–82.

Silver JR, Silver DD, Godfrey JJ. 1986. Trampoline injuries of the spine. *Injury* 17(2):117–124.

Silver RL, Garza J de la, Rang M. 1985. The myth of muscle balance. *Surgery* 67B(3):432–437.

Simon SR, Radin EL, Paul IL, Rose RM. 1972.The response of joints to impact loading, II: in vivo behavior of subchondral bone. *Journal of Biomechanics* 5:267–272.

Singh AK, Starkweather KD, Hollister AM, Jatana S, Lupichuk AG. 1992. Kinematics of the ankle: a hinge axis model. *Foot & Ankle* 13(8):439–446.

Skelton DA, Young A, Greig CA, Malbut KE. 1995. Effects of resistance training on strength, power and selected functional abilities of women aged 75 and older. *Journal of the American Geriatrics Society* 43:1081–1087.

Skinner HB, Wyatt MP, Stone ML, Hodgdon JA, Barrack RL. 1986. Exercise-related knee joint laxity. *American Journal of Sports Medicine* 14:30–34.

Skoglund LB, Miller JA. 1981. The length of the thoracolumbar spine in children with idiopathic scoliosis. *Acta Orthopaedica Scandinavica* 52:77–185.

Smith EL.1982. Exercise for prevention of osteoporosis: a review. *Physician and Sportsmedicine* 10(3):72–83.

Smith M, Fryer G. 2008. A comparison of two muscle energy techniques for increasing flexibility of the hamstring muscle group. *Journal of Bodywork and Movement Therapies* 12(4):312–317.

Smith RKW, Birch HL, Goodman S, Heinegård D, Goodship AE. 2002. The influence of ageing and exercise on tendon growth and degeneration: hypotheses for the initiation and prevention of strain-induced tendinopathies. *Comparative Biochemistry and Physiology Part A* 133:1039–1050.

Soler T, Calderon C. 2000. The presence of spondylolisthesis in the elite Spanish athlete. *American Journal of Sports Medicine* 28:57–62.

Soligard T, Myklebust G, Steffen K, Holme I, Silvers H, Bizzini M, Junge A, Dvorak J, Bahr R, Andersen TE. 2008. Comprehensive warm-up programme to prevent injuries in young female footballers: cluster randomised controlled trial. *British Medical Journal* 337:a2469.

Sommer HM. 1988. Patellar chondropathy and apicitis: muscle imbalances of the lower extremities in competitive sports. *Sports Medicine* 5:386–394.

Sonne-Holm S, Jacobsen S, Rovsing HC, Monrad H, Gebuhr P. 2007. Lumbar spondylolysis: a life long dynamic condition? A cross sectional survey of 4151 adults. *European Spine Journal* 16(6):821–828.

Spanjaard M, Reeves RD, van Dieën JH, Baltzopoulos V, Maganaris CN. 2006. Gastrocnemius muscle fascicle behavior during stair negotiation in humans. *Journal of Applied Physiology* 102:1618–1623.

Sparto PJ, Parnianpour M, Reinsel TE, Simon S. 1997. The effect of fatigue on multijoint kinematics, coordination, and postural stability during a repetitive lifting task. *Journal of Orthopaedic and Sports Physical Therapy* 25:3–12.

Speer DP, Braun JK. 1985. The biomechanical basis of growth plate injuries. *Physician and Sportsmedicine* 13(2):72–78.

Sperryn PN. 1988. Safety and hygiene in sport. In *The Olympic book of sports medicine*, eds Ditrix A, Knuttgen HG, Tittel K. Oxford, UK: Blackwell Scientific.

Stacoff A, Kalin X, Stussi E. 1991. The effects of shoes on the torsion and rearfoot motion in running. *Medicine and Science in Sports and Exercise* 23:482–490.

Standaertand CJ, Herring SA. 2000. Spondylolysis: a critical review. *British Journal of Sports Medicine* 34:415–422.

Standring S. 2004. *Gray's anatomy: the anatomical basis of clinical practice*. 39th ed. Edinburgh, UK: Churchill Livingstone.

Steinbach HL, Russell W. 1964. Measurement of the heel pad as an aid to diagnosis of acromegaly. *Radiology* 82:418–423.

Steindler A. 1955. *Kinesiology of the human body*. Springfield, IL: Charles C Thomas.

Steiner ME. 1987. Hypermobility and knee injuries. *Physician and Sportsmedicine* 15(6):159–165.

Steiner ME, Grana WA, Chillag K, Schelberg-Karnes E. 1986. The effect of exercise on anterior-posterior knee laxity. *American Journal of Sports Medicine* 14:24–29.

Steyers CM, Schelkun PH. 1995. Practical management of carpal tunnel syndrome. *Physician and Sportsmedicine* 23(1):83–88.

Stillman RJ, Lohman TG, Slaughter MH, Massey BH. 1986. Physical activity and bone mineral content in women aged 30 to 85 years. *Medicine and Science in Sports and Exercise* 18:576–580.

Stovitz SD, Pardee PE, Vazquez G, Duval S, Schwimmer JB. 2008. Musculoskeletal pain in obese children and adolescents. *Acta Paediatrica* 97:489-493.

Taber LA. 1995. Biomechanics of growth, remodeling, and morphogenesis. *Applied Mechanical Review* 48(8):487–545.

Taft TN, Wilson FC, Oglesby JW. 1987. Dislocation of the acromioclavicular joint: an end result study. *Journal of Bone and Joint Surgery* 69:1045–1051.

Tall RL, Devault W. 1993. Spinal injury in sport: epidemiological considerations in clinical sports medicine. *Clinics in Sports Medicine* 12(3):441–448.

Talmage RV, Anderson JJB. 1984. Bone density loss in women: effects of childhood activity, exercise, calcium intake, and oestrogen therapy. *Calcified Tissue International* 36:522–532.

Taunton JE, Ryan MB, Clement DB, McKenzie DC, Lloyd-Smith DR, Zumbo BD. 2003. A prospective study of running injuries: the Vancouver "in training" clinics. *British Journal of Sports Medicine* 37(3):239–244.

Taylor DC, Dalton JD, Seaber AV, Garrett WE. 1990. Viscoelastic properties of muscle-tendon units. *American Journal of Sports Medicine* 18:300–308.

Tesch PA, von Walden F, Gustafsson T, Linnehan RM, Trappe TA. 2008. Skeletal muscle proteolysis in response to short-term unloading in humans. *Journal of Applied Physiology* 105:902–906.

Thijs Y, De Clerc D, Roosen P, Witvrouw E. 2008. Gait-related intrinsic risk factors for patellofemoral pain in novice recreational runners. *British Journal of Sports Medicine* 42:466–471.

Torg JS, Vegso JJ, Sennet B, Dass M. 1985. The National Football Head and Neck Injury Registry: 14 year report on cervical quadriplegia, 1971-1984. *Journal of the American Medical Association* 254:3439–3443.

Tortora GJ. 2004. *Principles of human anatomy.* New York: Wiley.

Tortora GJ, Anagnostakos NP. 1984. *Principles of anatomy and physiology.* New York: Harper & Row.

Troup JDG. 1970. Functional anatomy of the spine. *British Journal of Sports Medicine* 5(1):27–33.

Tyler TF, Nicholas SJ, Campbell RJ, McHugh MP. 2001. The association of hip strength and flexibility with the incidence of adductor muscle strains in professional ice hockey players. *American Journal of Sports Medicine* 29:124–128.

Urso M, Clarkson P, Price T. 2006. Immobilization effects in young and older adults. *European Journal of Applied Physiology* 96(5):564–571.

Van der Wall H, Magee M, Reiter L, Frater CJ, Qurashi S, Loneragan R. 2006. Degenerative spondylolysis: a concise report of scintigraphic observations. *Rheumatology* 45:209–211.

van Gent R, van Middelkoop M, van Os A, Bierma-Zeinstra S, Koes B. 2007. Incidence and determinants of lower extremity running injuries in long distance runners: a systematic review. *British Journal of Sports Medicine* 41:469–480.

Van Ghent RN, Siem D, van Middelkoop M, van Os AG, Bierma-Zeinstra SMA, Koes BW. 2007. Incidence and determinants of lower extremity running injuries in long distance runners: a systematic review. *British Journal of Sports Medicine* 41:469–480.

Van Ingen Schenau GJ, Pratt CA, Macpherson JM. 1994. Differential use and control of monoand biarticular muscles. *Human Movement Science* 13:495–517.

Van Mechelen W, Hlobil H, Kemper HCG, Voorn WJ, de Jonge HR. 1993. Prevention of running injuries by warm-up, cool down, and stretching exercises. *American Journal of Sports Medicine* 21(5):711–719.

Villwock MR, Meyer EG, Powell JW, Fouty AJ, Haut RC. 2009. Football playing surface and shoe design affect rotational traction. *American Journal of Sports Medicine* 37:518–525.

Vizsolyi P, Taunton J, Robertson G, Filsinger L, Shannon HS, Whittingham D, Gleave M. 1987. Breaststroker's knee: an analysis of epidemiological and biomechanical factors. *American Journal of Sports Medicine* 15(1):63–71.

Voight ML, Hoogenboom BL, Prentice WE. 2006. *Musculoskeletal interventions: techniques for therapeutic exercise.* New York: McGraw-Hill Professional.

Voloshin A, Wosk J. 1981. Influence of artificial shock absorbers on human gait. *Clinical Orthopaedics and Related Research* 160:52–56.

Voloshin A, Wosk J. 1982. In-vivo study of low back pain and shock absorption in human locomotor system. *Journal of Biomechanics* 15:21-27.

Wackerhage H, Rennie MJ. 2006. How nutrition and exercise maintain the human musculokeletal mass. *Journal of Anatomy* 208:451–458.

Walker ML, Rothstein JM, Finucane SD, Lamb RL. 1987. Relationships between lumbar lordosis, pelvic tilt, and abdominal muscle performance. *Physical Therapy* 67(4):512–516.

Watkins J. 1982. Warming up and warming down. *New Zealand Journal of Sports Medicine* 10 (1):10–14.

Watkins J. 1987a. Qualitative movement analysis. *British Journal of Physical Education* 18(4):177–179.

Watkins J. 1987b. Quantitative movement analysis. *British Journal of Physical Education* 18(6):271–275.

Watkins J. 2007. *An introduction to biomechanics of sport and exercise.* Edinburgh, UK: Churchill Livingstone.

Watkins J, Green BN. 1992. Volleyball injuries: a survey of injuries of Scottish National League male players. *British Journal of Sports Medicine* 26(2):135–137.

Watkins J, Peabody P. 1996. Sports injuries in children and adolescents treated at a sports injury clinic. *Journal of Sports Medicine and Physical Fitness* 36(1):43–48.

Watson AWS. 1995. Sports injuries in footballers related to defects of posture and body mechanics. *Journal of Sports Medicine and Physical Fitness* 35 (4):289–294.

Watt DGD, Jones GM. 1971. Muscular control of landing from unexpected falls in man. *Journal of Physiology* 219:729–737.

Wearing SC, Hennig EM, Byrne NM, Steele JR, Hills AP. 2006. The impact of childhood obesity on musculoskeletal form. *Obesity Reviews* 7:209–218.

Weiker GG, Munnings F. 1994. Selected hip and pelvis injuries. *Physician and Sportsmedicine* 22(2):96, 98, 99, 103–106.

Weiss H-R, Negrini S, Rigo M, Kotwicki T, Hawes MC, Grivas TB, Maruyama T, Landauer F. 2006. Indications for conservative management of scoliosis (guidelines). *Scoliosis* 1:5–6.

Wen DY. 2007. Risk factors for overuse injuries in runners. *Current Sports Medicine Reports* 6(5):307–316.

Whedon GD. 1984. Disuse osteoporosis: physiological aspects. *Calcified Tissue International* 36:146–150.

Wickiewicz TL, Roy RR, Powell PL, Edgerton VR. 1983. Muscle architecture of the human lower limb. *Clinical Orthopaedics and Related Research* 179:275–283.

Wilder DG, Pope MH, Frymoyer JW. 1980. The functional topography of the sacroiliac joint. *Spine* 5(6):575–579.

Wilkerson GB, Nitz AJ. 1994. Dynamic ankle stability: mechanical and neuromuscular interrelationships. *Journal of Sport Rehabilitation* 3:43–57.

Willems TM, Witvrouw E, Delbaere K, Mahieu N, De Bourdeaudhuij I, De Clercq D. 2005. Intrinsic risk factors for inversion ankle sprains in male subjects: a prospective study. *American Journal of Sports Medicine* 33:415–423.

Williams JM. 2001. Psychology of injury risk and prevention. In *Handbook of sport psychology*, eds Singer RN, Hausenblas HA, Janelle CM. New York: Wiley, 766–786.

Williams P, Goldspink G. 1973. The effect of immobilization on the longitudinal growth of striated muscle fibers. *Journal of Anatomy* 116:45–55.

Williams P, Goldspink G. 1978. Changes in sarcomere length and physiological properties in immobilized muscle. *Journal of Anatomy* 127:459–468.

Wilmore JH, Costill DL, Kenny WL. 2008. *Physiology of sport and exercise*, 4th ed. Champaign, IL: Human Kinetics.

Windhorst U. 2007. Muscle proprioceptive feedback and spinal networks. *Brain Research Bulletin* 73:155–202.

Winslow J, Yoder E. 1995. Patellofemoral pain in female ballet dancers: correlation with iliotibial band tightness and tibial external rotation. *Journal of Orthopaedic and Sports Physical Therapy* 22(1):18–21.

Winter JA, Allen TJ, Proske U. 2005. Muscle spindle signals combine with the sense of effort to indicate limb position. *Journal of Physiology* 568:1035–1046.

Witvrouw E, Lysens R, Bellemans J, Cambier D, Vanderstraeten G. 2000. Intrinsic risk factors for the development of anterior knee pain in an athletic population: a two-year prospective study. *American Journal of Sports Medicine* 28(4):480–489.

Wolff J. 1988. *The law of bone modeling*, trans Maquet P, Furlong R. New York: Springer.

Woods V. 2005. Work-related musculoskeletal health and social support. *Occupational Medicine* 55(3):177-189.

Wosk J, Voloshin A. 1985. Low back pain: conservative treatment with artificial shock absorbers. *Archives of Physical Medicine and Rehabilitation* 66:145–148.

Wretenberg P, Hugo A, Broström E. 2008. Hip joint load in relation to leg length discrepancy. *Medical Devices: Evidence and Research* 1:13–18.

Yates JW, Jackson DW.1984.Current status of meniscus injury. *Physician and Sportsmedicine* 12(2):51–59.

Yoshikawa T, Mori S, Santiesteban AJ, Sun TC, Hafstad E, Chen J, Burr DB. 1994. The effects of muscle fatigue on bone strain. *Journal of Experimental Biology* 188:217–233.

Zachazewski JE, Geissler G. 1992. When to prescribe a knee brace. *Physician and Sportsmedicine* 20(1):91–99.

Zernicke RF, Loitz BJ. 1992. Exercise-related adaptations in connective tissue. In *Strength and power in sport*, ed Komi PV. Oxford, UK: Blackwell Scientific.

Zhang G, Young BB, Ezura Y, Favata M, Soslowsky LJ, Chakravarti S, Birl DE. 2005. Development of tendon structure and function: regulation of collagen fibrillogenesis. *Journal of Musculoskeletal & Neuronal Interactions* 5(1):5–21.

Zohar J. 1973. Preventive conditioning for maximum safety and performance. *Sports Coach* 42(9): 65, 113-115.

Index

Note: The italicized *f* and *t* following page numbers refer to figures and tables, respectively.

About the Author

James Watkins, PhD, is a professor of biomechanics in the School of Human Sciences and director of the Sport and Exercise Science Research Centre at the Swansea University in Wales. Watkins spent over 20 years in Glasgow, Scotland, as a lecturer and researcher, and he served as the head of the department of physical education, sport and outdoor education at Jordanhill College and later at the University of Strathclyde, both in Glasgow.

Watkins' main teaching and research specializations are musculoskeletal anatomy and the biomechanics of sport and exercise. He has authored over 80 publications, including three well-known textbooks (*An Introduction to the Mechanics of Human Movement*, 1983; this text, *Structure and Function of the Musculoskeletal System*; and *An Introduction to Biomechanics of Sport and Exercise*, 2007).

Watkins is a fellow of the British Association of Sport and Exercise Sciences (BASES), a fellow of the Physical Education Association of the United Kingdom (PEAUK), and an honorary member of the Association for Physical Education (afPE). He is an advisory board member of the *Journal of Sports Sciences* and a former chair of the Biomechanics Section of BASES.

In 1975 Watkins received his PhD in biomechanics from the University of Leeds in England. He resides in Swansea, where he enjoys walking, playing golf, and reading about the history of science.

You'll find
other outstanding
biomechanics and
anatomy resources at

www.HumanKinetics.com

In the U.S. call

1-800-747-4457

Australia..08 8372 0999
Canada 1-800-465-7301
Europe...+44 (0) 113 255 5665
New Zealand.......................................0064 9 448 1207

 HUMAN KINETICS
The Information Leader in Physical Activity
P.O. Box 5076 • Champaign, IL 61825-5076 USA